Electromagnetic Transients in Power Systems

ELECTRONIC & ELECTRICAL ENGINEERING RESEARCH STUDIES

HIGH-VOLTAGE POWER TRANSMISSION SERIES

Series Editor: **Dr V. T. Morgan**
CSIRO Division of Applied Physics, Australia

1. Electromagnetic Transients in Power Systems
   Pritindra Chowdhuri

Electromagnetic Transients in Power Systems

Pritindra Chowdhuri
Center for Electric Power
Tennessee Technological University
Cookeville, TN, USA

RESEARCH STUDIES PRESS LTD.
Taunton, Somerset, England

JOHN WILEY & SONS INC.
New York · Chichester · Toronto · Brisbane · Singapore

RESEARCH STUDIES PRESS LTD.
24 Belvedere Road, Taunton, Somerset, England TA1 1HD

Marketing and Distribution:

Australia and New Zealand:
Jacaranda Wiley Ltd.
GPO Box 859, Brisbane, Queensland 4001, Australia
Canada:
JOHN WILEY & SONS CANADA LIMITED
22 Worcester Road, Rexdale, Ontario, Canada

Europe, Africa, Middle East and Japan:
JOHN WILEY & SONS LIMITED
Baffins Lane, Chichester, West Sussex, UK, PO19 1UD

North and South America:
JOHN WILEY & SONS INC.
605 Third Avenue, New York, NY 10158, USA

South East Asia:
JOHN WILEY & SONS (SEA) PTE LTD.
37 Jalan Pemimpin 05-04
Block B Union Industrial Building, Singapore 2057

Library of Congress Cataloging-in-Publication Data
Available

British Library Cataloguing in Publication Data
A catalogue record for this book is available from the British Library.

ISBN 0 86380 180 3 (Research Studies Press Ltd.) *[Identifies the book for orders except in America.]*
ISBN 0 471 95746 1 (John Wiley & Sons Inc.) *[Identifies the book for orders in USA.]*

Printed in Great Britain by SRP Ltd., Exeter

In loving memories of

my parents

AHINDRA and SUDHIRA

CHOWDHURI

who

nurtured and inspired me

TABLE OF CONTENTS

EDITORIAL FOREWORD

This excellent monograph on Electromagnetic Transients in Power Systems by Prof. Pritindra Chowdhuri is the first book to be published in a new series on High-Voltage Power Transmission. The electricity supply industry is undergoing rapid change, with increasing pressure for the fullest possible utilisation of transmission and distribution assets. Hence, it is essential that the best technical and economic solutions are applied to such problems as overvoltages, power losses, fault currents, protection, electromagnetic interference, current-carrying capacities, insulation failures, excessive sags, corrosion, fatigue, wind-induced oscillations and constraints on line planning.

It is intended that this series will address the electrical, mechanical, civil, environmental and planning aspects of high-voltage power transmission, with authoritative reviews of recent research and developments by experts in their fields.

Sydney, Australia
June 1996

Vincent T. Morgan

PREFACE

This book has been in preparation during the last 10 years when I joined the Center for Electric Power at Tennessee Technological University. It is the outgrowth of the notes I prepared for two graduate courses on electromagnetic transients in power systems. I felt that graduate courses on this topic should deal not only with the basic mathematical and physical principles of electromagnetic transients and their effects on power systems but should also address the latest problems on electromagnetic transients which the electric power industry is trying to solve. Power-system grounding, lightning-induced voltages on overhead power lines, protection of substations and transients in low-voltage systems with emphasis on power-electronic systems are very important topics today.Although a wealth of knowledge exists on these topics in the published literature, there is no single depository where information on these topics can be found. Separate chapters have been devoted to these topics in this book. Each chapter contains extensive references for those who want to pursue the topic further.

I have drawn heavily on my own involvement during the last 40 years with electromagnetic transients in both high- and low-voltage electrical systems. I have also been benefitted by the knowledge and wisdom of my colleagues, particularly through my association with the Institute of Electrical and Electronics Engineers (IEEE). I hope that this book will be useful not only to the graduate students but also to the specialists who are practicing in this field.

I wish to thank the IEEE and the Electric Power Research Institute (EPRI) for permitting me to reproduce from their publications. I wish to thank ABB Power T&D Company, Cooper Power Systems, General Electric Company, Harris Corporation and Ohio Brass Company for making available to me information on surge protectors. Dr. K.L. Cummins of Global Atmospherics, Professor K. Feser of Stuttgart University and Dr. D.R. MacGorman of University of Oklahoma were very kind to permit me to reproduce several figures from their publications. My special thanks go to Mr. J. Autery, Dixons Mills, Alabama for providing us with the striking photo of lightning flash which is printed on the cover of the book with his permission.

I am grateful to Dr. C.E. Hickman, Director of the Center for Electric Power for making available the Center facilities and his constant encouragement for writing this book. I acknowledge Mr. K. A. Jones, R & D Engineer of the Center for scanning many of the figures in the book, and Ms. Helen Knott for help with word processing. I am thankful to my students for their valuable suggestions and help in weeding out errors in the manuscript. I am indebted to my daughter, Naomi Tyler for reading and correcting the manuscript in spite of her own heavy load in her Ph.D. program. My appreciation and gratitude go to my wife, Sharon for literally living through the vow, 'for better and for worse' that she once took and for her encouragement, understanding and patience in letting me wander into this venture.

Last but not the least, I am indebted to my editor, Dr. V.T. Morgan and to my publisher, Mrs. V.A. Wallace for their extraordinary patience and encouragement.

Cookeville, Tennessee, U.S.A. P. Chowdhuri
July 1996

CHAPTER 1
INTRODUCTION

1.1 Sources of Electrical Transients

Transients in an electrical system are generated by:

1. Lightning, and
2. Switching.

Lightning is associated with the discharge of large currents (sometimes exceeding 200 kA) to grounded objects. When lightning strikes a point, the voltage developed is the lightning discharge current multiplied by the impedance of the system as seen by the lightning current. This voltage can be in the order of several tens of megavolts. The taller a structure is, the higher is the probability of its being struck by lightning. High-voltage power lines and tall towers are susceptible to direct lightning strokes. However, even if a lightning stroke strikes the level ground, the electric and magnetic fields of the lightning channel can induce high enough voltage on a nearby low-voltage power line to cause system outage. Sometimes, the effects of lightning are felt through the rise in voltage of the ground caused by the flow of heavy current. The electrostatic field of the cloud has also caused malfunctions in electronic and communications equipment.

A lightning current rises to its peak in a time which varies from less than a microsecond to 10 or 20 microseconds, and then decays to a small value within a few hundred microseconds. A second low-current component persists sometimes for a few seconds. A lightning stroke is a random phenomenon. However, statistical data on the peak current, waveshape and the frequency of strike (i.e., ground flash density) around the world have been accumulated over the years. These data are used to design protective schemes for electrical systems.

Protection of power systems against lightning has been a serious concern of power engineers almost since the beginning of the twentieth century. Although much progress has been made, lightning-caused outages of power systems are still numerous. Computational techniques have made great strides since the introduction of the digital computers; yet much of the knowledge and its applications are still empirical. Nonetheless, the advances in the techniques of lightning protection of power systems during the past thirty years have been impressive in the face of the demand for better security of power system operation. It is expected that this pace in the progress will continue, and better techniques of lightning protection will evolve.

Many outstanding engineers have devoted their whole careers to unraveling the mysteries of lightning and its effects on the power system. However, it has not been easy for the practicing engineer to understand the complexities of the physical principles and their mathematical applications. Perhaps understanding is not essential if things work. And things have worked well indeed because of the ingenuity of the practicing engineer. This knowledge gained in the field by the engineers has given incentive to the researchers to further explore the physical principles of the interaction between lightning and the power system.

A switching surge is generated because of the inherent inductance associated with an electrical circuit. The theory of constant flux linkage states that an inductance opposes a change in the current by inducing a voltage. Therefore, when a switch (circuit breaker) tries to interrupt a current, a voltage is developed by the circuit inductance to oppose any change in the current. The faster a switch tries to interrupt a current, the higher is the resulting switching surge voltage. The energy stored in the inductance moves back and forth between the inductance and the capacitance of the electrical system, and finally is dissipated in the resistive elements of the system, giving rise to damped oscillatory voltage waves.

The switching surge, like lightning, is also a random phenomenon to some extent. The severity of a switching surge depends upon the instant of the power frequency voltage when the switch is opened. It depends also upon the dynamic characteristics of the arc which is drawn when the switch contacts separate. However, unlike lightning, the switching surge can be controlled at the source by proper design of the switch. Switching surges are generally associated with long front times (e.g., several hundred microseconds) and tail times (e.g., several milliseconds). However, certain types of switch (e.g., vacuum switches, SF_6 switches) are known to produce voltage transients in the nanoseconds regime.

The switching surge plays a prominent role in the operation of EHV (extra high voltage) and UHV (ultra high voltage) power systems whose operating voltage exceeds 345 kV. The basic impulse insulation level (BIL) of a power system at these voltage levels is high enough to withstand many direct lightning strokes. However, it has been found that the switching surge withstand level is dangerously low when the front time is within a particular band (between 100 to 300 microseconds) - sometimes very near the peak of the operating voltage. The development of new kinds of circuit breakers (e.g., vacuum and SF_6) has caused the switching surge problem to be more severe, although the fault-current interruption capability of the circuit breakers has improved significantly.

A third potential source of electrical transients is the detonation of nuclear devices at high altitudes (e.g., 40 km and higher) above the earth's surface. These transients are called HEMP (high-altitude electromagnetic pulse), or alternatively NEMP (nuclear electromagnetic pulse). Such high-altitude nuclear detonation does not produce the other effects with which nuclear explosions are associated, such as heat, shock waves or radioactive fallout. For analysis, HEMP can be broken into three parts [1]: E_1, E_2 and E_3. E_1 is a steep-front short-duration (SFSD) transient electric field which rises to its peak of tens of kV per meter in less than 10 nanoseconds, and decays to near zero in less than a microsecond. E_1 is followed by intermediate-type EMP or E_2, which has an amplitude on the order of a hundred volts per meter over a time period from about one microsecond to 0.1 second. A late-time EMP, called E_3 or magnetohydrodynamic EMP (MHD-EMP), results from geomagnetic perturbations caused by the high-altitude nuclear detonation. MHD-EMP is a low-magnitude EMP on the order of 10 volts per km which occurs about 0.1 second after the blast, and lasts from a few seconds to many tens of seconds. E_1 and E_3 are significant for the electric power systems because of their strong influence and large areas of coverage.

EMP, although biologically harmless, can play havoc with communications, as well as electronic and electrical systems, by inducing current and voltage surges through electrically conducting materials [2]. In some cases, the surges trip circuit breakers, shutting down a piece of equipment or power line. In other cases, individual components, such as semiconductor devices or circuits are destroyed. Results of extensive analyses on EMP are available [3]. However, actual test results are classified.

1.2 Basic Mathematical Concepts for Transient Analysis

1.2.1 Laplace transform

Laplace transform converts a set of differential equations into a set of algebraic equations. A solution is sought for the set of simultaneous algebraic equations. Then the final result is transformed back to the time and spatial coordinates by inverse transform. In most cases, it is relatively simple to invert the s-domain solution in the Laplace transform. In some cases, numerical techniques are required [4,5]. An inversion routine, called INLAP, is also available in the IMSL software package [6].

Laplace transform is a subject by itself. Therefore, it will not be discussed here; it will be assumed that the reader is familiar with the operation of the Laplace transform.

1.2.2 Representation of transient waveshape

A transient current or a voltage wave is generally represented by a double exponential function (Fig. 1.1a):

$$A(t) = A_o(e^{-a_1 t} - e^{-a_2 t}) = A_1(t) - A_2(t)\,, \tag{1.1}$$

where A(t) is the transient function (e.g., current or voltage), and A_o, a_1 and a_2 are constants which determine the front time, the time to half value and the peak magnitude.

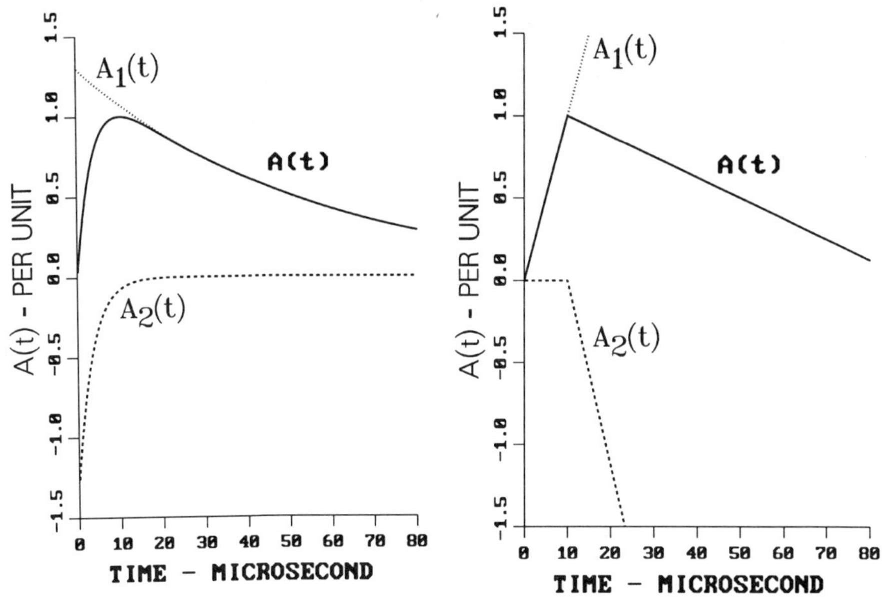

(a) (b)

Fig. 1.1 Representation of a double exponential transient function.

In Laplace transform: $$A(s) = A_o\left(\frac{1}{s+a_1} - \frac{1}{s+a_2}\right). \qquad (1.1a)$$

The double exponential function is sometimes simplified by a linearly rising and decaying function (Fig. 1.1b):

$$A(t) = \alpha_1 t u(t) - \alpha_2 (t-t_f) u(t-t_f) = A_1(t) + A_2(t) , \qquad (1.2)$$

$$\text{where} \quad \alpha_1 = \frac{A_o}{t_f} , \quad \alpha_2 = \frac{2t_H - t_f}{2t_f\,(t_H - t_f)} A_o , \qquad (1.3)$$

t_f = front time; t_H = time of decay to half value, and A_o = peak value.

$u(t) = 0$ for $t < 0$

$= 1$ for $t \geq 0$

$u(t - t_f) = 0$ for $t < t_f$

$= 1$ for $t \geq t_f$

In Laplace transform: $A(s) = \frac{\alpha_1}{s^2} - \frac{\alpha_2}{s^2} e^{-t_f s}$. (1.2a)

Sometimes the analysis is simplified greatly if the transient function is represented by a unit step function (Fig. 1.2):

$$A(t) = u(t) . \quad (1.4)$$

In Laplace transform: $A(s) = \frac{1}{s}$. (1.4a)

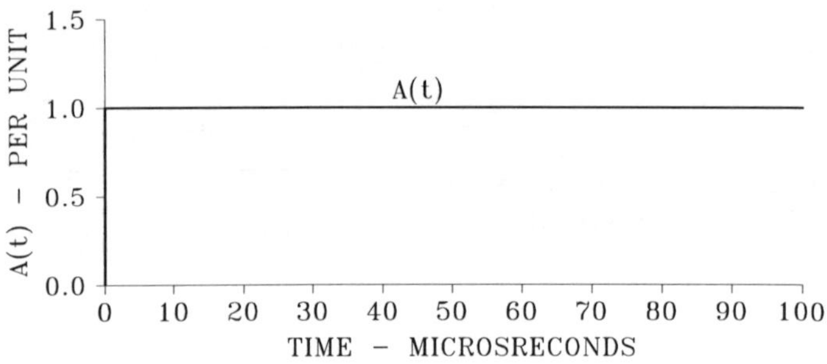

Fig. 1.2 Representation of a transient by step function.

The result obtained with the unit step-function input can be converted to a response corresponding to any other input function by the superposition principle (convolution or Duhamel integral).

1.2.3 Superposition principle - Duhamel integral [7-9]

In a linear system, the response is linearly proportional to the stimulus. In other words, in a linear system, if the stimulus S_1 produces a response R_1, and a stimulus S_2 produces a response R_2, then a stimulus $S = S_1 + S_2$ will produce a response $R = R_1 + R_2$.

If A(t) is a stimulus (voltage, current or whatever), it can be approximated by a series of step-function waves. The resultant response of the circuit will be the sum of the responses due to each of these input step functions.

For instance, if a unit step-function voltage, **1**, is applied to a circuit with an operational impedance Z(s), then the current will be

$$I(s) = \frac{1}{sZ(s)} \cdot \tag{1.5}$$

1 is the stimulus, and I(s) is the response. The response to a unit step function is called the step response, $u_r(t)$. In the example,

$$u_r(t) = \mathcal{L}^{-1}\frac{1}{sZ(s)} \cdot \tag{1.6}$$

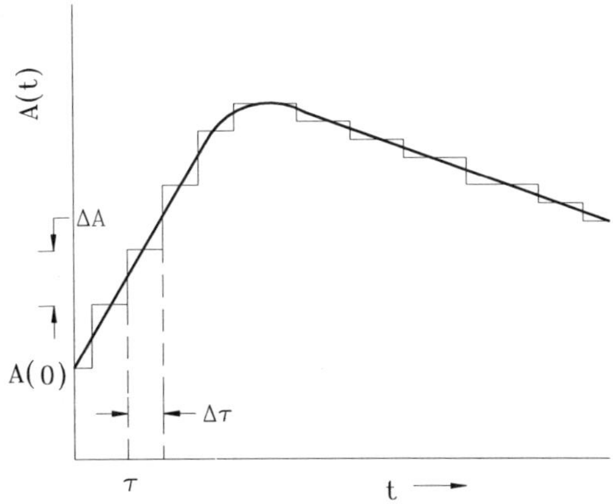

Fig. 1.3 Approximation of an arbitrary function by a series of delayed step functions.

For the function shown in Fig. 1.3, a step input of A(0) will have a response $A_0 \cdot u_r(t)$. To this, we must add the responses due to the series of step inputs as shown. If $\Delta\tau$ is the time interval chosen, then the height of each step input will be

$$\frac{\partial A}{\partial t}\Delta\tau = A'(\tau)\Delta\tau \tag{1.7}$$

at the instant τ. The response of this input will be $A'(\tau)\Delta\tau u_r(t - \tau)$ when the input was given at $t = \tau$. The resultant response, R(t), to A(t) will then be the summation of all the individual responses:

$$R(t) = A(0)u_r(t) + \int_0^t A'(\tau)u_r(t - \tau)d\tau. \tag{1.8}$$

This is the Duhamel integral, where $u_r(t)$ = unit step response, and R(t) = response due to the input A(t).

Duhamel integral has several forms in addition to that given in (1.8):

$$R(t) = A(0)u_r(t) + \int_0^t u_r(\tau)A'(t - \tau)d\tau, \tag{1.8a}$$

$$R(t) = u_r(0)A(t) + \int_0^t u_r'(\tau)A(t - \tau)d\tau, \tag{1.8b}$$

$$R(t) = u_r(0)A(t) + \int_0^t A(\tau)u_r'(t - \tau)d\tau, \tag{1.8c}$$

$$R(t) = \frac{d}{dt}\int_0^t A(\tau)u_r(t - \tau)d\tau, \tag{1.8d}$$

$$R(t) = \frac{d}{dt}\int_0^t u_r(\tau)A(t - \tau)d\tau. \tag{1.8e}$$

1.3 References to Chapter 1

1. P.R. Barnes, "EMP Research on Electric Power Systems Program Update", Report No. ORNL/M-1392, Oak Ridge National Laboratory, May 1991.

2. ***IEEE Guide for the Installation of Electrical Equipment to Minimize Electrical Noise Inputs to Controllers from External Sources.*** ANSI/IEEE Std 518-1982. The Institute of Electrical and Electronics Engineers, Inc.

3. K.S.H. Lee, Editor, ***EMP Interaction: Principles, Techniques, and Reference Data.*** Hemisphere Publishing Company, 1986.

4. R. Bellman, R.E. Kalaba and J.A. Lockett, ***Numerical Inversion of the Laplace Transform: Applications to Biology, Economics, Engineering, and Physics.*** American Elsevier Publishing Company, Inc., New York, 1966.

5. A. Ahmed, P. Johannet and P. Auriol, "Efficient Inverse Laplace Transform Algorithm for Transient Overvoltage Calculation", ***IEE Proc.-C***, Vol.139, No.2, pp.117-121, March 1992.

6. ***User's Manual FORTRAN subroutines for mathematical applications***, Math/Library, Version 2.0, September 1991.

7. M.F. Gardner and J.L. Barnes, ***Transients in Linear Systems***. John Wiley, New York, 1950.

8. L.V. Bewley, ***Traveling Waves on Transmission Systems***, Second Edition. John Wiley, New York, 1951.

9. A. Greenwood, ***Electrical Transients in Power Systems***, Second Edition. John Wiley, New York, 1991.

Problems for Chapter 1

1.

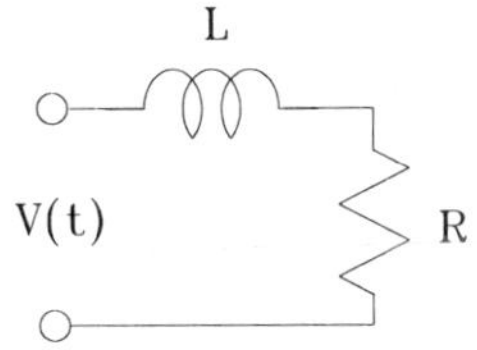

For the circuit shown above, find the voltage across the resistor, when the applied voltage, V(t), is a 1.2/50-μs wave given by:

$$\text{(a)}\quad V(t) = 1.0223\,(e^{-a_1 t} - e^{-a_2 t}),$$

where $a_1 = 1.46\text{x}10^4$, and $a_2 = 4.47\text{x}10^6$.

$$\text{(b)}\quad V(t) = \alpha_1 t u(t) - \alpha_2 (t - t_f) u(t - t_f),$$

where the peak value of the voltage is 1.0.
Solve by (i) using Duhamel integral, and (ii) applying the voltage V(t) directly.

2. For the circuit of problem 1, if L = 10 μH and R = 100 Ω, plot the voltage across the resistor for both the input waveforms upto t = 10 μs.

CHAPTER 2
WAVE PROPAGATION

Voltage and current propagate along a conductor with a finite velocity like waves on a water surface. Just like water waves, the voltage and current waves are reflected at and transmitted across a point of discontinuity. Under normal circumstances, the effects of the propagation are not felt in an electric circuit, because the travel time (at a speed of 300,000 kilometers, or 186,000 miles, per second) around the circuit is negligible compared with the rise time (time to crest) of the voltage or current waves. However, where the conducting path is more than several hundred meters long, such as overhead lines or underground cables in an electric power system, the disastrous effects of voltage build-up from short-duration transients (i.e. microseconds regime), such as those caused by lightning, can be felt. It is essential, therefore, to understand the principles of propagation of transient voltage and current waves along power lines (overhead or underground) in order to assess the severity of voltage buildup and to mitigate their effects on terminal apparatus, such as insulators, transformers, circuit breakers, motors and generators. Such understanding can easily be achieved by means of the wave equation.

2.1 Wave Equation

A lossless and uniform transmission line can be represented by a uniformly distributed inductance, shunted by a uniformly distributed capacitance, as shown in Fig. 2.1.

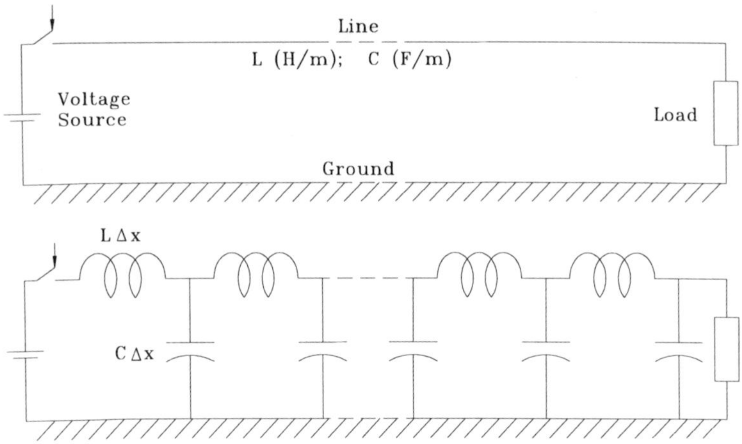

Fig. 2.1 Schematic of a lossless transmission line.
L = self inductance of line per meter
C = capacitance to ground of line per meter

The voltage drop across an elemental length **Δ**x of the line is given by:

$$\Delta V = -L\,\Delta x \frac{\partial I}{\partial t}, \tag{2.1}$$

$$\text{For } \Delta x \to 0, \quad \frac{\partial V}{\partial x} = -L \frac{\partial I}{\partial t}, \tag{2.2}$$

$$\text{Differentiating (2.2),} \quad \frac{\partial^2 V}{\partial x^2} = -L \frac{\partial^2 I}{\partial x\,\partial t}. \tag{2.3}$$

The current through the capacitance to ground for an electrical length **Δ**x of the line is given by:

$$-\Delta I = C\,\Delta x \frac{\partial V}{\partial t}. \tag{2.4}$$

$$\text{This gives:} \quad \frac{\partial^2 I}{\partial x\,\partial t} = -C \frac{\partial^2 V}{\partial t^2}. \tag{2.5}$$

Combining (2.3) and (2.5): $$\frac{\partial^2 V}{\partial t^2} = v^2 \frac{\partial^2 V}{\partial x^2}\ . \tag{2.6}$$

It can be similarly shown that:

$$\frac{\partial^2 I}{\partial t^2} = v^2 \frac{\partial^2 I}{\partial x^2}\ , \tag{2.7}$$

where $$v^2 = \frac{1}{LC}, \tag{2.8}$$

$$L = 2\times 10^{-7}\ell n\frac{2h}{r},\ (\text{H/m}),\ C = \frac{k\,10^{-9}}{18\,\ell n\dfrac{2h}{r}},\ (\text{F/m}), \tag{2.9}$$

h = height of line above ground, m,
r = radius of line conductor, m, and
k = dielectric constant of the medium (k=1 for air).

For a bundle-conductor line, r in (2.9) is replaced by the equivalent radius of the line conductor:

$$r_{eq} = [n r_{sc} A^{\,n-1}]^{1/n}\ , \tag{2.10}$$

where n=number of subconductor per phase, r_{sc}=radius of a subconductor, and A=bundle radius. The inductance and the capacitance of a coaxial cable are given by:

$$L = 2\text{x}10^{-7}\ell n\frac{b}{a},\ (\text{H/m})\ \text{and}\ C = \frac{k\,10^{-9}}{18\,\ell n\dfrac{b}{a}},\ (\text{F/m})\ , \tag{2.11}$$

where a,b = outer radius of the inner conductor and the inner radius of the outer coaxial conductor. Equations (2.6) and (2.7) are called the wave equations. The solution of (2.6) takes the following alternate forms.

$$V(x,t) = f_1(t-\frac{x}{v})+f_2(t+\frac{x}{v}) = V_f + V_r\ . \tag{2.12a}$$

$$\text{Or,}\ V(x,t) = f_1(x - vt) + f_2(x + vt) \ = V_f + V_r. \tag{2.12b}$$

The functions f_1 and f_2 are called the forward-moving and the backward-moving (reverse) voltage waves which travel with a velocity, v. f_1 and f_2 can be evaluated from the initial and boundary conditions of the line. The wave equation for the current in (2.7) can be similarly solved.

2.2 Velocity of Traveling Waves

From (2.8) and (2.12), it is evident that the velocity of the traveling wave is given by:

$$v = \frac{1}{\sqrt{LC}}. \tag{2.13}$$

Replacing L and C from (2.9),

$$v = \frac{3\times10^8}{\sqrt{k}} \quad \text{meters/second.} \tag{2.14}$$

In air (k=1), v=300,000 km/s, which is the velocity of light in free space. In underground cables, where k is greater than one, the traveling waves have lower velocities.

2.3 Relation between the Voltage and Current Waves

From (2.2) and (2.12),

$$\frac{\partial I}{\partial t} = -\frac{1}{L}\frac{\partial V}{\partial x} = -\frac{1}{L}\frac{\partial}{\partial x}[f_1(t-\frac{x}{v})+f_2(t+\frac{x}{v})].$$

$$\text{Or,}\ \frac{\partial I}{\partial t} = \frac{1}{Lv}[f_1'(t-\frac{x}{v})-f_2'(t+\frac{x}{v})]$$

$$= \sqrt{\frac{C}{L}}[f_1'(t-\frac{x}{v})-f_2'(t+\frac{x}{v})], \tag{2.15}$$

where the 'primed' functions are differentiation with respect to x.
Integrating with respect to t,

$$I=\sqrt{\frac{C}{L}}\left[f_1(t-\frac{x}{v})-f_2(t+\frac{x}{v})\right]=\frac{V_f}{Z}-\frac{V_r}{Z}=I_f+I_r\,, \qquad (2.16)$$

$$\text{where}\quad Z=\sqrt{\frac{L}{C}}\quad \Omega. \qquad (2.17a)$$

Putting the values of L and C from (2.9),

$$Z=\frac{60}{\sqrt{k}}\ell n\frac{2h}{r}\quad \Omega. \qquad (2.17b)$$

For the coaxial cable:

$$Z=\frac{60}{\sqrt{k}}\ell n\frac{b}{a}\quad \Omega. \qquad (2.17c)$$

Z is called the surge (or characteristic) impedance of the line. Equation (2.16) shows that the forward-moving current, I_f, is directly proportional to V_f. The relationship between V_r and I_r is similar, but ***with the sign reversed***. Figure 2.2 shows the schematic relationships between the forward and reverse voltage and current waves.

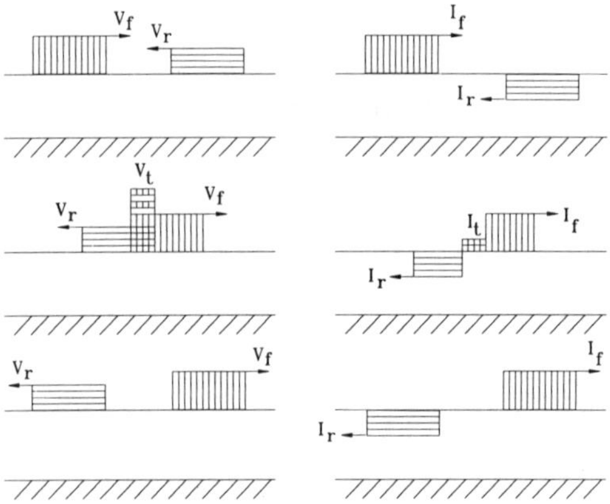

Fig. 2.2 Schematic representation of the relationships between forward and reverse voltage and current waves.

2.4 Line of Finite Length: Point of Discontinuity

A power line may terminate to a resistor, reactor, transformer, cable, capacitor, an open circuit breaker, or a bus which may be connected to several other lines. Under these conditions, a part of the forward-moving voltage and current waves will be reflected back at the point of discontinuity, and a part of these waves will be transmitted forward through the point of discontinuity.

In general then, the voltage and current waves at any point of a line of finite length are the algebraic sum of the respective forward and reflected waves. In other words,

$$V_t = V_f + V_r \,, \quad I_t = I_f + I_r \,, \quad \text{and} \tag{2.18}$$

$$I_t = \frac{V_f}{Z} - \frac{V_r}{Z} \,, \tag{2.19}$$

where the subscripts f, r and t denote the forward, reflected and transmitted (or total) waves. Equations (2.18) and (2.19) corroborate (2.12) and (2.16). From (2.19):

$$ZI_t = V_f - V_r \,. \tag{2.20}$$

From (2.18) and (2.20),

$$V_t + ZI_t = 2V_f \,. \tag{2.21}$$

Once the terminal conditions are known, then the reflected and the transmitted components can be determined in terms of the forward component. For instance, if the line is terminated by a resistor R (Fig. 2.3), then at the terminal, V_t=RI_t. Putting this value of I_t in (2.21),

$$V_t = \frac{2R}{R+Z} V_f = a_t V_f \,, \tag{2.22}$$

where a_t=2R/(R+Z) is called the coefficient of voltage transmission.

From (2.18) and (2.22),

$$V_r = \frac{R-Z}{R+Z} V_f = a_r V_f \,, \tag{2.23}$$

where a_r=(R-Z)/(R+Z) is the coefficient of voltage reflection.
Similarly, for the current waves,

$$I_t = \frac{2Z}{R+Z} I_f = b_t I_f \,, \quad \text{and} \tag{2.24}$$

$$I_r = -\frac{R-Z}{R+Z} I_f = b_r I_f \,. \tag{2.25}$$

Figure 2.3 shows the various voltage and current waves on a line terminated by a resistor. Figure 2.4 shows the same but for a line terminated by a second line.

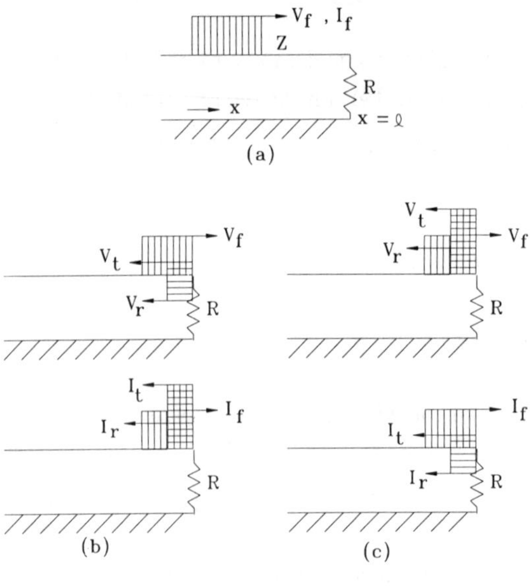

Fig. 2.3 Voltage and current waves on a line terminated by a resistor.
(a) schematic of the line; (b) Z > R; (c) Z < R.

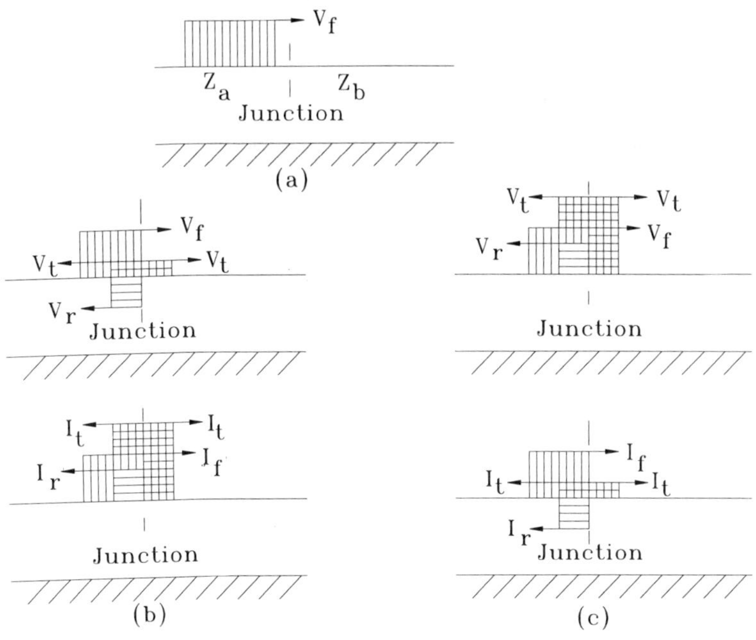

Fig. 2.4 Voltage and current waves on a line terminated by a second line.
(a) schematic of the lines; (b) $Z_a > Z_b$; (c) $Z_a < Z_b$.

2.5 Examples of Line Terminations

2.5.1 Line terminated by its surge impedance: R = Z

If the line is terminated by its surge impedance, then, from (2.22) to (2.25),

$a_t = 1,\ a_r = 0,\ b_t = 1,\ b_r = 0.$

This means that there is no reflection at the line terminal; the line behaves like a line of infinite length.

2.5.2 Open-circuited line

Putting $R = \infty$ in (2.22) to (2.25),

$a_t = 2$, $a_r = 1$, $b_t = 0$, $b_r = -1$.

This means that the voltage at the line end becomes double the forward voltage wave V_f, and the current there, as expected, becomes zero. If a line is terminated by a resistor which is significantly higher than its surge impedance, then the voltage at the termination will be almost double the forward voltage wave. As an example, the voltage at the junction between an overhead line and an underground cable will be almost double if the forward voltage wave is moving along the cable towards the overhead line, because the surge impedance of an overhead line is much higher than that of an underground cable. The voltage at the junction will be reduced significantly if the forward voltage propagates along the overhead line towards the underground cable. This will be clear later.

2.5.3 Short-circuited line

Putting $R = 0$ in (2.22) to (2.25),

$a_t = 0$, $a_r = -1$, $b_t = 2$, $b_r = 1$.

The line terminal voltage will be zero, as expected of a short-circuited line. A negative reflected voltage wave will start to travel back to the line, thus neutralizing the forward voltage wave. The terminal current will, however, double and a positive reflected current wave will travel back to the line, thus increasing the line current. If the overhead line is terminated by an underground cable, then the voltage transmitted along the cable from the overhead line will be significantly reduced. If the surge impedances of the overhead line and the underground cable are 500 and 50 ohms respectively (Z_{ol}=500 Ω, Z_{uc}=50 Ω), then a_t=0.182. If a 1000-kV transient voltage were propagating along the overhead line towards the cable, then the voltage transmitted along the cable would be only 182 kV. However, if the cable is short and is terminated at the other end by a high impedance, then the successive reflections at the two cable terminals will raise the voltage level to almost double the original forward voltage on the overhead line. The effect of the cable will be only to reduce the steepness of the oncoming traveling voltage wave. This will have an important bearing in the protection of cable terminal equipment. This will be dealt with later.

2.6 Line Terminated by a Complex Impedance

If the line is terminated by a complex impedance, such as an RLC network, then the propagation characteristics of the oncoming voltage and current waves can be determined by the use of Laplace transform. If the Laplace transform of the terminal impedance is Z_s, then (2.22) and (2.23), in Laplace transform, can be written as

$$V_t(s) = \frac{2Z_s}{Z_s+Z} V_f = a_t(s)V_f(s) \text{ , and} \tag{2.26}$$

$$V_r(s) = \frac{Z_s-Z}{Z_s+Z} V_f(s) = -a_r(s)V_f(s) \text{ .} \tag{2.27}$$

If, for instance, the line is terminated by an inductor, then Z_s=Ls. Further, if the oncoming voltage wave is a step function, then $V_f(s)=V_f/s$,

$$V_t = 2V_f\, e^{\frac{-Zt}{L}} \text{ , and} \tag{2.28}$$

$$V_r = V_f\,(-1+2e^{\frac{-Zt}{L}}) \text{ .} \tag{2.29}$$

The voltage across the terminal inductor will be $2V_f$ at t=0, and will approach zero exponentially. In other words, the inductor will act like an open circuit initially, but will behave as a short circuit at infinite time.

Similarly, for a line terminated by a capacitor, V_t and V_r are given by

$$V_t = 2V_f\,(1-e^{\frac{-t}{ZC}}) \text{, and} \tag{2.30}$$

$$V_r = V_f\,(1-2e^{\frac{-t}{ZC}}) \text{.} \tag{2.31}$$

The capacitor will behave like a short circuit at t=0, and will act like an open circuit at infinite time.

2.7 Application of Thevenin's Theorem

Thevenin's theorem can also be applied to determine the reflected and the transmitted waves at a transition point. Sometimes it may even be convenient. According to Thevenin's theorem, an active network, containing voltage sources, connected to an external passive impedance network can be represented by a voltage source, V_n, in series with an internal impedance, Z, (Fig. 2.5a). V_n is determined by disconnecting the external passive impedance network, Z_ℓ, from the active network, and measuring the voltage across the open-circuited terminals of the active network. The internal impedance, Z, is determined by short-circuiting the internal voltage sources and measuring the impedance across the open-circuited terminals of the active network. The current through and the voltage across Z_ℓ can then be calculated.

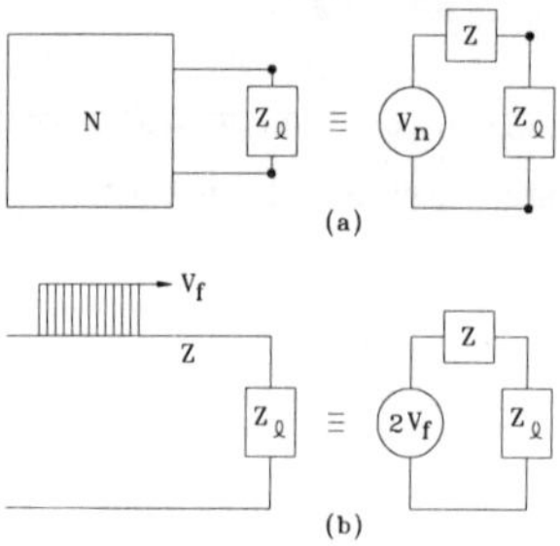

Fig. 2.5 Thevenin's equivalent of a terminated line.

Applying Thevenin's theorem to a line terminated by a complex impedance, Z_ℓ (Fig. 2.5b), disconnecting Z_ℓ from the line will produce a voltage $2V_f$ across the open-circuited line terminals. As the surge impedance of the line, Z, is known, the current through Z_ℓ is given by

$$I_t = \frac{2V_f}{Z+Z_\ell}. \tag{2.32}$$

Terminal voltage across Z_ℓ, $V_t = I_t Z_\ell = \dfrac{2Z_\ell}{Z+Z_\ell} V_f$. (2.33)

Reflected voltage from Z_ℓ, $V_r = V_t - V_f = \dfrac{Z_\ell - Z}{Z_\ell + Z} V_f$. (2.34)

Equations (2.33) and (2.34) are the same as (2.26) and (2.27).

2.8 General Transition Point

Figure 2.6 shows the case for a general transition point. The impedances $z_1(s)$,.....$z_n(s)$ are the lumped impedances, expressed in Laplace transform, connected in series with the lines of surge impedances Z_1,.....Z_n. For instance, these lumped impedances could be current-limiting reactors connected in series with the various lines. The impedance $z_g(s)$ is connected from the transition point to the ground. An example of $z_g(s)$ is a capacitor bank or a surge suppressor (lightning arrester).

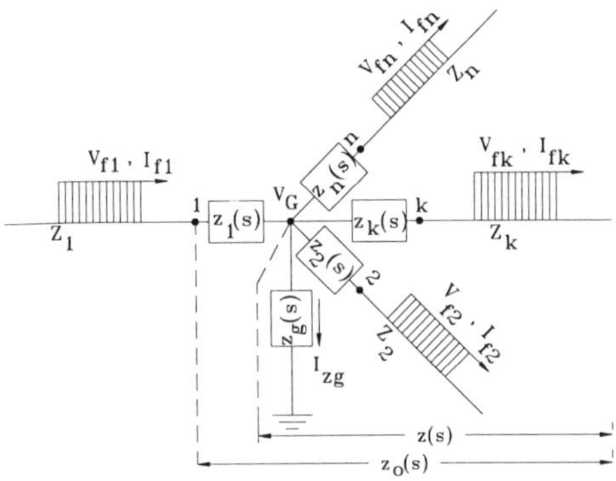

Fig. 2.6 General transition point.

The following analysis will be performed in Laplace transform. However, the Laplace transform operator, s, will be omitted in the notations for the sake of simplicity. From Fig. 2.6,

$$V_{f1} = Z_1 I_{f1} , \tag{2.35}$$

$$V_{r1} = -Z_1 I_{r1} , \text{ and} \tag{2.36}$$

$$V_{fk} = Z_k I_{fk} . \tag{2.37}$$

Recalling (2.18) and (2.19), the total voltage and current at the transition point 1 are:

$$V_{t1} = V_{f1} + V_{r1} = z_o I_{t1} , \text{ and} \tag{2.38}$$

$$I_{t1} = I_{f1} + I_{r1} . \tag{2.39}$$

From (2.38) and (2.39),

$$V_{r1} = \frac{z_o - Z_1}{z_o + Z_1} V_{f1} \ ; \ V_{t1} = \frac{2z_o}{z_o + Z_1} V_{f1} \ . \tag{2.40}$$

$$I_{r1} = -\frac{z_o - Z_1}{z_o + Z_1} I_{f1} \ ; \ I_{t1} \frac{2Z_1}{z_o + Z_1} \ . \tag{2.41}$$

Now, $z_o = z_1 + z$

$$\text{or,} \quad z_o = z_1 + \frac{1}{\dfrac{1}{z_g} + \sum \dfrac{1}{z_k + Z_k}} \ . \tag{2.42}$$

$$\text{Therefore,} \quad V_G = zI_{t1} = \frac{2z}{z_o + Z_1} V_{f1} \ , \tag{2.43}$$

$$I_{zg} = \frac{V_G}{z_g} = \frac{2z}{z_g(z_o + Z_1)} V_{f1} \ , \tag{2.44}$$

$$I_{fk} = \frac{V_G}{z_k + Z_k} = \frac{2z}{(z_o + Z_1)(z_k + Z_k)} V_{f1} \ , \ \text{and} \tag{2.45}$$

$$V_{fk} = Z_k I_{fk} = \frac{2z}{z_o + Z_1} \frac{Z_k}{z_k + Z_k} V_{f1} \ . \tag{2.46}$$

The potential drops across the lumped series impedances are:

$$V_1 = z_1 I_{t1} = \frac{2z_1}{z_o + Z_1} V_{f1} \ , \ \text{and} \tag{2.47}$$

$$V_k = z_k I_{fk} = \frac{2z}{z_o + Z_1} \frac{z_k}{z_k + Z_k} V_{f1} \ . \tag{2.48}$$

2.9 Multiple Reflections: Lattice Diagram

Thus far, only one junction between lines has been discussed. The lines were considered to be semi-infinite. However, in actual electrical networks, both ends of a line are separated by a finite distance. In that case, waves reflected from one end will travel back to the other terminal of the line and be reflected back from the second terminal, creating successive reflections at the two terminals. At each junction, the waves will also be repeatedly transmitted across the junction to the contiguous lines and, in turn, will produce successive reflections on these contiguous lines. Such multiple reflections may build up dangerous overvoltages at certain points of the network. Sometimes it is exceedingly difficult to keep track of the multiplicity of these successive reflections. Bewley [1] developed a lattice or space-time diagram which shows the position and direction of motion of every incident, reflected and transmitted wave on the system at every instant of time.

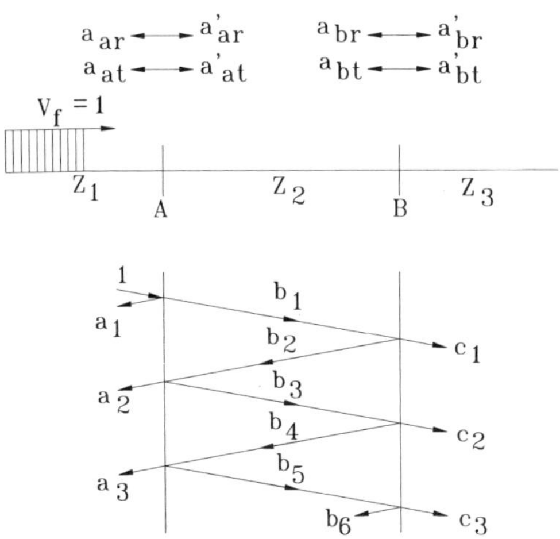

Fig. 2.7 The lattice diagram showing multiple reflections.

$a_1=a_{ar}$; $a_2=a_{at}b_2$; $a_3=a_{at}b_4$

$b_1=a'_{at}$; $b_2=a_{br}b_1$; $b_3=a'_{ar}b_2$; $b_4=a_{br}b_3$; $b_5=a'_{ar}b_4$

$c_1=a'_{bt}b_1$; $c_2=a'_{bt}b_3$; $c_3=a'_{bt}b_5$

The principle of the lattice diagram is shown in Fig. 2.7 where three lines with two junctions are shown. Lines 1 and 3 are assumed to be semi-infinite, so that the multiple reflections and transmissions only on line 2 are considered. A voltage wave travels along line 1, and goes over to line 3 through line 2. The lattice diagram shows the multiple reflections at the two junctions. To construct the lattice, the junctions are laid off to scale at intervals equal to the times of passage of the wave on each line between junctions. The circuits between junctions may either be overhead lines or underground cables having, in general, different surge impedances, velocities of propagation and attenuation factors. The advantage of laying off the junctions at intervals equal to the times of wave passage instead of to the actual lengths between the junctions is that all lines connecting the junctions have the same slope. A suitable time scale is chosen for the vertical axis. The coefficients of reflection and transmission at each junction are placed above the junction with the proper arrow to show the direction of travel.

It is convenient to assume the peak of the incident wave to the left of junction A to be unity. Starting at the origin of the initial incident wave at the upper left-hand corner of the lattice, the reflected and the transmitted components of the wave at each junction are obtained by multiplying the incident wave at that junction, arriving there from both the left and the right, by the appropriate coefficients of reflection and transmission. This is continued until the lattice is completed.

It should be observed that:

1. All waves travel downhill.

2. The position of any wave at any time is given by the time scale at the left of the lattice.

3. The resultant voltage at any point at any instant of time is the superposition of all the waves which have arrived at that point up until that instant of time, displaced in position from each other by intervals equal to the differences in their time of arrival.

4. The previous history of any wave is easily traced. In other words, one can find where the wave came from, and what other waves went into its composition.

The lattice diagram for the current waves can similarly be constructed.

2.10 Examples of Multiple Reflections

2.10.1 Charging an open-circuited line from a dc voltage source

A simple example of multiple reflections is shown in Fig. 2.8 where a voltage source of negligible internal impedance is switched on to an open-circuited line. The coefficients of reflection and transmission at the open end can be calculated from eqs. (2.22) to (2.25) by putting $R=\infty$. The coefficients at the source end are similarly computed by putting $R=0$. As shown in Fig. 2.8, the coefficient of voltage reflection at the source end, a_{r1} is -1 because the source voltage is held constant at V_o. Therefore, any voltage wave returning to the source must be neutralized by sending back a wave of opposite polarity. A voltage wave V_o starts to travel from the source towards the open end of the line at the instant the switch is closed ($t=0$). It travels with a velocity $\upsilon=1/\sqrt{(LC)}$. The voltage at the open end is zero until the time, $\tau=\ell/\upsilon$, where ℓ is the length of the line. At that instant, the open-end voltage jumps to $2V_o$, because $a_{t2}=2$. A reflected voltage V_o travels towards the source. It reaches the source at $t=2\tau$. As was mentioned above, a reflected voltage wave $-V_o$ travels towards the open end, and reaches it at $t=3\tau$. The voltage at the open end jumps by a step of $-2V_o$ at this instant. The voltage profile at the open end is shown in Fig. 2.8d. The voltages at the midpoint ($x=\ell/2$) and at the open end ($x=\ell$) are given by

$$V(x=\frac{\ell}{2}) = V_o u(t-\frac{\tau}{2}) + V_o u(t-\frac{3\tau}{2}) - V_o u(t-\frac{5\tau}{2}) - V_o u(t-\frac{7\tau}{2})\ldots \tag{2.49}$$

$$V(x=\ell) = 2V_o u(t-\tau) - 2V_o u(t-3\tau) + 2V_o u(t-5\tau) - 2V_o u(t-7\tau)\ldots \tag{2.50}$$

where the functions u(t-a) are shifted unit step functions, signifying that

$$u(t-a)=0 \text{ for } t<a, \text{ and} \tag{2.51}$$

$$u(t-a)=1 \text{ for } t\geq a. \tag{2.52}$$

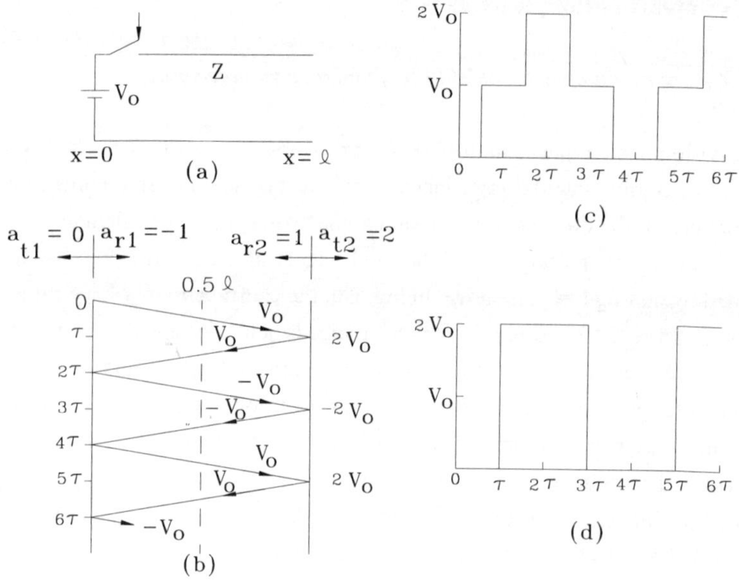

Fig. 2.8 Open-circuited line suddenly energized by dc voltage.
(a) schematic of line; (b) voltage lattice diagram;
(c) voltage at mid-point of line; (d) voltage at open end.

The voltage at the open end oscillates between $2V_o$ and zero, maintaining an average value of V_o, which a voltmeter will measure. Similarly, the current at the source end oscillates between $I(=V_o/Z)$ and $-I$, with an average value of zero which an ammeter will measure.

2.10.2 Short-circuited line suddenly switched to a dc voltage source

This case is shown in Fig. 2.9. At the instant the switch closes (t=0), a current $I_o=V_o/Z$ starts to travel from the zero-impedance voltage source towards the shorted end with a velocity υ. At a time $\tau=\ell/\upsilon$, the current wave reaches the shorted end. Until this time the current at the shorted end was zero. At this instant, the current at the shorted end jumps to $2I_o$, and a reflected current I_o travels back towards the source. This reflected current wave reaches the source at $t=2\tau$. The current at the source is I_o during the time interval $t=0$, 2τ. At $t=2\tau$, the current at the source jumps by a step of $2I_o$, making the total current there equal to $3I_o$. A reflected current, I_o, travels towards the

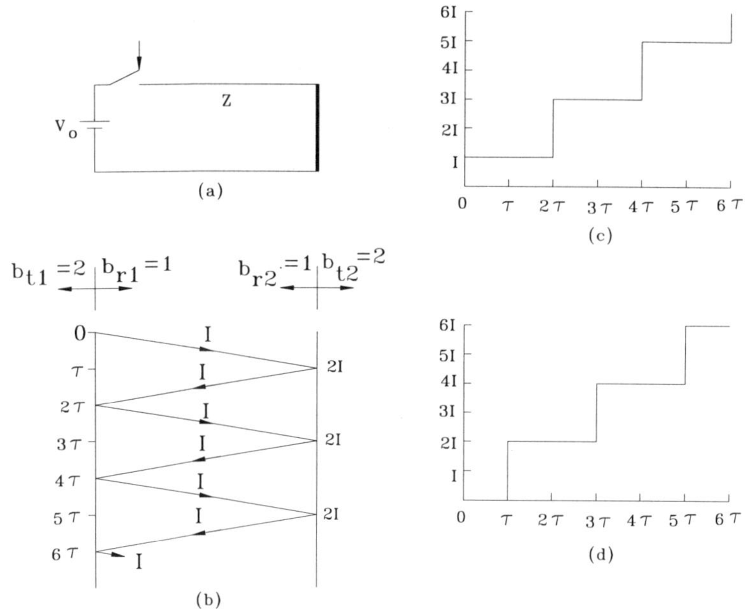

Fig. 2.9 Short-circuited line suddenly switched to dc voltage.
(a) schematic of line; (b) current lattice diagram;
(c) current at the source; (d) current at shorted end.

shorted end. Every time a current wave reaches either end of the line, the current there jumps by a step of $2I_o$. The currents at the source (at x=0) and at the shorted end (at $x=\ell$) are given by

$$I(x=0)=I_o u(t)+2I_o u(t-2\tau)+2I_o u(t-4\tau)+\ldots \qquad (2.53)$$

$$I(x=\ell)=2I_o u(t-\tau)+2I_o u(t-3\tau)+2I_o u(t-5\tau)+\ldots \qquad (2.54)$$

2.11 References to Chapter 2

1. L.V. Bewley, ***Traveling Waves on Transmission Systems***, Second Edition. John Wiley, New York, 1951.

Problems for Chapter 2

1. Solve the wave equation:

$$\frac{\partial^2 V}{\partial t^2} = v^2 \frac{\partial^2 V}{\partial x^2} ,$$

when V(0,t) = F(t) and V(x,0) = 0. The solution should be in the following forms:

$$1.\ V(x,t) = f_1(t - \frac{x}{v}) + f_2(t + \frac{x}{v}) , \quad \text{or}$$

$$2.\ V(x,t) = f_1(x - vt) + f_2(x + vt) .$$

2. For a 300-m long line suddenly excited by a unit step-function voltage, plot

   (a) the voltages at the midpoint and the far end of the line when the line is open at the far end.

   (b) the currents at the source end and the far end of the line when the line is shorted at the far end.

3.

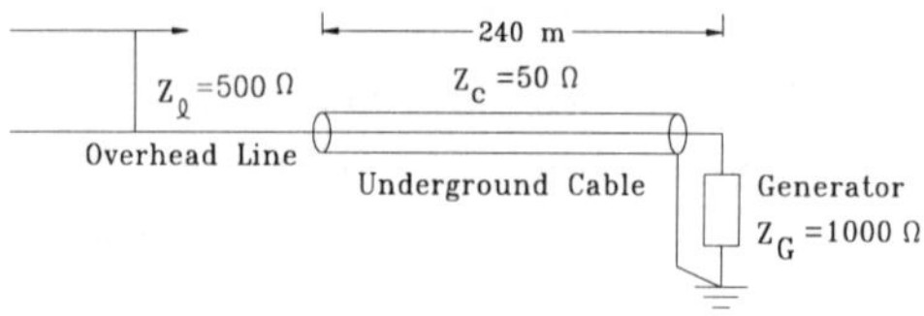

A generator is connected to an overhead line through a 240-m long underground cable, as shown above. The velocity of the traveling wave through the cable is 80 percent of the velocity on the overhead line.

   (a) Plot the voltage across the generator terminal for a time interval of 20 μs when a 1000-kV rectangular voltage wave is traveling along the overhead line towards the generator.

   (b) Calculate the final peak voltage across the generator terminal.

4.

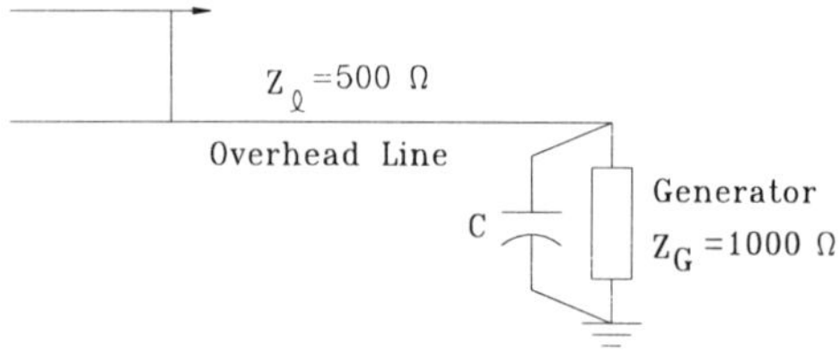

If the underground cable of problem 3 is replaced by a discrete capacitor of the same value as the total capacitance of the replaced cable across the generator terminal:

(a) Find the expression for the voltage across the generator terminal.

(b) Plot the generator terminal voltage against time.

(c) What will be the final peak generator voltage ?

5.

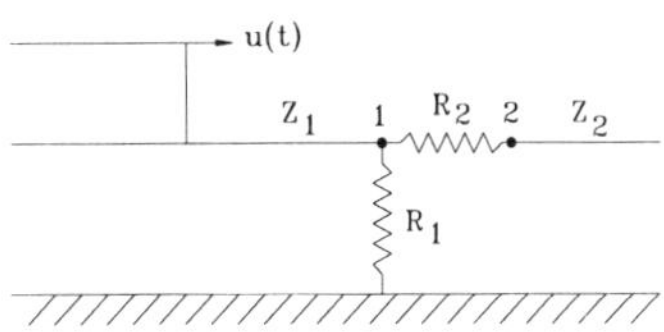

For the line configuration shown above, what are the voltages V_1 and V_2 at the junctions 1 and 2 ?

6.

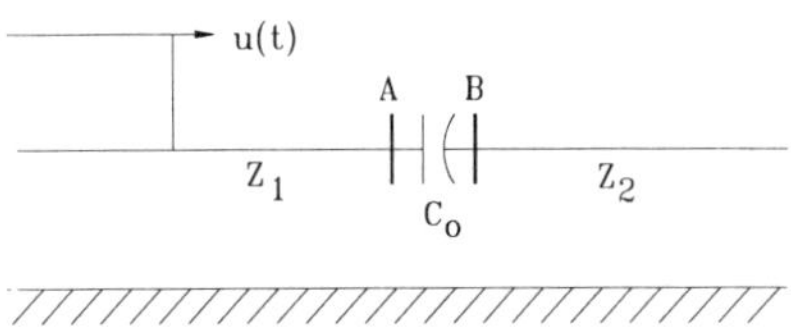

Two overhead power lines are connected to each other through a series capacitor bank, C_o, as shown above. If a unit step-function voltage is traveling on line # 1 towards line # 2, find the voltages, V_A and V_B, on the busses A and B.

7.

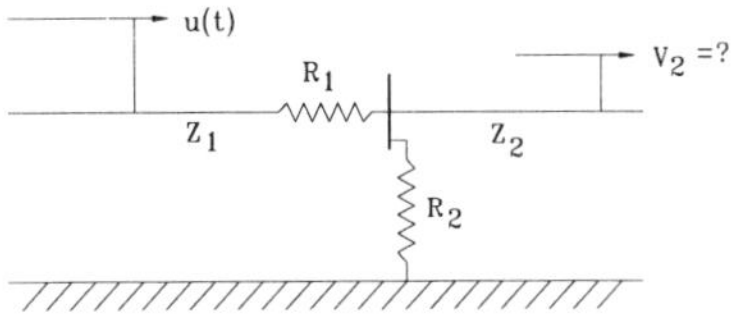

For the system shown above, find an expression for the voltage, V_2, which will be transmitted along line #2 when a unit step-function voltage, V_1 is injected on to line #1.

CHAPTER 3

WAVE PROPAGATION ON MULTICONDUCTOR SYSTEMS

3.1 Group of Conductors: Coefficients of Potential, Capacitance and Induction

By the principle of superposition, the potentials set up when charges q_1, q_2,....,q_n are simultaneously deposited on conductors 1,2,...,n are obtained by addition, e.g.,

$$
\begin{aligned}
V_1 &= p_{11}q_1 + p_{12}q_2 +..+ p_{1r}q_r +..+ p_{1n}q_n ,\\
V_2 &= p_{21}q_1 + p_{22}q_2 +..+ p_{2r}q_r +..+ p_{2n}q_n ,\\
&\cdot\\
V_r &= p_{r1}q_1 + p_{r2}q_2 +..+ p_{rr}q_r +..+ p_{rn}q_n ,\\
&\cdot\\
V_n &= p_{n1}q_1 + p_{n2}q_2 +..+ p_{nr}q_r +..+ p_{nn}q_n .
\end{aligned}
\tag{3.1}
$$

The constants p's are called the ***coefficients of potential***. In matrix form, (3.1) becomes

$$[V] = [p][q] \tag{3.2}$$

or,
$$[q] = [p]^{-1}[V]. \tag{3.3}$$

Putting, $[p]^{-1} = [C]$, (3.4)

$$
\begin{aligned}
q_1 &= C_{11}V_1 + C_{12}V_2 +..+ C_{1r}V_r +..+ C_{1n}V_n ,\\
q_2 &= C_{21}V_2 + C_{22}V_2 +..+ C_{2r}V_r +..+ C_{2n}V_n ,\\
&\cdot\\
q_r &= C_{r1}V_1 + C_{r2}V_2 +..+ C_{rr}V_r +..+ C_{rn}V_n ,\\
&\cdot\\
q_n &= C_{n1}V_1 + C_{n2}V_2 +..+ C_{nr}V_r +..+ C_{nn}V_n .
\end{aligned}
\tag{3.5}
$$

The self terms, C_{rr} (r=1 to n), are called the ***coefficients of capacitance***, and the mutual terms, C_{rs} ($r \neq s$, r,s=1 to n), are called the ***coefficients of induction***.

The important properties of the coefficients of potential, capacitance and induction are:

1. Reciprocity: $C_{rs}=C_{sr}$, and $p_{rs}=p_{sr}$.
2. All p's are positive or zero.
3. C_{rr} are positive or zero.
4. C_{rs} are negative or zero.
5. The sum $C_{r1}+C_{r2}+....+C_{rn}$ is positive or zero.

3.2 Multiconductor Line above Earth

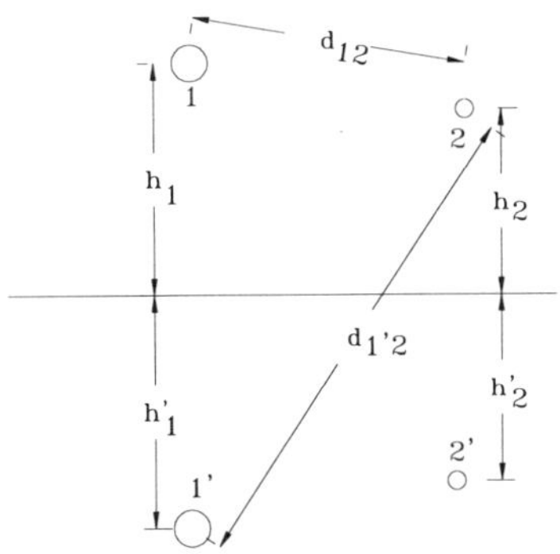

Fig. 3.1 Schematic of two line conductors above earth.

When the conductors carry charges $q_1,q_2,....$per unit length (Fig. 3.1), the electric field between the conductors and earth is the same as would be produced by line charges $q_1,q_2,....$practically at the center of the conductors 1,2,... in conjunction with charges $-q_1,-q_2,....$ along the image lines 1',2'.... . The image conductors will have the same height below the earth's surface as the height of the actual conductors ($h_1=h'_1$, $h_2=h'_2$, etc.) if the earth is assumed to be perfect, *i.e.,* earth resistivity = 0. The coefficients of potential can, then, be evaluated as follows:

$p_{11}q_1$ = potential on 1 due to charge q_1 per unit length on 1 ***plus*** $-q_1$ on its image line 1' = $\dfrac{q_1}{2\pi\epsilon_o}\ell n\dfrac{2h_1}{r_1}$.

As $\epsilon_o = 10^{-9}/36\pi$ F/m, then

$$p_{11}=18\times10^9\,\ell n\frac{2h_1}{r_1}, \tag{3.6}$$

where h_1,r_1 = height and radius, in meters, of line conductor 1.

Similarly, $p_{21}q_1$ = potential on 2 due to charge q_1 on 1 ***and*** $-q_1$ on 1'

$$= \frac{q_1}{2\pi\epsilon_o}\ell n\frac{d_{1'2}}{d_{12}}.$$

$$\text{Then,}\quad p_{21} = 18\times10^9\,\ell n\frac{d_{1'2}}{d_{12}}, \tag{3.7}$$

where d_{12} = distance between centers of 1 and 2, and
$d_{1'2}$ = distance between centers of 1' and 2.

In general, ***in air***

$$p_{rr} = 18\times10^9\,\ell n\frac{2h_r}{r_r}, \text{ and} \tag{3.8}$$

$$p_{rs} = 18\times10^9\,\ell n\frac{d_{r's}}{d_{rs}}. \tag{3.9}$$

3.3 Wave Equations

A lossless distributed-constant multiconductor line can be represented by its self inductances, L_{rr} in H/m, capacitances to ground, C_{rg} in F/m, mutual inductances, L_{rs} in H/m and mutual capacitances, C_{rs} in F/m, where the subscripts r and s stand for the rth and the sth conductors. A 2-conductor line sectionalized into Δx segments is shown schematically in Fig. 3.2.

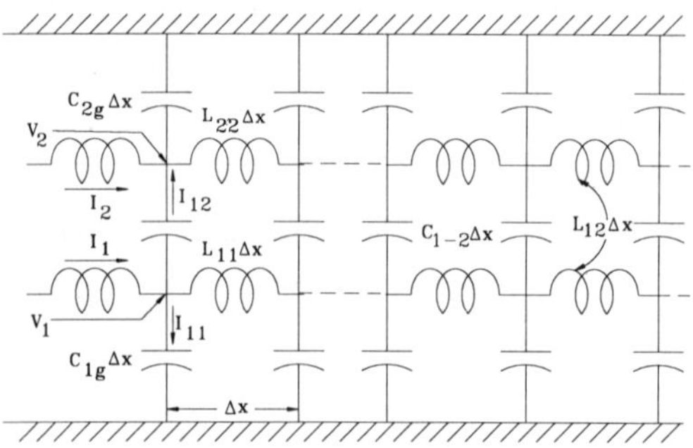

Fig. 3.2 Equivalent circuit of a two-conductor line.

The voltage drop across a section Δx of a multiconductor line is given by:

$$\Delta V_1 = -L_{11}\Delta x\frac{\partial I_1}{\partial t} - L_{12}\Delta x\frac{\partial I_2}{\partial t} - \ldots - L_{1n}\Delta x\frac{\partial I_n}{\partial t}$$

. . . .

. . . .

$$\Delta V_n = -L_{n1}\Delta x\frac{\partial I_1}{\partial t} - L_{n2}\Delta x\frac{\partial I_2}{\partial t} - \ldots - L_{nn}\Delta x\frac{\partial I_n}{\partial t}. \tag{3.10}$$

The partial differential equations, in matrix form, are given by

$$\frac{\partial[V]}{\partial x} = -[L]\frac{\partial[I]}{\partial t}. \tag{3.11}$$

For the 2-conductor system in Fig. 3.2, $I_1=I_{11}+I_{12}$. For an n-conductor system, $I_1=I_{11}+I_{12}+\cdots+I_{1n}$; $I_r=I_{r1}+I_{r2}+\cdots+I_{rn}$. Then:

$$-\frac{\partial I_1}{\partial x} = -[C_{1g}\frac{\partial V_1}{\partial t}+C_{1-2}\frac{\partial}{\partial t}(V_1-V_2)+\cdots+C_{1-n}\frac{\partial}{\partial t}(V_1-V_n)]$$

$$=(C_{1g}+C_{1-2}+\cdots+C_{1-n})\frac{\partial V_1}{\partial t}-C_{1-2}\frac{\partial V_2}{\partial t}-\cdots-C_{1-n}\frac{\partial V_n}{\partial t}$$

$$=C_{11}\frac{\partial V_1}{\partial t}+C_{12}\frac{\partial V_2}{\partial t}+\cdots+C_{1n}\frac{\partial V_n}{\partial t}, \tag{3.12}$$

where $C_{1r}=-C_{1-r}$, and $C_{11}=C_{1g}+C_{1-2}+\cdots+C_{1-n}$. In matrix form:

$$\frac{\partial[I]}{\partial x} = -[C]\frac{\partial[V]}{\partial t}. \tag{3.13}$$

Combining (3.11) and (3.13): $$\frac{\partial^2[V]}{\partial x^2} = [L][C]\frac{\partial^2[V]}{\partial t^2}, \tag{3.14}$$

where, in air: $$L_{rr}=\frac{\mu_o}{2\pi}\ell n\frac{2h_r}{r_r}=2\times10^{-7}\ell n\frac{2h_r}{r_r}\ \text{H/m}. \tag{3.15}$$

$$L_{rs} = \frac{\mu_o}{2\pi}\ell n\frac{d_{r's}}{d_{rs}} = 2\times10^{-7}\ell n\frac{d_{r's}}{d_{rs}}\ \text{H/m}. \tag{3.16}$$

From (3.8) and (3.9), the L's can also be expressed as:

$$L_{rr} = \mu_o \epsilon_o p_{rr} = \frac{p_{rr}}{v_c^2}, \text{ and} \tag{3.17}$$

$$L_{rs} = \frac{p_{rs}}{v_c^2}, \tag{3.18}$$

$$\text{where } \mu_o = 4\pi\times 10^{-7}, \text{and } v_c = \frac{1}{\sqrt{\mu_o \epsilon_o}} = 3\times 10^8 \text{ m/s}. \tag{3.19}$$

$$\text{In matrix form, } [L] = \frac{[p]}{v_c^2}. \tag{3.20}$$

$$\text{Incorporating (3.4): } [L][C] = \frac{[p][p]^{-1}}{v_c^2} = \frac{[1]}{v_c^2}, \tag{3.21}$$

where [1] = identity matrix. Therefore, ***for a perfect earth*:**

$$\frac{\partial^2 [V]}{\partial x^2} = \frac{[1]}{v_c^2}\frac{\partial^2 [V]}{\partial t^2}. \tag{3.22}$$

$$\text{For the k-th conductor: } \frac{\partial^2 V_k}{\partial x^2} = \frac{1}{v_c^2}\frac{\partial^2 V_k}{\partial t^2}. \tag{3.23}$$

This equation is similar to (2.8). Therefore, its solution is

$$V_k = f_{k1}(t - \frac{x}{v_c}) + f_{k2}(t + \frac{x}{v_c}) = V_{kf} + V_{kr}. \tag{3.24}$$

$$\text{And, } I_k = \frac{V_{kf}}{Z} - \frac{V_{kr}}{Z} = I_{kf} + I_{kr}. \tag{3.25}$$

$$\text{As } I = \frac{dQ}{dt} = \left(\frac{\partial Q}{\partial x}\right)\left(\frac{\partial x}{\partial t}\right) = q v_c, \text{ then } q = \frac{I}{v_c}. \tag{3.26}$$

$$\text{Invoking (3.2), (3.8) and (3.9),}\quad [V] = \frac{[p]}{v_c}[I] \tag{3.27}$$

$$= \frac{\sqrt{\dfrac{\mu_o}{\epsilon_o}}}{2\pi}\begin{bmatrix} \ell n\dfrac{2h_1}{r_1} & \cdot\;\cdot\;\cdot & \ell n\dfrac{d_{1'n}}{d_{1n}} \\ \cdot & \cdot\;\cdot\;\cdot & \cdot \\ \cdot & \cdot\;\cdot\;\cdot & \cdot \\ \ell n\dfrac{d_{n'1}}{d_{n1}} & \cdot\;\cdot\;\cdot & \ell n\dfrac{2h_n}{r_n} \end{bmatrix}[I]$$

$$= 60\times\begin{bmatrix} \ell n\dfrac{2h_1}{r_1} & \cdot\;\cdot\;\cdot & \ell n\dfrac{d_{1'n}}{d_{1n}} \\ \cdot & \cdot\;\cdot\;\cdot & \cdot \\ \cdot & \cdot\;\cdot\;\cdot & \cdot \\ \ell n\dfrac{d_{n'1}}{d_{n1}} & \cdot\;\cdot\;\cdot & \ell n\dfrac{2h_n}{r_n} \end{bmatrix}[I]$$

$$= \begin{bmatrix} Z_{11} & \cdot\;\cdot\;\cdot & Z_{1n} \\ Z_{21} & \cdot\;\cdot\;\cdot & Z_{2n} \\ \cdot & \cdot\;\cdot\;\cdot & \cdot \\ \cdot & \cdot\;\cdot\;\cdot & \cdot \\ Z_{n1} & \cdot\;\cdot\;\cdot & Z_{nn} \end{bmatrix}[I] = [Z][I]. \tag{3.28}$$

This means that

$$\begin{aligned} V_1 &= Z_{11}I_1+Z_{12}I_2+....+Z_{1n}I_n \\ V_2 &= Z_{21}I_1+Z_{22}I_2+....+Z_{2n}I_n \\ &\;\;.. \\ V_n &= Z_{n1}I_1+Z_{n2}I_2+....+Z_{nn}I_n \end{aligned} \tag{3.29}$$

From (3.28), $[I] = [Z]^{-1}[V] = [Y][V]$ (3.30)

An alternative way to derive (3.28) is to invoke (3.11). In Laplace transform, (3.11) becomes:

$$\frac{\partial[V(s)]}{\partial x} = -[L]\,s\,[I(s)] = -\frac{[p]}{v_c^2}\,s\,[I(s)]. \tag{3.27a}$$

Recalling that the Laplace transform of the forward voltage wave, $V(t - x/v_c)$ is $V[\exp(-sx/v_c)]$, (3.27a) is converted into:

$$-\frac{s}{v_c}[V(s)] = -\frac{s}{v_c^2}[p][I(s)], \tag{3.27b}$$

$$\text{or,}\quad [V(s)] = \frac{[p][I(s)]}{v_c} = [Z][I(s)]. \tag{3.27c}$$

By taking inverse Laplace transform:

$$[V]=[Z][I]. \tag{3.28}$$

3.4 General Solution of the Wave Equation

The wave equation for a multiconductor line, in matrix form, is given by

$$\frac{\partial^2[V]}{\partial x^2} = [L][C]\frac{\partial^2[V]}{\partial t^2} = [M]\frac{\partial^2[V]}{\partial t^2}. \tag{3.31}$$

When the earth resistivity is zero, e.g., for a perfect earth, (according to (3.21)):

$$[M] = [L][C] = \frac{[1]}{v_c^2}, \tag{3.32}$$

where v_c is the velocity of the traveling wave. This is a fortunate situation, since the wave equation for each conductor is independent of the mutual influence of the other conductors in the system, i.e., the individual wave equations are uncoupled. That is to say that the wave equation for the k-th conductor is

$$\frac{\partial^2 V_k}{\partial x^2} = \frac{1}{v_c^2}\frac{\partial^2 V_k}{\partial t^2}. \tag{3.33}$$

Equation (3.32) may not always be true. For instance, when the earth has a finite resistivity (i.e. for an imperfect earth), the image of the current goes deeper into the earth, while the image of the line charge stays where it was for a perfect earth, and [M] $\neq [1]/v_c^2$. However, for very high-frequency excitations, the depth of the current image almost coincides with that for a perfect earth. This will be true for lightning which contains high-frequency components (> 1 MHz). This may not be true for switching surges which may contain lower frequency components. Similarly, if the air breaks down in partial discharges (corona) under high transient voltage, the corona radius is used to compute the C-matrix whereas the actual conductor radius is used to compute the L-matrix, and (3.32) is again invalid.

Considering the general case when [M] is not a diagonal matrix, the evaluation of the voltage on each conductor of the multiconductor system from the mutiplicity of equations is tedious because of the coupling of each conductor with the others. One method to solve the wave equations is to uncouple the component equations by the well-known method of diagonalization of matrix. This is called the modal analysis.

3.5 Modal Analysis [1]

If we transform the conductor voltage matrix [V] by

$$[V] = [T][V_m], \tag{3.34}$$

where we define $[V_m]$ as the matrix of the modal voltages, and [T] as the transformation matrix, then (3.31) is transformed into

$$\frac{\partial^2 [T][V_m]}{\partial x^2} = [M]\frac{\partial^2 [T][V_m]}{\partial t^2}.$$

$$\text{Or,} \quad \frac{\partial^2 [V_m]}{\partial x^2} = [T]^{-1}[M][T]\frac{\partial^2 [V_m]}{\partial t^2}. \tag{3.35}$$

If the component equations are to be uncoupled, then $[T]^{-1}[M][T]$ must be diagonal, i.e.,

$$[T]^{-1}[M][T] = [\lambda], \tag{3.36}$$

where $[\lambda]$ is a diagonal matrix. The procedure is then to diagonalize [M] by a proper choice of a transformation matrix [T], such that (3.36) is true.

This is done by determining the eigenvalues of [M] from the solution of its characteristic equation, i.e.,

$$\det.\left|[M] - \lambda_i[1]\right| = 0. \tag{3.37}$$

The number of eigenvalues λ_i will be equal to the order of the matrix [M] if the roots of the characteristic equation are all distinct. Once the eigenvalues are determined, its eigenvectors can be found from the set of equations given by

$$\left[[M] - \lambda_i[1]\right][T_i] = 0, \tag{3.38}$$

where $[T_i]$ is the eigenvector corresponding to the eigenvalue λ_i. If [M] is an nxn matrix, and it has n distinct eigenvalues λ_i (i=1,2,...n), then each λ_i has an eigenvector$[T_i]_{n,1}$. The columns of the transformation (or modal) matrix [T] are the eigenvectors of [M].

The following are the four steps to solve the general wave equation of a multiconductor line:

Step 1: Find the eigenvalues λ_i of [M] from

$$\det.\left|[M] - \lambda_i[1]\right| = 0.$$

Step 2: Solve for the modal voltages from

$$\frac{\partial^2[V_m]}{\partial x^2} = [\lambda]\frac{\partial^2[V_m]}{\partial t^2},$$

where $[\lambda]$ is a diagonal matrix of the eigenvalues.

Step 3: Find the eigenvectors corresponding to each λ_i, and hence the modal matrix [T].

Step 4: Find the actual conductor-voltage matrix [V] from $[V] = [T][V_m]$.

3.6 A Simple Example of Modal Analysis

Find the eigenvalues and the eigenvectors of the 2x2 matrix

$$[M] = \begin{bmatrix} 2 & 1 \\ 1 & 2 \end{bmatrix}.$$

Set up the characteristic equation:

$$\begin{vmatrix} \lambda-2 & -1 \\ -1 & \lambda-2 \end{vmatrix} = \lambda^2 - 4\lambda + 3 = 0 .$$

$$\lambda_1, \lambda_2 = 3.0, 1.0 .$$

Set up the eigenvector equations:

$$\begin{bmatrix} \lambda-2 & -1 \\ -1 & \lambda-2 \end{bmatrix} \begin{bmatrix} T_1 \\ T_2 \end{bmatrix} = [0] .$$

$$(\lambda - 2)T_1 - 1T_2 = 0 \; ; \; - 1T_1 + (\lambda - 2)T_2 = 0 .$$

For $\lambda_1 = 3.0$, setting $T_{11} = 1.0$, then $T_{21} = 1.0$.

$$[T_1] = \frac{1}{\sqrt{T_{11}^2 + T_{21}^2}} \begin{bmatrix} T_{11} \\ T_{21} \end{bmatrix} = \frac{1}{\sqrt{2}} \begin{bmatrix} 1.0 \\ 1.0 \end{bmatrix} = \begin{bmatrix} 0.707 \\ 0.707 \end{bmatrix}.$$

Similarly, for $\lambda_2 = 1.0$, setting $T_{12} = 1.0$, then $T_{22} = -1.0$.

$$[T_2] = \frac{1}{\sqrt{T_{12}^2 + T_{22}^2}} \begin{bmatrix} T_{12} \\ T_{22} \end{bmatrix} = \frac{1}{\sqrt{2}} \begin{bmatrix} 1.0 \\ -1.0 \end{bmatrix} = \begin{bmatrix} 0.707 \\ -0.707 \end{bmatrix}.$$

Combining the two eigenvectors, the modal matrix, [T], is given by

$$[T] = \begin{bmatrix} T_{11} & T_{12} \\ T_{21} & T_{22} \end{bmatrix} = \begin{bmatrix} 0.707 & 0.707 \\ 0.707 & -0.707 \end{bmatrix}.$$

$$[T]^{-1} = \frac{\text{Adj.}[T]}{|T|} = \begin{bmatrix} 0.707 & 0.707 \\ 0.707 & -0.707 \end{bmatrix}.$$

$$[T]^{-1}[M][T] = \begin{bmatrix} 0.707 & 0.707 \\ 0.707 & -0.707 \end{bmatrix} \begin{bmatrix} 2 & 1 \\ 1 & 2 \end{bmatrix} \begin{bmatrix} 0.707 & 0.707 \\ 0.707 & -0.707 \end{bmatrix}$$

$$= \begin{bmatrix} 3.0 & 0 \\ 0 & 1.0 \end{bmatrix} = \begin{bmatrix} \lambda_1 & 0 \\ 0 & \lambda_2 \end{bmatrix}.$$

3.7 Physical Significance of Modal Analysis

The modal wave equation is given by

$$\frac{\partial^2 [V_m]}{\partial x^2} = [T]^{-1}[M][T] \frac{\partial^2 [V_m]}{\partial t^2} = [\lambda] \frac{\partial^2 [V_m]}{\partial t^2}. \tag{3.39}$$

For an n-conductor system, the matrices are of order n. Equation (3.39) signifies that for an n-conductor system, n number of modal voltages are generated which are independent of each other. Each wave travels with a velocity $\upsilon_k = 1/\sqrt{\lambda_k}$, where k=1,2,...n. The actual voltages on the conductors are given by

$$[V] = [T][V_m].$$

For a two-conductor system, only two modes exist, as shown in Fig. 3.3. For the line-to-line mode, the modal quantity (voltage, current, charge, etc.) will travel over one conductor and return through the second conductor; however, for the ground mode, the modal quantities travel over both the conductors in the same direction and return through the ground. For a three-conductor system, two line-to-line modes and a ground mode exist (Fig. 3.4).

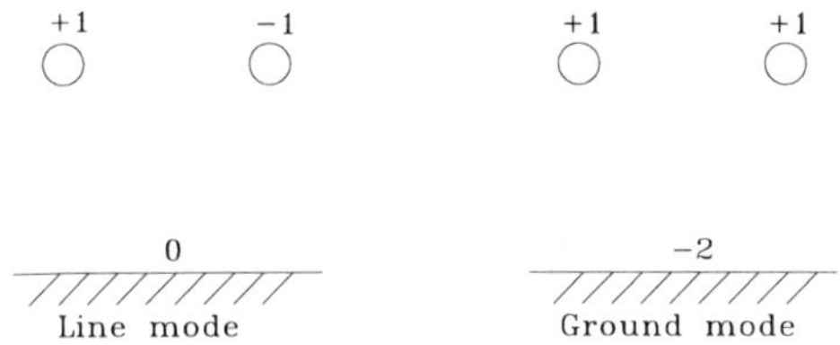

Fig. 3.3 Modes of propagation in a two-conductor system.

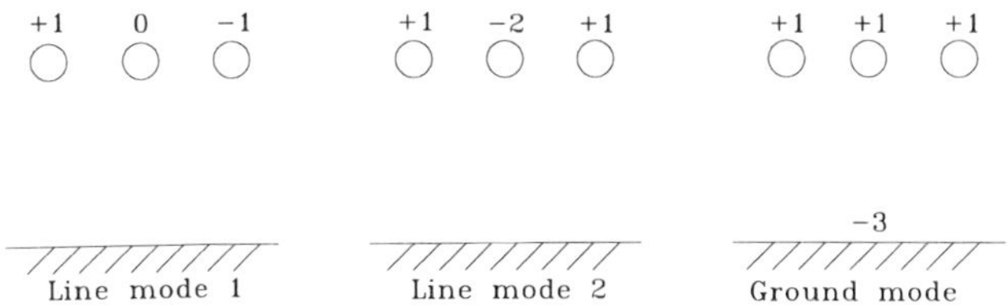

Fig. 3.4 Modes of propagation in a three-conductor system.

Similarly, for an n-conductor system, (n-1) line-to-line and one ground mode would exist. This applies to transposed systems where the impedance matrix is symmetrical and the diagonal terms are equal. Propagation through the ground would exist in more than one mode for untransposed systems.

Generally, the line-to-line modes of propagation travel at velocities very nearly equal to that of light; they also encounter less loss, and hence less attenuation and distortion than the ground mode as they travel. However, as the ground has considerably more resistance than that of the conductors, the ground mode is more attenuated and distorted, and its velocity of propagation is considerably lower than that of light.

The resistances of the conductors and the earth have been neglected in the present analysis. However, in practical systems, these resistances may play a significant role, especially for the ground-mode propagation. These resistances are not constant. They are frequency-dependent in frequency-domain phenomena, and are time-dependent in time-dependent phenomena, such as electromagnetic transients. No satisfactory solution exists, at the present time, which can include time-varying resistances. If it existed, then the wave equation would be modified. In Laplace transform:

$$\frac{\partial^2[V(s)]}{\partial x^2} = (s^2[L][C] + [R(s)][C]s)[V(s)], \tag{3.40}$$

where R(s) are the Laplace transforms of the time-varying resistances.

Although the application of modal analysis is shown here for transient phenomena, modal analysis can be applied universally where it is desired to uncouple a set of equations from each other [1]: steady-state and transient phenomena in engineering and scientific disciplines.

One application of the modal analysis is in the method of symmetrical components.

$$\begin{bmatrix} V_a \\ V_b \\ V_c \end{bmatrix} = \begin{bmatrix} 1 & 1 & 1 \\ 1 & a^2 & a \\ 1 & a & a^2 \end{bmatrix} \begin{bmatrix} V_o \\ V_1 \\ V_2 \end{bmatrix}, \quad a = e^{j120^\circ} \tag{3.41}$$

$$\text{Here, } \begin{bmatrix} 1 & 1 & 1 \\ 1 & a^2 & a \\ 1 & a & a^2 \end{bmatrix} = [T]. \tag{3.42}$$

Clarke's components (α,β and zero components) and Park's components for rotating machines (d-,q- and zero-axis components) are some of the other applications.

3.8 Transition Points in Multiconductor Systems [2,3]

3.8.1 Conductors in parallel

A transient voltage may travel along a phase conductor of a three-phase line and come across a bus from where several lines may emanate. Fig. 3.5 shows the case where a single conductor from the left terminates on a bus and two conductors originate from the bus. A transient voltage traveling along conductor 0 will continue to travel along the

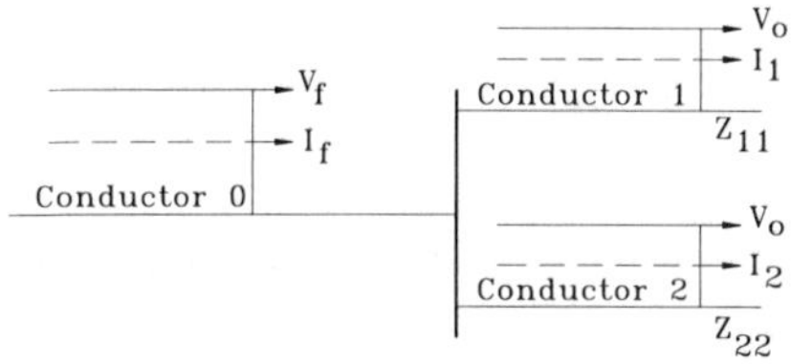

Fig. 3.5 Example of a transient traveling along parallel conductors.

conductors 1 and 2 after reaching the bus from the left. The voltages on conductors 1 and 2 will generally be the same; however, the currents in the two conductors will be different, depending on their surge impedances.

Assuming that the voltages on conductors 1 and 2 are equal to V_o, and invoking (3.29), we get

$$I_1 = V_o \frac{Z_{22}-Z_{12}}{Z_{11}Z_{22}-Z_{12}^2} = V_o \frac{\begin{vmatrix} 1 & Z_{12} \\ 1 & Z_{22} \end{vmatrix}}{\begin{vmatrix} Z_{11} & Z_{12} \\ Z_{12} & Z_{22} \end{vmatrix}} . \tag{3.43}$$

$$I_2 = V_o \frac{\begin{vmatrix} Z_{11} & 1 \\ Z_{12} & 1 \end{vmatrix}}{\begin{vmatrix} Z_{11} & Z_{12} \\ Z_{12} & Z_{22} \end{vmatrix}} \tag{3.44}$$

For n conductors in parallel,

$$I_1 = V_o \frac{A_1}{D} , \; I_2 = V_o \frac{A_2}{D} , \; I_r = V_o \frac{A_r}{D} , \; I_n = V_o \frac{A_n}{D} , \tag{3.45}$$

where D is the determinant of the impedance matrix, and A_r is the determinant of the modified impedance matrix with the r-th column replaced by a column of 1. If the total current is

$$I_o = \sum_{r=1}^{n} I_r , \quad \text{then, } V_o = Z_{eq} I_o ,$$

$$\text{where } Z_{eq} = \frac{D}{A_1 + A_2 + \cdots + A_n}. \tag{3.46}$$

If the self surge impedances are equal and the mutual surge impedances are equal for all the n lines, then it can be shown from (3.29) that

$$Z_{eq} = \frac{Z_s + (n-1)Z_m}{n}, \tag{3.47}$$

where Z_s and Z_m are the self and the mutual surge impedances, respectively.

3.8.2 Second conductor uniformly grounded

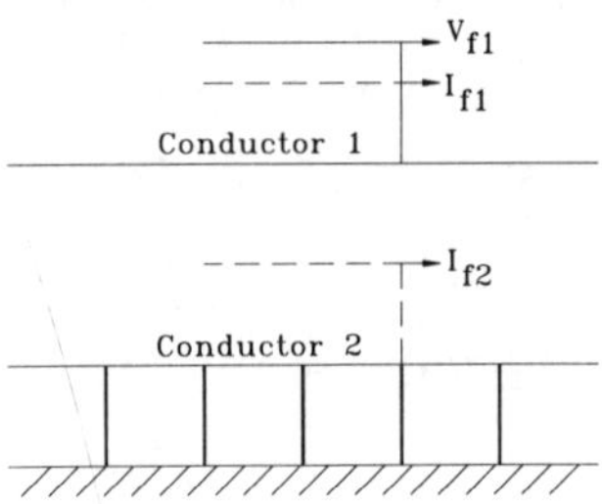

Fig. 3.6 Second conductor uniformly grounded.

In this case (Fig. 3.6) V_{f2}=0. Then,

$$I_{f2} = -\frac{Z_{12}}{Z_{22}} I_{f1}, \tag{3.48}$$

$$\text{and, } V_{f1} = Z_{11}I_{f1} + Z_{12}I_{f2} = (Z_{11} - \frac{Z_{12}^2}{Z_{22}})I_{f1}. \tag{3.49}$$

The effect of the grounded conductor is to reduce the effective surge impedance of conductor 1. This is one of the advantages of an overhead shield wire. This will be clear later.

3.8.3 Second conductor of finite length and open-circuited

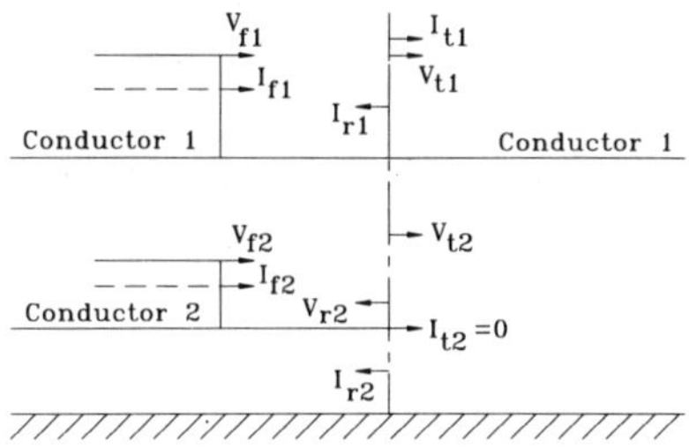

Fig. 3.7 Second conductor of finite length and open-circuited.

The following conditions hold at the transition point for this case (Fig. 3.7):

$$V_{f1} + V_{r1} = V_{t1} \tag{3.50}$$

$$V_{t1} = Z_{11} I_{t1} \tag{3.51}$$

$$I_{f1} + I_{r1} = I_{t1} \tag{3.52}$$

$$I_{f2} + I_{r2} = 0. \tag{3.53}$$

If V_{f1} and V_{f2} are known, then I_{f1} and I_{f2} can be determined from (3.29):

$$I_{f1} = \frac{\begin{vmatrix} V_{f1} & Z_{12} \\ V_{f2} & Z_{22} \end{vmatrix}}{\begin{vmatrix} Z_{11} & Z_{12} \\ Z_{12} & Z_{22} \end{vmatrix}} = \frac{Z_{22}V_{f1} - Z_{12}V_{f2}}{Z_{11}Z_{22} - Z_{12}^2}. \tag{3.54}$$

Similarly,

$$I_{f2} = -I_{r2} = \frac{\begin{vmatrix} Z_{11} & V_{f1} \\ Z_{12} & V_{f2} \end{vmatrix}}{\begin{vmatrix} Z_{11} & Z_{12} \\ Z_{12} & Z_{22} \end{vmatrix}} = \frac{-Z_{12}V_{f1} + Z_{11}V_{f2}}{Z_{11}Z_{22} - Z_{12}^2}. \tag{3.55}$$

From (3.50) to (3.52),

$$V_{f1} + V_{r1} = Z_{11}(I_{f1} + I_{r1}). \tag{3.56}$$

However,

$$V_{f1} = Z_{11}I_{f1} + Z_{12}I_{f2}\ , \tag{3.57}$$

$$V_{r1} = -Z_{11}I_{r1} - Z_{12}I_{r2}\ . \tag{3.58}$$

Similarly,

$$V_{f2} = Z_{12}I_{f1} + Z_{22}I_{f2}\ , \text{ and} \tag{3.59}$$

$$V_{r2} = -Z_{12}I_{r1} - Z_{22}I_{r2}\ . \tag{3.60}$$

Substituting these values of V_{f1} and V_{r1} in (3.56), one gets

$$Z_{12}I_{f2} - 2Z_{11}I_{r1} - Z_{12}I_{r2} = 0. \tag{3.61}$$

Substituting for $I_{r2} = -I_{f2}$, from (3.53),

$$I_{r1} = \frac{Z_{12}}{Z_{11}}I_{f2}\ . \tag{3.62}$$

Substituting the values of I_{r1} from (3.62) and I_{r2} from (3.53) into (3.58):

$$V_{r1} = -Z_{11}(\frac{Z_{12}}{Z_{11}})I_{f2} - Z_{12}(-I_{f2}) = 0. \tag{3.63}$$

Although I_{r1} exists, there is no reflected voltage on conductor 1 (V_{r1}=0).

$$\text{Therefore, } V_{t1} = V_{f1}\ , \text{ and} \tag{3.64}$$

$$I_{t1} = \frac{V_{t1}}{Z_{11}} = \frac{V_{f1}}{Z_{11}}. \tag{3.65}$$

Similarly, from (3.60):

$$V_{r2} = -Z_{12}(\frac{Z_{12}}{Z_{11}})I_{f2} - Z_{22}(-I_{f2}) = \frac{Z_{22}Z_{11} - Z_{12}^2}{Z_{11}} I_{f2} \ . \tag{3.66}$$

Substituting I_{f2} from (3.55),

$$V_{r2} = -\frac{Z_{12}}{Z_{11}} V_{f1} + V_{f2} \ . \tag{3.67}$$

The open-circuit voltage of conductor 2 is:

$$V_{t2} = V_{f2} + V_{r2} = -\frac{Z_{12}}{Z_{11}} V_{f1} + 2V_{f2} \ . \tag{3.68}$$

3.8.4 Second conductor of finite length and shorted to ground

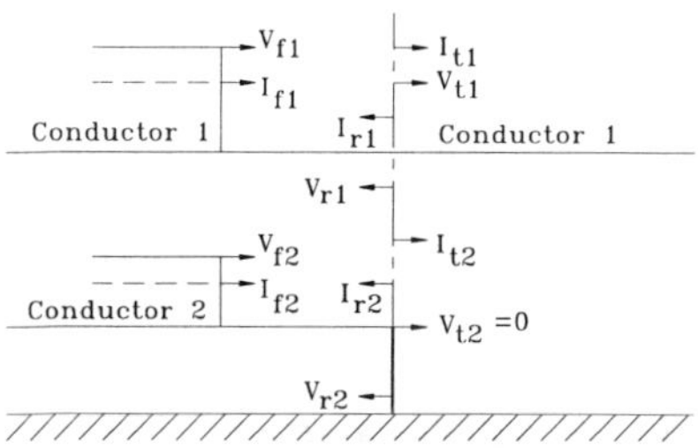

Fig. 3.8 Second conductor of finite length and shorted to ground.

The new boundary conditions (Fig. 3.8) are the following:

$$V_{f1} + V_{r1} = V_{t1} \ , \tag{3.69}$$

$$V_{t1} = Z_{11} I_{t1} \ , \tag{3.70}$$

$$I_{f1} + I_{r1} = I_{t1} \ , \text{ and} \tag{3.71}$$

$$V_{t2} = V_{f2} + V_{r2} = 0 \ . \tag{3.72}$$

From these four equations, in addition to (3.57) to (3.60):

$$V_{r1} = \frac{Z_{12}^2}{2Z_{11}Z_{22} - Z_{12}^2} V_{f1} - \frac{2Z_{11}Z_{12}}{2Z_{11}Z_{22} - Z_{12}^2} V_{f2} \ , \tag{3.73}$$

$$V_{t1} = \frac{2Z_{11}Z_{22}}{2Z_{11}Z_{22} - Z_{12}^2}V_{f1} - \frac{2Z_{11}Z_{12}}{2Z_{11}Z_{22} - Z_{12}^2}V_{f2}\,, \tag{3.74}$$

$$V_{r2} = -V_{f2}\,, \tag{3.75}$$

$$V_{t2} = 0\,. \tag{3.76}$$

I_{f1} and I_{f2} can be found from (3.54) and (3.55). I_{t1} is found from (3.74) and (3.70). From (3.71):

$$I_{r1} = I_{t1} - I_{f1}\,. \tag{3.77}$$

From (3.58) and (3.60):

$$I_{r2} = \frac{\begin{vmatrix} Z_{11} & -V_{r1} \\ Z_{12} & -V_{r2} \end{vmatrix}}{\begin{vmatrix} Z_{11} & Z_{12} \\ Z_{12} & Z_{22} \end{vmatrix}}\,. \tag{3.78}$$

3.8.5 Conditions at the beginning of a parallel

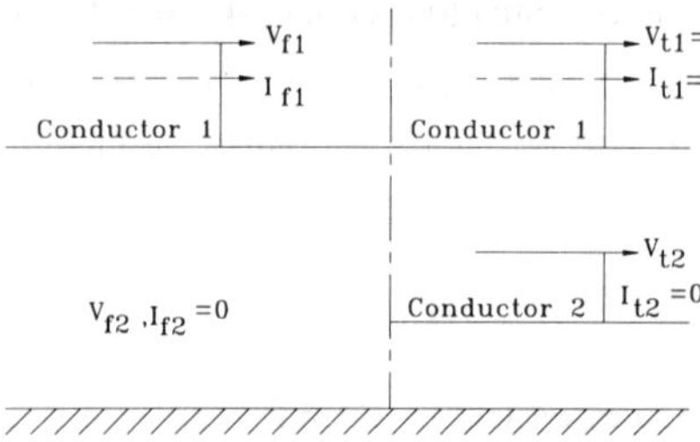

Fig. 3.9 Conditions at the beginning of a parallel.

This situation is shown in Fig. 3.9. As no current to the left of conductor 2 can flow, then:

$$I_{f2} = I_{r2} = 0\,. \tag{3.79}$$

$$\text{Therefore,} \quad I_{t2} = I_{f2} + I_{r2} = 0. \tag{3.80}$$

For conductor 1:

$$V_{f1} + V_{r1} = V_{t1} = Z_{11}(I_{f1} + I_{r1})$$

or, $\quad Z_{11}I_{f1} + Z_{12}I_{f2} - Z_{11}I_{r1} - Z_{12}I_{r2} = Z_{11}I_{f1} + Z_{11}I_{r1}$

or, $\quad 2Z_{11}I_{r1} = Z_{12}I_{f2} - Z_{12}I_{r2} = 0 \text{ (as } I_{f2} = I_{r2} = 0)$

Therefore, $I_{r1} = 0$. Then, $I_{f1} = I_{t1}$. However,

$$V_{t2} = Z_{12}I_{f1} + Z_{22}I_{t2} = Z_{12}\frac{V_{f1}}{Z_{11}} = K_{12}V_{f1}, \tag{3.81}$$

where $K_{12} = Z_{12}/Z_{11}$ is called the coupling factor. Because of this coupling factor provided by an overhead ground wire, the voltage across an insulator string of an overhead line will be lower than that without a ground wire. This is another advantage of the overhead ground wire. This will be further discussed later.

3.9 Effects of Dissipation on Wave Propagation

The effects of dissipation have so far been neglected in the analysis of wave propagation on overhead lines. Dissipation causes attenuation and distortion of the wave as it travels along the line. The sources of dissipation are (i) conductor losses, (ii) ground losses due to finite soil conductivity, and (iii) impulse corona on the line. The effect of conductor losses is the least of the three effects, and is generally neglected. The effects of the other two sources of losses are discussed below.

3.9.1 Effect of finite soil conductivity

Time-dependent ground impedances are connected in series with the line inductances in Fig. 3.2. The voltage and the current equations are then given by:

$$\frac{\partial[V]}{\partial x}\Delta x = \big([L]s + [Z_g(s)]\big)[I]\Delta x, \text{ and} \tag{3.82}$$

$$-\frac{\partial[I]}{\partial x}\Delta x = [C]s[V]\Delta x, \tag{3.83}$$

where $[Z_g(s)]$ is the time-dependent ground impedance, expressed in Laplace transform. Differentiating (3.82) with respect to x, and eliminating [I]:

$$\frac{\partial^2}{\partial x^2}[V(s)] = \left([L][C]s^2 + [Z_g(s)][C]s\right)[V(s)]$$

$$= [k]^2[V(s)]. \tag{3.84}$$

The solution of (3.84) is given by:

$$[V(s)] = e^{-[k]x}[V_o(s)]. \tag{3.85}$$

The matrix $[V_o(s)]$ can be evaluated from the boundary conditions. Differentiating (3.85) with respect to x, and equating with (3.82):

$$\frac{\partial}{\partial x}[V(s)] = -[k][V(s)] = -([L]s + [Z_g(s)])[I(s)]. \tag{3.86}$$

From (3.86):

$$[V(s)] = [k]^{-1}([L]s + [Z_g(s)])[I(s)] = [Z(s)][I(s)], \tag{3.87}$$

where Z(s) is the time-dependent surge impedance matrix of the multiconductor overhead line.

3.9.2 Modal analysis

The matrix [V(s)] can be evaluated from (3.85) by invoking Sylvester's theorem. An alternate solution will be by modal analysis which also imparts a physical meaning to the solution. Transforming (3.85) by (3.34), the modal wave equation is:

$$\frac{\partial^2}{\partial x^2}[V_m(s)] = \frac{s^2}{v_c^2}[V_m(s)] + s[T]^{-1}[Z_g(s)][C][T][V_m(s)], \tag{3.88}$$

$$\text{where } \frac{[1]}{v_c^2} = [L][C], \text{ and} \tag{3.89}$$

$$[Z_g(s)][C] = \begin{bmatrix} Z_{g1}(s) & . & . & Z_{gn}(s) \\ . & . & . & . \\ . & . & . & . \\ Z_{g1}(s) & . & . & Z_{gn}(s) \end{bmatrix} \begin{bmatrix} C_{11} & . & . & C_{1n} \\ . & . & . & . \\ . & . & . & . \\ C_{n1} & . & . & C_{nn} \end{bmatrix}. \tag{3.90}$$

For a transposed line, $Z_{g1}(s)=Z_{g2}(s)=\cdots=Z_{gn}(s)=Z_g(s)$; also, all diagonal elements are equal, and all off-diagonal elements in the C-matrix are equal. Equation (3.90) is then simplified to:

$$[Z_g(s)][C]=Z_g(s)C_g[1_{nn}], \tag{3.91}$$

$$\text{where } C_g=\sum_{m,n=1}^{n} C_{mn}, \text{ and} \tag{3.92}$$

$[1_{nn}]$ is an nxn matrix with all its elements equal to 1. Equation (3.88) can be rearranged as:

$$\frac{\partial^2}{\partial x^2}[V_m(s)]=\frac{s^2}{v_c^2}[V_m(s)]+sZ_g(s)C_g[T]^{-1}[1_{nn}][T][V_m(s)]$$

$$=\frac{s^2}{v_c^2}[V_m(s)]+sZ_g(s)C_g[\lambda][V_m(s)], \tag{3.93}$$

where $[\lambda]$ is the eigenvalue matrix of $[1_{nn}]$, given by

$$[\lambda]=\begin{bmatrix} \lambda_1 & 0 & . & . & . & 0 \\ 0 & \lambda_2 & . & . & . & 0 \\ . & . & . & . & . & . \\ 0 & . & . & \lambda_r & . & 0 \\ . & . & . & . & . & \\ . & . & . & . & . & \lambda_n \end{bmatrix}. \tag{3.94}$$

Equation (3.93) is composed of n decoupled equations which can be solved individually like a single-conductor case. For instance, the nth equation is given by:

$$\frac{\partial^2 V_{mn}(s)}{\partial x^2}=\frac{s^2}{v_c^2}V_{mn}(s)+sZ_g(s)C_g\lambda_n V_{mn}(s)=k_n^2 V_{mn}(s), \tag{3.95}$$

$$\text{where } k_n^2=\frac{s^2}{v_c^2}+sZ_g(s)C_g\lambda_n\ . \tag{3.96}$$

The solution of (3.95) is:

$$V_{mn}(x,s)=V_{mn0}(s)\,e^{-\frac{sx}{v_c}\sqrt{1.0+\frac{\lambda_n v_c^2 Z_g(s)C_g}{s}}}, \tag{3.97}$$

where $V_{mn0}(s)$ can be determined from the boundary conditions at x=0.

3.9.3 Three-phase lines

For a 3-conductor line, for instance, the eigenvalues of $[1_{33}]$ are: $\lambda_1=0$, $\lambda_2=0$, and $\lambda_3=3$. Then, the eigenvector matrix, [T], is given by:

$$[T]=\begin{bmatrix} 1 & 1 & 1 \\ -1 & 0 & 1 \\ 0 & -1 & 1 \end{bmatrix}. \tag{3.98}$$

$$\text{Putting, } g_1=\sqrt{1.0+\frac{3\,v_c^2 Z_g(s)\,C_g}{s}}, \tag{3.99}$$

$$V_{m1}(x,s)=V_{m10}(s)\,e^{-\frac{s}{v_c}x}, \tag{3.100}$$

$$V_{m2}(x,s)=V_{m20}(s)\,e^{-\frac{s}{v_c}x}, \text{ and} \tag{3.101}$$

$$V_{m3}(x,s)=V_{m30}\,e^{-\frac{sg_1}{v_c}x}. \tag{3.102}$$

The actual voltages on the three conductors are found by applying (3.33):

$$V_1=V_{m10}(s)\,e^{-\frac{s}{v_c}x}+V_{m20}(s)\,e^{-\frac{s}{v_c}x}+V_{m30}(s)\,e^{-\frac{sg_1}{v_c}x}, \tag{3.103}$$

$$V_2(x,s)=-V_{m10}(s)\,e^{-\frac{s}{v_c}x}+0+V_{m30}\,e^{-\frac{sg_1}{v_c}x}, \text{ and} \tag{3.104}$$

$$V_3(x,s)=0-V_{m20}\,e^{-\frac{s}{v_c}x}+V_{m30}\,e^{-\frac{sg_1}{v_c}x}. \tag{3.105}$$

It should be noticed in (3.103) to (3.105) that all three components of the modal voltages exist on the struck conductor (conductor 1). One line-mode voltage is missing on the voltages induced on conductors 2 and 3. The line-mode voltage which is induced on conductors 2 and 3 is of equal magnitude but of opposite sign to that on the struck conductor. The ground-mode voltage is the same in all three conductors. The two line-mode voltages travel with a velocity, v_c, which is about the same as the velocity of electromagnetic waves in free space. However, the ground-mode voltage travels with a time-dependent velocity which is lower than v_c.

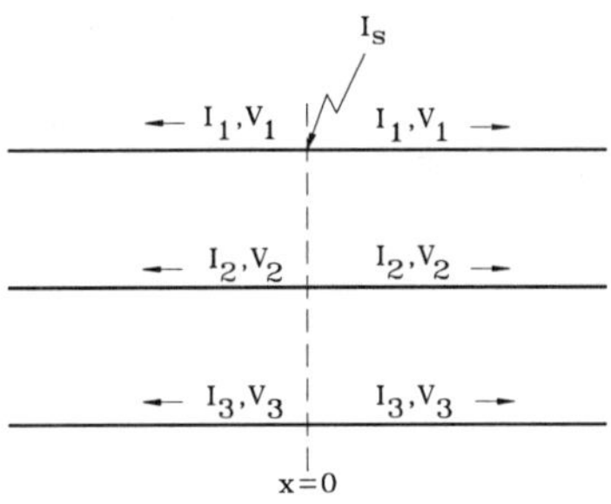

Fig. 3.10 Midspan strike on a three-conductor line.

Figure 3.10 shows an example of a midspan strike on a three-phase overhead line. If lightning struck conductor 1, then $I_1(0,s)=I_s(s)/2$, $I_2(0,s)=I_3(0,s)=0$ at x=0 for a return-stroke current, $I_s(s)$.

Differentiating (3.103) to (3.105) with respect to x, and equating with (3.82):

$$V_{m10}(s)=\left(Z_{11}-2Z_{21}+Z_{31}\right)\frac{I_1(0,s)}{3}, \tag{3.106}$$

$$V_{m20}(s)=\left(Z_{11}+Z_{21}-2Z_{31}\right)\frac{I_1(0,s)}{3}, \text{ and} \tag{3.107}$$

$$V_{m30}(s)=\left(Z_{11}+Z_{21}+Z_{31}+\frac{3\,Z_g(s)\,v_c}{s}\right)\frac{I_1(0,s)}{3\,g_1}. \tag{3.108}$$

For transposed lines, $Z_{21}=Z_{31}$; therefore, the two line-mode voltages are equal ($V_{m10}=V_{m20}$).

Because of the lower velocity of the ground-mode wave, it arrives at a distant point on the line at a later time than the two line-mode waves. As a result, the waveshape is 'sheared' (elongated). The peak of the ground-mode voltage falls behind the line-mode voltages more as they travel farther, producing attenuation of the resultant voltage. For the neighboring conductors (e.g., conductors 2 and 3), a bipolar induced voltage will appear at a distant point, because the line-mode induced voltage is of opposite polarity and the ground-mode voltage arrives later at the distant point. Multivelocity voltages have been reported during field tests [2]. Figure 3.11 shows the propagation of the modal voltages on a three-conductor overhead line.

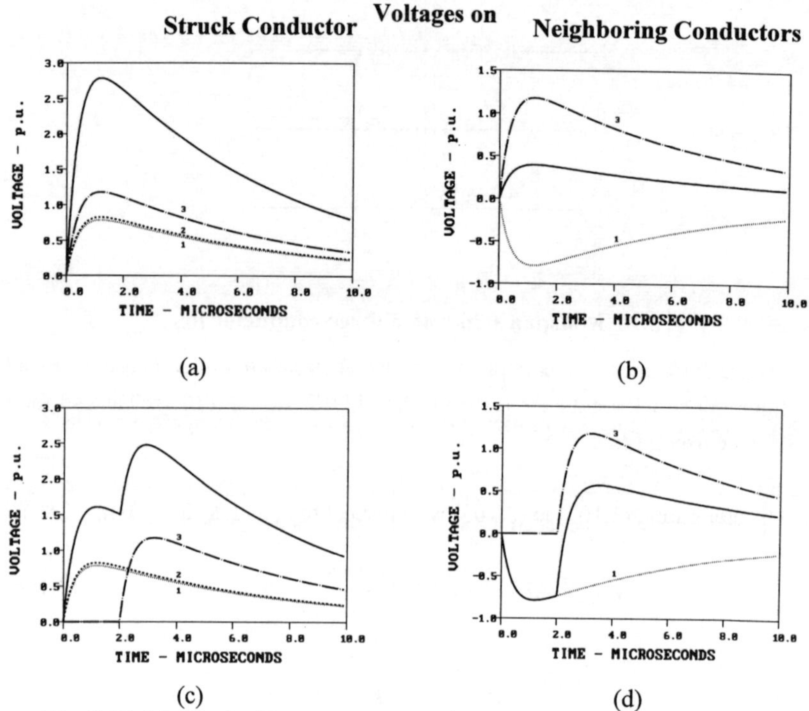

Fig. 3.11 Multivelocity wave propagation on a three-phase overhead line over imperfect earth.

(a) and (b) at x=0; (c) and (d) at a distant point, x=ℓ km.

Solid curve: total voltage; numbered curves: modal components.

3.9.4 Ground impedance

In (3.106) to (3.108), all parameters are known except the ground impedance, $Z_g(s)$. Rudenberg represented an overhead line by laying it on the ground plane at the center of a groove of radius h from which the earth has been scooped out, h being the height of the overhead line above earth [4]. This is shown in Fig. 3.12. The equation for the current distribution in earth is given by [4]:

$$\frac{\partial^2 J_g}{\partial r^2} + \frac{1}{r}\frac{\partial J_g}{\partial r} - \frac{\mu_o}{2\rho_g}\frac{\partial J_g}{\partial t} = 0, \tag{3.109}$$

where r =distance of the field point from the center of the conductor, m, J_g=current density in earth, A/m^2, ρ_g=resistivity of soil, Ω-m, and μ_o=permeability of free space, H/m.

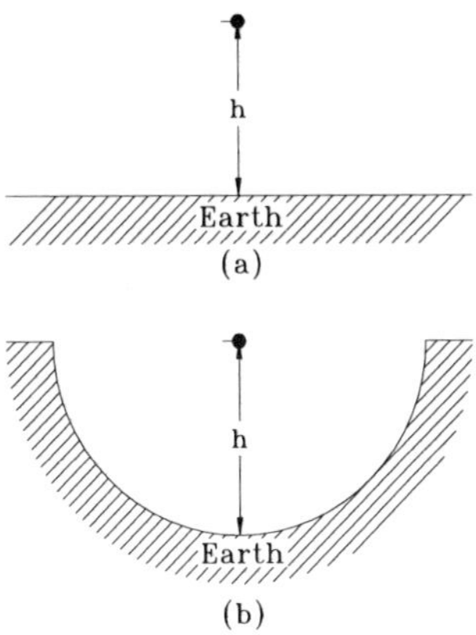

Fig. 3.12 Rudenberg's model of overhead line above imperfect earth.

Taking the Laplace transform of (3.109), assuming the initial conditions to be zero, and the current density in the earth to be zero at infinity, one gets [5]:

$$J_g(s) = A\,j\,H_o^{(1)}(jK'r), \tag{3.110}$$

where $K'=(\mu_o s/2\rho_g)$,
s = Laplace transform operator,
$H_o^{(1)}$ = Hankel function of order zero, and $j=\sqrt{(-1)}$.

The constant, A, in (3.110) is determined from the condition that the total current through the ground must be equal and opposite to the current, I, in the conductor of the overhead line. Applying this condition, the current density at the earth's surface is:

$$J_{ge}(s) = \frac{jK'H_o^{(1)}(jK'h)}{\pi h H_1^{(1)}(jK'h)} I(s). \tag{3.111}$$

The Laplace transform of the voltage gradient, E_g, on the surface of the earth is:

$$E_g(s) = \rho_g J_{ge}(s). \tag{3.112}$$

Hence, the impedance function (per meter length of the ground-return circuit) is:

$$Z_g(s) = \frac{E_g(s)}{-I(s)}. \tag{3.113}$$

Combining (3.111) and (3.113),

$$Z_g(s) = 0.5 R_g K'h \frac{K_0(K'h)}{K_1(K'h)}, \tag{3.114}$$

where $R_g = 2\rho_g/\pi h^2$,
K_0=modified Bessel function of order zero and second kind, and
K_1=modified Bessel function of order one and second kind.

The voltage equations can be expressed in the time domain with the help of an inversion routine by the digital computer, such as the INLAP routine in the IMSL software package.

3.9.5 Effect of impulse corona

Field tests have shown that lightning overvoltages produce intense corona around the struck line, causing attenuation and distortion of the traveling voltage waves [6-13]. Boehne, in discussing [8], suggested that the ionized air surrounding the struck conductor effectively increases the diameter of the conductor during corona, thus increasing its capacitance. As the velocity of the traveling wave is $v_c = 1/\sqrt{(LC)}$, the more intense the corona is the slower will be the velocity of propagation of the wave. A voltage wave of finite wavefront can be subdivided into layers of voltage elements of increasing voltage magnitudes. Because of the more intense corona for the higher voltage levels, these different voltage elements will travel at different velocities, the higher voltage elements traveling slower. Thus, the the top of the wave will be continually 'toppling' over to the rear because of its slower velocity. This will result in an 'apparent' attenuation of the wave. Because positive corona is more severe, the attenuation and distortion of the positive-polarity wave will be higher than that of the negative-polarity wave. This has been observed in the field tests.

Skilling and Dykes suggested an analysis for the attenuation and distortion due to corona which has been widely applied [10]. Combining the Skilling-Dykes analysis with Peek's equation for corona loss, Bewley [2] suggested the following equation for the velocity of the traveling voltage wave on an ovehead line under impulse corona:

$$v_c' = \frac{v_c}{1 + \frac{k}{C}\frac{V - V_c}{V}}, \tag{3.115}$$

where v_c= velocity of the wave below corona level, m/s

V=voltage magnitude,

V_c=corona inception voltage,

k=corona energy/half cycle/meter (from Peek's equation)=$k_o\sqrt{(r/2h)}$,

k_o=1.2x10^{-11} for positive voltages,

=0.52x10^{-11} for negative voltages,

r,h=conductor radius, height above ground, respectively, and

C=conductor capacitance, F/m.

At time t=x/v, the ordinate, V, lags behind its original position in time by:

$$\tau = \frac{x}{v_c'} - \frac{x}{v_c} = \frac{x}{v_c}\frac{k}{C}(1 - \frac{V_c}{V}). \tag{3.116}$$

Each point on the wavefront is retarded by τ until the final point intersects the tail. As τ is a function of both x (distance traveled) and the instantaneous voltage, V, the longer the distance traveled and the higher the instantaneous voltage is, the longer will be the retardation effect. This is shown in Fig. 3.13.

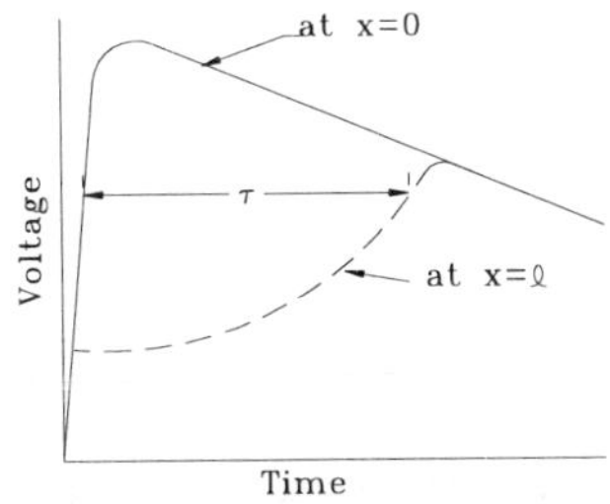

Fig. 3.13 Attenuation and distortion according to Skilling-Dykes [2,10].

The corona inception voltage, V_c, is given by:

$$V_c = \frac{2\pi r \epsilon_0 E_c}{C}, \quad \text{kV} \tag{3.117a}$$

where r is the radius of the conductor in m, ϵ_o is the permittivity of free space, E_c is the corona inception electric field (kV/m) on the conductor surface, and C is the capacitance of the overhead line (F/m). Using Peek's equation for E_c, V_c can be expressed as:

$$V_c = 3\times10^3 m\,\delta\left(1 + \frac{0.03}{\sqrt{\delta r}}\right)\ell n\frac{2h}{r}, \tag{3.117b}$$

where m is the surface roughness factor, δ is the relative air density and h (in m) is the conductor height above ground.

Peek's equation for corona loss under power-frequency voltage may not be applicable to impulse corona. Wagner and Lloyd [14] proposed that the distortion in the traveling voltage wave is caused by the increase in the capacitance of the conductor as Boehne had suggested earlier. They suggested that the increase in the conductor capacitance can be estimated from the charge-voltage (q-V) characteristics of the line under impulse corona. Since then, many researchers have based their work on this principle; a few of them [14-21] are cited in the reference. The q-V characteristic and the attendant voltage distribution are shown in Fig. 3.14.

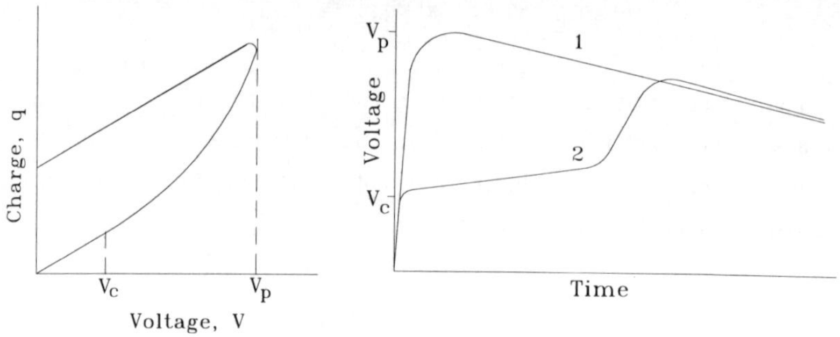

(a) (b)

Fig. 3.14 Voltage distortion by impulse corona according to q-V characteristic.

The charge on a conductor is related to its voltage by the equation, q=CV, where C is the capacitance of the conductor. As the voltage on the line conductor is increased from zero, the charge on the conductor increases linearly, signifying that the capacitance is the constant slope of the q-V curve (Fig. 3.14a). This capacitance, C_n is called the natural [14] or the geometric capacitance [20] of the line, which is given by:

$$C_n = \frac{10^{-9}}{18\,\ell n \frac{2h}{r_o}}, \ \text{F/m} \tag{3.118}$$

where h is the conductor height above ground, and r_o is the conductor radius. The slope of the q-V curve increases when the voltage is increased beyond the corona inception voltage level, V_c. The slope of the q-V curve is called the dynamic capacitance of the line [20], C_d=dq/dV. The slope increases until the peak of the voltage, V_p, is reached because of increased ionization at the higher voltage levels. In other words, C_d is a variable capacitance which is a function of the applied voltage above the corona inception level. During the subsequent fall of the voltage, the slope of the q-V curve is about the same as that during the preionization level. During the preionization period, the voltage travels with a velocity, v_c, given by:

$$v_c = \frac{1}{\sqrt{L C_n}}. \tag{3.119}$$

On the same basis, the velocity of the voltage wave above corona inception level would be:

$$v_d = \frac{1}{\sqrt{L C_d}}. \tag{3.120}$$

As $C_d > C_n$, then $v_d < v_c$. Because C_d increases with voltage, the voltage elements above the corona inception voltage level will be increasingly retarded behind the voltage elements at the lower voltage levels. This will naturally flatten the wavefront. If the voltage wave has traveled a distance, ℓ, from the point of origin, then a point on the voltage wave (above corona inception level) will be retarded by:

$$\tau = \ell\left(\frac{1}{v_d} - \frac{1}{v_c}\right) = \frac{\ell}{v_c}\left(\sqrt{\frac{C_d}{C_n}} - 1\right). \tag{3.121}$$

Figure 3.14b shows two voltage profiles, where the curve 1 is the voltage at the point of origin, and curve 2 is the voltage profile at a distane ℓ from the point of origin. This is similar to Fig.3.13.

C_n can be computed by conventional means. The q-V characteristic of the line needs to be known to compute C_d. However, based on parabolic relationship between q and V in the corona region, empirical equation for q is given in [20]:

$$q = V_c C_n \left(\frac{V}{V_c}\right)^B, \tag{3.122}$$

where B is an empirical parameter shown in the table below [20].

Conductor Type	B	
	Positive Polarity	Negative Polarity
Single	0.22r+1.2	0.07r+1.12
Bundle	1.52-0.15ℓn (n)	1.28-0.08ℓn (n)

Note: r (in cm) is conductor radius; n is number of subconductors

On this basis, C_d is given by [20]:

$$C_d = C_n B \left(\frac{V}{V_c} \right)^{B-1} . \qquad (3.123)$$

Wagner and Lloyd [14] concluded that the charge on the front of the voltage wave is a function of the voltage magnitude, and is independent of the rate of rise. However, Gary et al. [20] found that C_d is higher for steeper wavefronts even though the slope effect is small and can be neglected. The effect of the polarity of the voltage wave is pronounced, the effect of the positive-polarity wave being significantly higher than that of the negative-polarity wave. This observation is reflected in the value of the parameter, B. The effects of polarity and distance traveled are shown in Fig. 3.15.

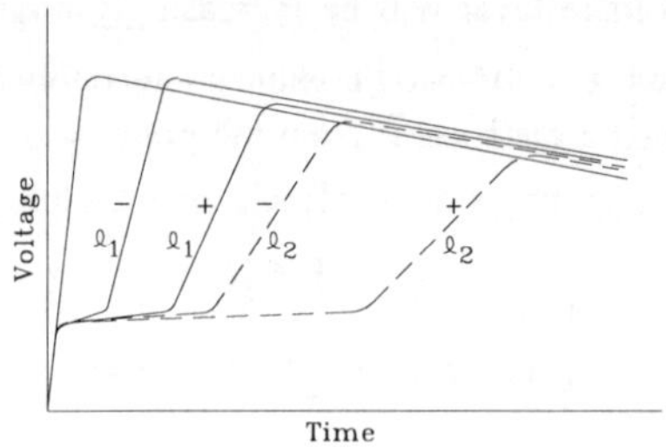

Fig. 3.15 Effects of polarity and distance traveled on wave distortion by impulse corona: $\ell_2 > \ell_1$.

Figure 3.16 shows the analog model of an overhead line under impulse corona, proposed by Wagner and Lloyd [14]. The effect of the impulse corona is simulated by a capacitance, ΔC (=C_d-C_n), in series with a diode and a voltage source, V_c, the corona inception voltage. This network is connected across the natural capacitance, C_n, of the line. Below the corona inception voltage level, ΔC is prevented from charging by the diode, maintaining the total line capacitance of C_n. When the line voltage exceeds V_c, ΔC is charged, and the total line capacitance is C_d (=C_n+ΔC). This analog model of an overhead line under impulse corona has been universally applied.

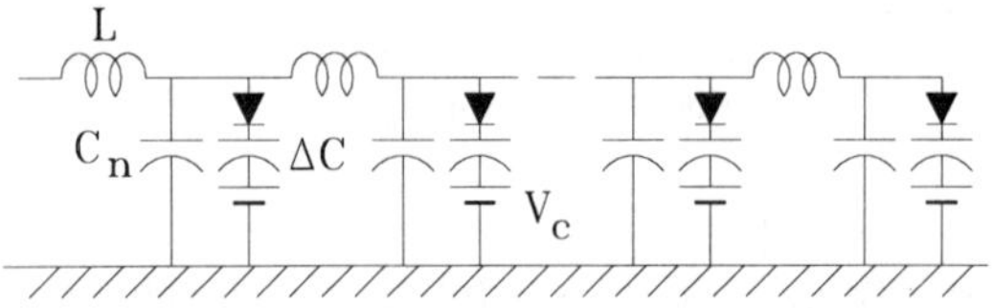

Fig. 3.16 Analog simulation of an overhead line under impulse corona.

3.10 References to Chapter 3

1. L.A. Pipes, ***Matrix Methods in Engineering***. Prentice-Hall, Englewood Cliffs, NJ, 1963.

2. L.V. Bewley, ***Traveling Waves on Transmission Systems***, Second Edition. John Wiley, New York, NY, 1951.

3. ***Electrical Transmission and Distribution Reference Book***, Fourth Edition. Westinghouse Electric Corporation, East Pittsburgh, PA, 1964.

4. R. Rudenberg, ***Transient Performance of Electrical Power Systems***. McGraw-Hill, New York, N.Y., 1950.

5. S. Hayashi, ***Surges on Transmission Systems***. Denki-Shoin, Kyoto, Japan, 1955.

6. R.N. Conwell and C.L. Fortescue, "Lightning Laboratory at Stillwater, New Jersey", ***AIEE Trans.***, Vol. 49, pp. 872-876, 1930.

7. K.B. McEachron, J.G. Hemstreet and W.J. Rudge, "Traveling Waves on Transmission Lines with Artificial Lightning", ***AIEE Trans.***, Vol. 49, pp. 885-894, 1930.

8. J.A. Dowell, "Attenuation and Successive Reflections of Traveling Waves", ***AIEE Trans.***, Vol. 50, pp.551-557, 1931.

9. O. Brune and J.R. Eaton, "Experimental Studies in the Propagation of Lightning Surges on Transmission Lines", ***AIEE Trans.***, Vol. 50, pp. 1132-1138, 1931.

10. H.H. Skilling and P. de K. Dykes, "Distortion of Traveling Waves by Corona", ***Electrical Engineering***, Vol. 56, pp. 850-857, 1937.

11. ***Surge Phenomena*** (book), Electrical Research Association. The British Electrical and Allied Industries Research Association, London, England, 1941.

12. M. Bőckman, N. Hylten-Cavallius and S. Rusck, "Propagation of Surge Generator Waves up to 850 kV on a 132-kV Line", ***Proc. CIGRE***, Paper No. 314, 1950.

13. C.F. Wagner, I.W. Gross and B.L. Lloyd, "High Voltage Impulse Tests on Transmission Lines", ***AIEE Trans.***, Vol. 73, Pt. III-A, pp. 196-210, 1954.

14. C.F. Wagner and B.L. Lloyd, "Effects of Corona on Traveling Waves", ***AIEE Trans.***, Vol. 74, Pt. III, pp. 858-872, 1955.

15. H.M. Kudyan and C.H. Shih, "A Nonlinear Model for Transmission Lines in Corona", ***IEEE Trans. on Power Apparatus and Systems***, Vol. PAS-100, pp. 1420-1430, 1981.

16. A. Inoue, "Propagation Analysis of Overvoltage Surges with Corona Based upon Charge Versus Voltage Curve", ***IEEE Trans. on Power Apparatus and Systems***, Vol. PAS-104, pp. 655-662, 1985.

17. A. Semlyen and H. Wei-Gang, "Corona Modelling for the Calculation of Transients on Transmission Lines", ***IEEE Trans. on Power Delivery***, Vol. PWRD-1, pp. 228-239, 1986.

18. M.A. Al-Tai, H.S.B. Elayyan, D.M. German, A. Haddad, N. Harid and R.T. Waters, "The Simulation of Surge Corona on Transmission Lines", ***IEEE Trans. on Power Delivery***, Vol. 4, pp. 1360-1368, 1989.

19. P. S. Maruvada, D.H. Nguyen and H. Hamadani-Zadeh, "Studies on Modeling Corona Attenuation of Dynamic Overvoltages", ***IEEE Trans. on Power Delivery***, Vol. 4, pp. 1441-1449, 1989.

20. C. Gary, D. Cristescu and G. Dragan, "Distortion and Attenuation of Travelling Waves Caused by Transient Corona", ***CIGRE Study Committee 33 Report***, 1989.

21. C. de Jesus and M.T. Correia de Barros, "Modelling of Corona Dynamics for Surge Propagation Studies", ***IEEE Trans. on Power Delivery***, Vol. 9, pp.1564-1569, 1994.

Problems for Chapter 3

1.

$$\text{Given} \quad [M] = \begin{bmatrix} -1 & -3 \\ 2 & 4 \end{bmatrix},$$

a. Find its eigenvalue matrix, $[\lambda]$ and the eigenvector (modal) matrix, $[T]$.
b. Show that $[T]^{-1}[M][T] = [\lambda]$.

2. A three-phase, completely transposed overhead power line has the following dimensions:

Average height above ground, h = 10 m.
Equivalent distance between phase conductors = 3 m.
Conductor radius = 1.5 cm.

The boundary conditions at x = 0 are:

a. $V_1(0,t) = 1.0$ (unit step-function voltage); b. $V_2(0,t) = V_3(0,t) = 0$.

Plot the voltages on each conductor at x = 1 km and 5 km for a time period of 10 μs when the conductor resistances under transient conditions are

i. $R_c = 5\text{x}10^{-4}\ \Omega/\text{m}$,

ii. $R_c = 5\text{x}10^{-3}\ \Omega/\text{m}$.

3. The inductance and the capacitance matrices of a completely transposed two-conductor overhead power line has the following form:

$$[L] = \begin{bmatrix} L_s & L_m \\ L_m & L_s \end{bmatrix}; \text{ and } [C] = \begin{bmatrix} C_s & C_m \\ C_m & C_s \end{bmatrix}.$$

The matrix of the conductor resistances is given in the following form:

$$[R_c] = \begin{bmatrix} R_c & 0 \\ 0 & R_c \end{bmatrix}.$$

Find the expressions for the voltages, $V_1(x,t)$ and $V_2(x,t)$, on conductors 1 and 2 at a distance x from the origin, if a unit step-function voltage is applied to conductor 1 at x = 0 while conductor 2 is shorted to ground at x = 0.

4. A 13-m high shield wire is used to shield a 10-m high phase conductor. The shielding angle is 30 degrees. The radii of the shield wire and the phase conductor are 1 cm and 2 cm, respectively. For a 100-kA lightning stroke to the shield wire, what will be the voltage across the shield wire and the phase conductor ?

5. For the problem 4, in case of a shielding failure, the lightning stroke will hit the phase conductor. What will be the voltage of the phase conductor? What would be the phase-conductor voltage if the shield wire did not exist?

6. A 1000-kV, 0.1/10-μs voltage is injected to an overhead line by a backflashover at a tower. If the line height is 10 m, and the conductor radius is 1.5 cm, plot the voltage ***vs.*** time at a distance (a) 1 km and (b) 5 km from the struck point for a (i) positive-polarity, and (ii) negative-polarity voltage wave.

CHAPTER 4
LIGHTNING PHENOMENA

It is believed that lightning sparked the breath of life on earth. Lightning or strong electrical discharges have been detected in the atmospheres of Jupiter, Neptune, Saturn and Uranus [1]. Lightning in Jupiter is believed to be 100 times more powerful than our terrestrial lightning. Recently, the space probe, Galileo, has detected strong electrical discharges in Venus [1,2]. Strong lightning-like discharges have been found in volcanoes and sand storms in deserts. However, the usual lightning is associated solely with thunderclouds.

4.1 Formation of a Thundercloud

A thundercloud consists of a dipole of electrical charges - usually layers of positive charge at the top and layers of negative charge at the bottom. A layer of weak positive charge is also found at the bottom of the thundercloud (Fig. 4.1).

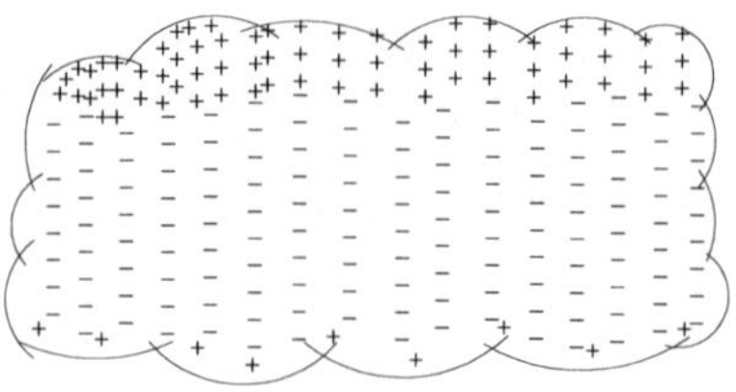

Fig. 4.1 Structure of a thundercloud.

Several theories exist which try to explain the formation of the thundercloud [3-7]. However, none has been able to completely explain all the phenomena. Of the many theories postulated, those of C.T.R. Wilson [8,9] and of G.C. Simpson [10-12] are widely known.

Wilson's theory is mainly based on two natural phenomena. These are:

1. Large numbers of ions, both positive and negative, exist in the atmosphere. These are mostly caused by the cosmic rays from outer space. Many of these ions attach themselves to dust particles and to extremely small drops of water. The mobility of these ions is in the range of 0.0003 to 0.0005 cm/sec.

2. An electrostatic field exists in the earth's atmosphere, with the earth being negative with respect to the upper atmosphere. This electric field is in the order of 1 V/cm at the earth's surface, gradually decreasing to 20 mV/cm at an altitude of about 9 km.

According to Wilson's theory, a relatively large drop of water (e.g., 1 mm radius) will be polarized by the atmospheric electric field, with a negative charge at the upper part and a positive charge at the lower region. The velocity of the fall of such a water drop under the influence of gravity is about 5.9 m/sec. The positive charge at the lower region of the water drop attracts negative ions from the atmosphere as the water drop falls under the influence of gravity. The atmosphere is then left with an abundance of positive charges. The slower moving, smaller water drops attract the atmospheric positive charges. The charges are thus separated - the heavier water drops accumulate the negative charges at the lower region of the cloud, and the smaller and lighter water drops accumulate the positive charges at the upper region of the cloud. Wilson's proposition that the thundercloud produces a dipolar electric field with the upper region positive has been proven by field measurements for approaching and receding thunderclouds.

Simpson's theory on the formation of thunderclouds is based on the following experimental evidences:

1. A water drop of diameter greater than 0.25 cm will break up into smaller drops when it falls through still or ascending air.

2. The steady-state velocity of falling water drops of 0.25-cm diameter is 8 m/sec. This is thus the limiting velocity of rain drops. No rain drops can fall to earth in an updraft of air exceeding 8 m/sec.

3. When a rain drop breaks up, the resulting droplets acquire positive charge, and the neighboring air becomes negatively charged.

The meteorological conditions within a thundercloud, according to Simpson, are shown in Fig. 4.2. The air enters the storm from the right, and passes under the forward end of the cloud where it takes an upward direction. Within the darkened region, C, the vertical velocity of the updraft of air exceeds 8 m/sec. Therefore, no water drops can enter into this region, and there is no accumulation of electric charge in region C. Above region C (in region B), the velocity is still very high, and only the larger drops can enter here. They are blown apart upon entering region B, forming positively charged smaller droplets. The air becomes negatively charged, and rises to higher altitudes. The smaller water drops recombine and fall in the region B, to be blown apart again, producing more charge separation.

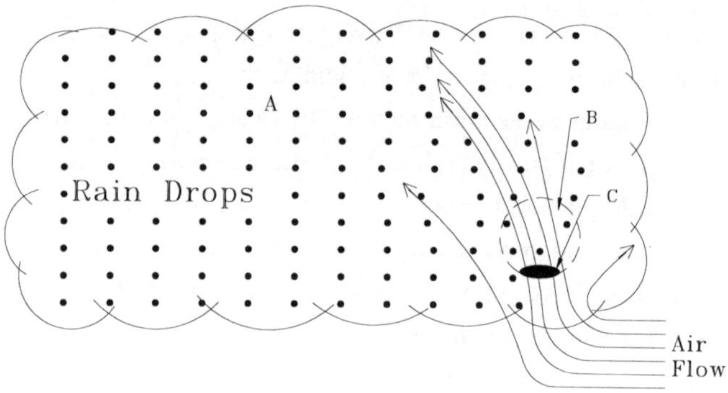

Fig. 4.2 Simpson's model of a thundercloud.

Wilson's theory suggests that a positive layer of charge resides on the upper part of a thundercloud, and a negatively charged layer at the bottom. Simpson's theory suggests that the main body of a thundercloud is negatively charged, and that the base of the cloud contains a region of high-density positive charges. Experiments with balloons equipped to measure electric fields confirmed that the upper region of the thundercloud indeed contains positive charges and the lower region contains negative charges, as Wilson predicted. However, a concentrated region of positive charge frequently exists at the base of the cloud, as Simpson had predicted. This has also been verified by ground-based instruments from observations on approaching and receding thunderclouds, and from observations directly beneath the cloud.

Simpson subsequently revised his theory. According to the revised theory, the temperature of the upper region of the positively charged layer is below -10°C, where water drops cannot exist. From observations on blizzards, Simpson concluded that the impact of the ice crystals in the upper region of the thundercloud would result in the ice crystals becoming negatively charged and the air becoming positively charged. The negatively charged ice crystals sink to the lower region, leaving the upper region of the thundercloud positively charged. More recent investigations tend to give ice crystals a more significant role in the electrification of thunderclouds. Field tests on overhead power lines have shown that the majority of lightning strokes to earth lowers negative charge. However, the frequency of positive lightning strokes increases during lightning activities in the winter.

4.2 Mechanism of the Lightning Flash

The systematic study of the mechanism of the lightning stroke was started by Schonland and his associates [13-21]. With the help of the Boys rotating camera, and later with the cathode ray oscilloscope, they found that a single flash of lightning is composed of a number of separate discharges and that the majority of the lightning flashes lowers negative charges from the cloud to the earth.

4.2.1 Pilot streamer and stepped leader

Because the base of the cloud is a few kilometers above the ground, the critical breakdown electrical field of air at the cloud base is lower than that at the ground level. Therefore, for strokes to level ground or to low objects, the discharge always starts at the cloud end. The discharge from the cloud consists of a series of streamers moving downwards in a step-like manner, and hence is called the 'stepped leader'. The length of each step is about 50 m. After completing a step, the tip of the streamer appears to pause for a time in the order of 100 microseconds; the streamer then extends still further, the new step being much brighter than the rest of the streamers. The branching of the stepped leader in open country appears to be affected by the nature of the terrain. In flat, bare country the discharges show little or no branching; whereas, in hilly, wooded country profuse branching is usually observed [22].

While the average velocity of advance of the stepped leader is about 0.15 m/μs, it actually advances in steps which proceed at a velocity of about 20 m/μs. Schonland postulated that the stepped leader is preceded by a 'pilot' streamer, carrying a current of several amperes and traveling continuously with a velocity equal to the effective rate of advance of the leader stroke (i.e., 0.15 m/μs). Because of the weak luminosity, it has not been possible to photograph it; therefore, its existence is circumstantial. The usual velocity of advance of the pilot streamer, which is 0.15 m/μs, is nearly equal to the observed velocity of electrons in the minimum ionizing field in air. The much higher velocity of about 20 m/μs, which is attained by the stepped leader, may be explained by the presence of a pre-ionized path ahead of its tip in which appreciable ionization remains from the passage of the preceding pilot streamer.

Cravath and Loeb [23], and Meek [24] refined Schonland's hypothesis. Their proposed mechanism of the pilot streamer-cum-stepped leader model is explained as follows. When the electric field at the base of the cloud exceeds the breakdown strength of air, the pilot streamer starts to advance into the virgin air by photoionization at the constant rate of about 0.15 m/μs. The pilot streamer proceeds until the electric field at its tip is insufficient to further ionize the air. In the meantime, the stem of the pilot streamer loses electrons by recombination. This results in an effective resistance along the channel. When the voltage drop caused by the current flow through this resistance is high enough, a new discharge flows along the channel. This is the stepped leader.

Bruce [25,26] objected to the Schonland-Loeb-Meek theory by stating that the experimental data on current and voltage show a discrepancy by a factor of 1000 from those postulated in the proposed theory. Bruce suggested that as the leader stroke progresses from the cloud towards the ground, the potential difference between the discharge track and the surrounding space causes lateral corona current to flow. The total corona current increases with the length of the channel, and when the current is about 1 A, the character of the channel changes suddenly from a glow to an arc. This, according to Bruce, accounts for the progress of the stepped leader.

4.2.2 Ground streamer and return stroke

As the stepped leader approaches the ground, the electric field on the ground increases. When the tip of the stepped leader is at a certain height from the ground, the electric field on the ground directly below the tip (for level terrain) exceeds the critical breakdown field of air. An upward-moving streamer starts from the ground reaching towards the downward-moving stepped leader. The velocity of the upward-moving ground streamer increases as it approaches the stepped leader. The height of the ground streamer has been measured by Schonland to be between 20 m to 150 m, and estimated by Wagner and McCann [27] to be a few feet (and definitely not over 100 feet) in length for strokes to open ground or to relatively low objects.

An intensely illuminated discharge, called the return stroke, propagates upwards when the ground streamer and the stepped leader meet. The velocity of the return stroke has been measured by Schonland to be between 20 to 144 m/μs, with the most frequent value being 35 m/μs. The velocity of the return stroke decreases as it travels upwards, the velocity of the lower part being on the order of 100 m/μs. The luminosity of the return stroke also decreases with height. Therefore, it can be concluded that the higher velocities are associated with higher luminosities. Stroke current measurements show the same general trend, so that the evidence is quite strong that higher return-stroke velocities are associated with higher return-stroke currents. Figure 4.3 depicts the progression of the stepped leader followed by the return stroke in a lightning flash.

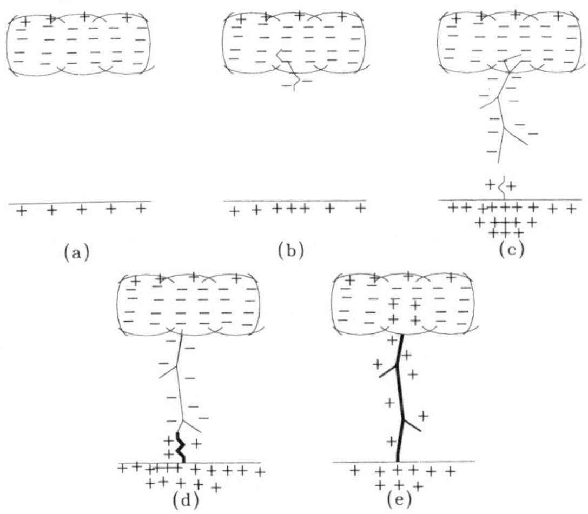

Fig. 4.3 Progression of a lightning flash.

(a) charged cloud; (b) stepped leader leader initiated from cloud base; (c) initiation of ground streamer; (d) initiation of return stroke; (e) completion of the first stroke.

4.2.3 Subsequent strokes

After the return stroke has reached the cloud, a new discharge may start from the cloud. This new discharge, called the 'dart leader' by Schonland, is generally not stepped, and its velocity is generally higher than that of the stepped leader. Once the dart leader reaches the ground, a return stroke from the ground travels towards the cloud with higher velocity, but lower current and luminosity, than the first return stroke. Schonland has measured as many as 27 component strokes in a single lightning flash, each stroke consisting of a stepped or dart leader and a return stroke. However, the first stroke is always much more intense than the succeeding ones.

4.2.4 Charge distribution along a stepped leader

Schonland concluded that the charge distribution along the stepped leader is uniform on the basis of comparison of the final leader field change with the net electrostatic field change caused by the return stroke. This conclusion was based on 18 observations out of 72 strokes. However, the other 54 observations showed nonuniform distribution of charge.

Bruce and Golde [28] assumed an exponential distribution of the charge:

$$q_h = q_o \exp(-\beta h) \, , \tag{4.1}$$

where, h = height above ground, q_o = charge per unit length near the ground end (leader tip), and β = a constant. Bruce and Golde assumed that q_h depends upon the capacitance and the potential of the channel at height h. The distribution of capacitance along the leader-stroke channel and a fairly uniform potential drop along this channel produced by the charge distribution, should, according to Bruce and Golde, lead to a distribution of charge per unit length decreasing markedly from earth to cloud, which agrees with the general form of the tail of the lightning current wave.

4.2.5 Velocity of stroke propagation

Schonland and his associates showed that the velocities of the stepped as well as the dart leaders decrease as they approach the ground. As for the return stroke, they found in their earlier study that the velocity did not follow a given pattern. For example, for a particular discharge, one earlier return stroke had a higher velocity up to 1.2 km from ground; then it decreased during the last part of its propagation to the cloud. During the same discharge, however, for a subsequent return stroke, the velocity was higher initially, then it decreased and finally it increased again [13]. In their later study, they gave two examples where the velocity of the return stroke increased initially before it fell to lower values. However, it is suspected that the initial lower value of the velocity may be caused by inadvertent inclusion of the upward-moving ground streamer.

Schonland has shown that the velocity of a streamer is dependent on the electron density on the trail [16]. The free electrons or ions produced during the previous ionizing process would tend to disappear by processes of capture and recombination, and the electron or ion density would be an inverse function of the age of the trail. As the charge density of the stepped leader decreases with its height above ground, the velocity of the return stroke should also decrease as it propagates towards the cloud. The decreased velocity of the stepped leader as it approaches the ground can be explained by the assumption that the density of free electrons in the atmosphere is higher at a greater height above ground.

By studying Schonland's experimental data, Bruce and Golde suggested that the return-stroke velocity can be represented by an expression of the form:

$$v(t) = v_0 \exp(-\gamma t) , \tag{4.2}$$

where v_0 = return-stroke velocity at the ground level,
t = time, and
γ = a constant.

Rudenberg found that the velocity of a streamer is dependent upon the current in the streamer as follows [29]:

$$v = \frac{18 I 10^6}{r k E_0} \text{ km/s} , \tag{4.3}$$

where v = velocity of the streamer,
I = current in the streamer, A,
r = radius of the streamer, cm,
k = dielectric constant of the medium, and
E_0 = critical breakdown field, V/cm.

The Rudenberg equation for the velocity of a streamer applies to the stepped leader in the case of the lightning stroke.

Lundholm calculated the velocity of the return stroke [30]:

$$v = \frac{c}{\sqrt{1 + \frac{A}{I}}} , \tag{4.4}$$

where c = velocity of light in free space,
A = a constant, varying between 300 to 900, and
I = return-stroke current, kA.

Rusck [31] modified Lundholm's equation by giving the constant, A, a definite value, namely A=500.

4.2.6 Waveshape of return-stroke current

To facilitate computations, the shape of the current wave in the return stroke is assumed to be rectangular in most analytical studies. Then by Duhamel's theorem, the results can be transformed to those corresponding to any specified current waveshape. Assuming uniform distribution of the charge, q_o, along the stepped leader, and a uniform velocity, v_o, of the return stroke, the current of the rectangular waveshape in the return stroke is

$$I_o = q_o v_o \; . \tag{4.5}$$

Bruce and Golde suggested a double-exponential waveshape for the return-stroke current:

$$I(t) = I_o[\exp(-\alpha t) - \exp(-\beta t)] \; . \tag{4.6}$$

This type of current wave rises to its peak fairly rapidly and decays rather slowly by proper choice of α and β. The decaying part of the wavetail can be explained by assuming a charge density which decreases with height, and a return-stroke velocity which decreases likewise with the height of the channel above ground. However, there seems to be no agreement on the physical reasoning for the front (i.e., the rising part) of the current wave. It is known from Schonland's observations that the diameter of the stepped leader is much larger than that of the return stroke. On this basis, Bruce and Golde proposed that it takes time for the charge in the return stroke at any height to neutralize the charge of the stepped leader, which is distributed over a much larger area. This should account for the finite front time of the current wave. Assuming an exponential function as the neutralizing time, Bruce and Golde were able to reproduce analytically a current waveshape of the double exponential form of (4.6).

Wagner proposed that the finite front time is caused by the time taken by the upward-moving ground streamer to meet the downward-moving stepped leader [32]. As the upward-moving ground streamer is not capable of carrying the heavy currents associated with the return stroke, this explanation does not seem to be feasible.

Rusck suggested that the upward-moving return stroke may not carry as much charge per unit length as the stepped leader [31]. This implies that the lightning channel cannot be neutralized instantaneously, and thus the finite time for the wave front of the return-stroke current can be explained. Rusck's proposition suggests that the charge density in the return stroke should depend upon the availability of charges on the earth, which in turn should depend upon the earth's resistivity. Therefore, the ratio of the charge density in the return stroke to that in the stepped leader should depend upon the earth's resistivity. In other words, the duration of the wavefront of the return-stroke current should depend upon the earth's resistivity; the lower the earth's resistivity, the steeper the wavefront of the return-stroke current will be. This implies that lightning strokes to the open sea would have very short wavefronts. However, no data are available to prove or disprove this thesis.

Berger, in his extensive research on lightning at the lightning research station on top of Monte San Salvatore, showed that the return-stroke current waveshape is anything but double exponential [33-35]. There seems to be a preponderance of return-stroke current waveshapes with a concave wavefront where the current rises very slowly at first, then changes the slope and rises rapidly to its crest. It was suggested by Chowdhuri [36] that the slow-rising part of the wavefront is caused by the upward-moving ground streamer. As neither the velocity nor the charge density of the ground streamer is high the low value of the current can thus be explained. Once the ground streamer meets the stepped leader and contact with the ground has been made, both the velocity and the charge density may increase so that the current increases rapidly. It should be mentioned here that the upward-moving ground streamers play a prominent role in the mechanism of lightning discharge on high-altitude objects, such as mountaintops and tall towers. Contrary to Berger's experience with concave wavefronts of the return-stroke currents, Yokoyama's data on the lightning discharges on a 200-m tall tower do not show the concave wavefronts [37-39].

4.2.7 *Other forms of lightning discharges*

The cloud-to-ground discharge just explained was first described by Schonland. However, there are several other types of cloud-to-ground discharges which though infrequent also occur. Berger classified eight types of cloud-to-ground lightning discharges [35]. These are shown schematically in Fig. 4.4. The type 1b, shown in Fig. 4.4, is the most frequent cloud-to-ground discharge.

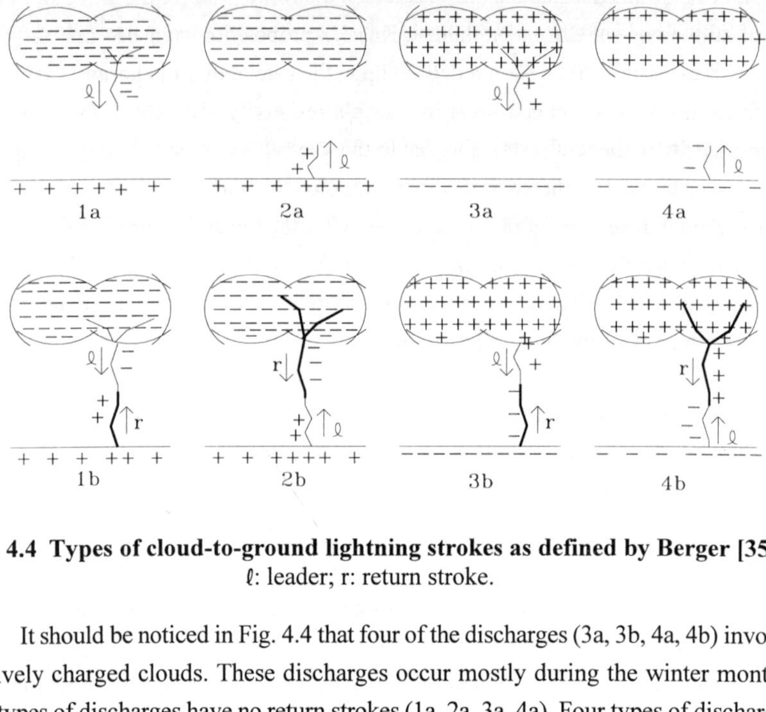

Fig. 4.4 Types of cloud-to-ground lightning strokes as defined by Berger [35].
ℓ: leader; r: return stroke.

It should be noticed in Fig. 4.4 that four of the discharges (3a, 3b, 4a, 4b) involve positively charged clouds. These discharges occur mostly during the winter months. Four types of discharges have no return strokes (1a, 2a, 3a, 4a). Four types of discharges are initiated from the ground (2a, 2b, 4a, 4b). These occur mostly from tall objects, such as mountaintops (San Salvatore) and tall structures (Empire State Building).

Apart from the cloud-to-ground lightning discharges, cloud-to-cloud discharges do exist. In fact, inter- and intra-cloud discharges are more frequent than the cloud-to-ground discharges. We will be concerned solely with the cloud-to-ground discharges because of their devastating effects to objects on the earth.

4.3 Statistical Variation of Lightning Parameters

For the protection of electrical power systems, the following characteristics of lightning are important:

1. frequency of occurrence,
2. magnitude of the return-stroke current, and
3. waveshape of the return-stroke current.

Lightning is a random phenomenon. No two lightning discharges are the same. Because of the random nature of lightning, measurement of its characteristics is extremely difficult. Therefore, statistical inferences are made from many years of measurements by many investigators around the world.

4.3.1 Frequency of occurrence

Traditionally, meteorologists have defined the severity of lightning storms in a given region by the number of days thunder was heard in that region per year. This is called the keraunic level of that region. As the severity of thunderstorms varies from year to year in a given region, data averaged over many years for a given region are published as its keraunic level. The isokeraunic maps of the world and the U.S.A. are shown in Figs. 4.5 and 4.6. The keraunic level does not distinguish between the number of times thunder was heard in a given day. For instance, if thunder was heard once in region A, and five times in region B in a given day, the keraunic levels for the two regions would still be the same, even though the damage done in region B would perhaps be higher. The keraunic level also does not take into account the duration of the lightning activity, nor does it distinguish between cloud-to-cloud and cloud-to-ground discharges. The frequency of cloud-to-ground discharges (i.e., ground flashes) is the parameter which is of significance for the protection of power systems against lightning. The keraunic level has been used by power engineers to design lightning protection because no other data on lightning severity were available. This resulted in the unreliable design of lightning protection. The empirical equations used to correlate the keraunic level with the ground flash density in different countries are shown in Table 4.1 [41].

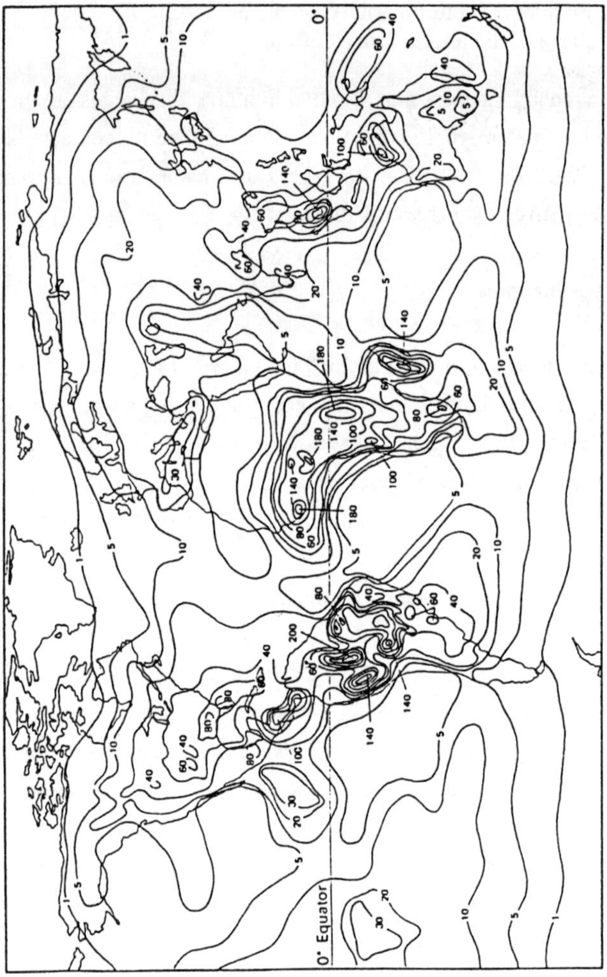

Fig. 4.5 Isokeraunic map of the world [40].
Numbers show the average number of thunderstorm days per year.

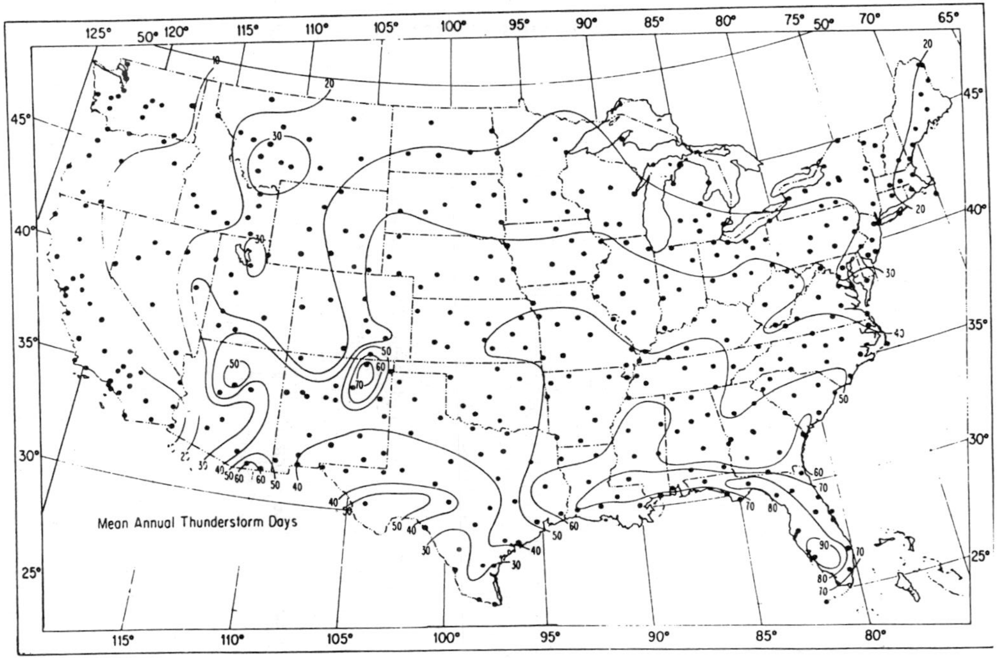

Fig. 4.6 Isokeraunic map of the U.S.A. [42].

Table 4.1: Empirical Relationship between Ground-Flash Density and Isokeraunic Level [41]

Location	Ground-Flash Density (Flashes/km^2/year)
India	0.1T
Rhodesia	0.14T
Sweden	$0.004T^2$
U.K.*	aT^b
U.S.A. (north)	0.11T
U.S.A. (south)	0.17T
U.S.A.	0.1T
U.S.A.	0.15T
U.S.S.R.	$0.036T^{1.3}$
World (temperate climate)	0.19T
World (temperate climate)	0.15T
World (tropical climate)	0.13T

T = IKL

*: $a = (2.6 \pm 0.2)10^{-3}$; $b = 1.9 \pm 0.1$

The true index of lightning severity would be the ground-flash density of a region. Ground-flash density is defined as the number of lightning flashes which struck a ground area of 1 km^2 per year. The ground-flash density is an extremely important parameter in the design of lightning protection. The outage rate of a power line from lightning is directly proportional to the ground-flash density. A 50 percent error in the ground-flash density (0.1T vs. 0.15T) would produce a 50 percent error in the estimation of the outage rate of a power line.

From lightning strike location data collected in Florida and Oklahoma, MacGorman and associates [42] found that the ground-flash density could be estimated by the equation:

$$n_g = 0.054\,H^{1.1}\,, \qquad (4.7)$$

where n_g is the ground-flash density and H is the thunderstorm duration in hours. They compiled the thunderstorm duration data from 450 aviation stations nationwide for a period between 8 and 30 years (Fig. 4.7). Using the above relation between n_g and H, they developed ground-flash density map of the U.S.A. This is shown in Fig. 4.8.

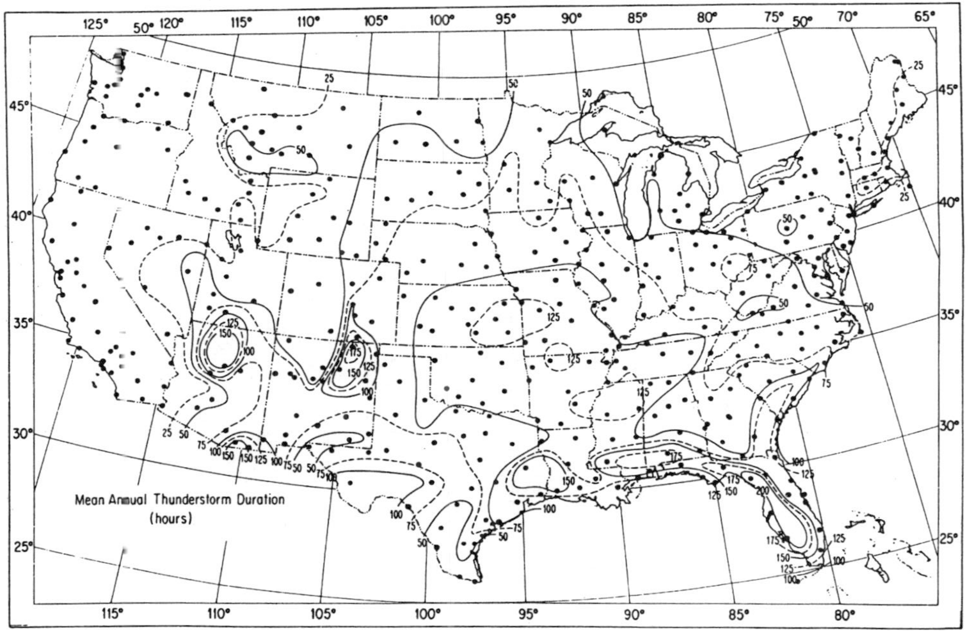

Fig. 4.7 Contour map of mean annual thunderstorm duration in the U.S.A. [42].

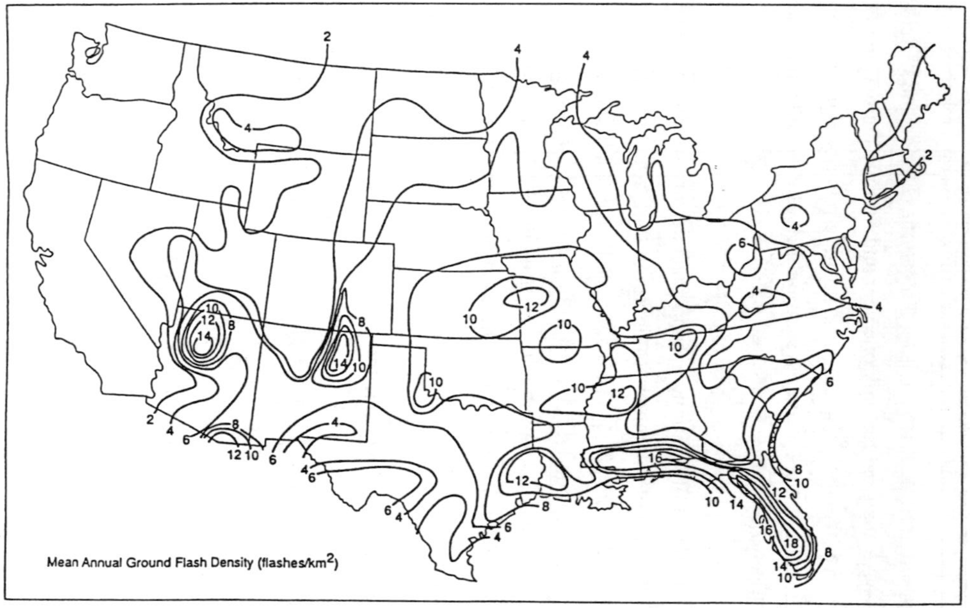

Fig. 4.8 Contour map of mean annual lightning ground-flash density in the U.S.A. [42].

Estimates of the lightning ground-flash density (Fig. 4.8) from thunderstorm duration data (Fig. 4.7) are an improvement on the previous estimates which were based on the keraunic levels. The contour map of Fig. 4.7 was developed from data collected by 450 aviation weather stations over many years across the United States. However, the limitations of Fig. 4.8 should be borne in mind. The ground-flash densities, n_g, in Fig. 4.8 were derived from the thunderstorm durations, H, in Fig. 4.7 by applying (4.7). As was concluded in [42], (4.7) is an empirical equation, based on a relatively small data base, and the data were collected from only two locations (Florida and Oklahoma). In each location, data were analyzed for only one year and for only the season of the year when thunderstorms were most frequent. The positive-polarity lightning strokes to ground were not detected by the locating system in [42]. This would tend to lower the actual ground-flash density.

Direct measurement of lightning ground-flash density would obviously be the best method. Research and development on lightning flash counters started almost 50 years ago, and is still continuing [43]. The biggest challenge of a lightning flash counter is to distinguish cloud-to-ground discharges from the cloud-to-cloud discharges, and ignore the latter. The latest development in the measurement of ground-flash density is discussed in [44,45]. The contour map of the lightning ground-flash density in the U.S.A. by direct measurement is shown in Fig. 4.9. The gridded data on the upper right-hand corner of Fig. 4.9 for Vermont (VT) show significant additional information beyond what can be determined from the contour plot because of higher spatial resolution. There is also no need for the smoothing which is necessary to produce contour lines. Such high-resolution gridded data should be used to estimate thunderstorm severity on a regional basis.

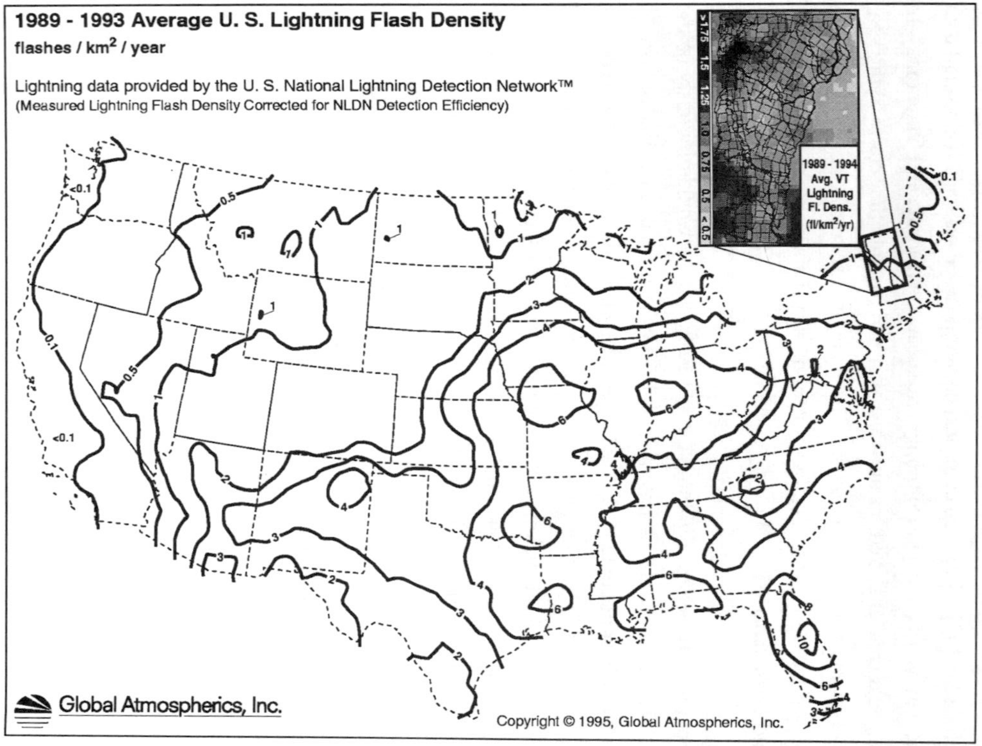

Fig. 4.9 Contour of lightning ground-flash density in the U.S.A. by direct measurement.

(Courtesy of Global Atmospherics, Inc.)

4.3.2 Magnitude of return-stroke current

Analysis of field data from around the world shows that the statistical variations of the peak return-stroke current, I_p, fit the log-normal distribution [46]. The cumulative frequency distribution of the peak return-stroke current is shown in Fig. 4.10 [47]. Assuming I_p to be a random variable, it will have a log-normal distribution when ℓnI_p has a normal distribution. The probability density function $p(I_p)$ of I_p then can be expressed as

$$p(I_p)=\frac{1}{\sqrt{2\pi}I_p\sigma_{\ell nI_p}}\exp[-0.5(\frac{\ell nI_p-\ell nI_{pm}}{\sigma_{\ell nI_p}})^2], \qquad (4.8)$$

where, $\sigma_{\ell nIp}$ = standard deviation of ℓnI_p, and
I_{pm} = median value of return-stroke current, I_p.

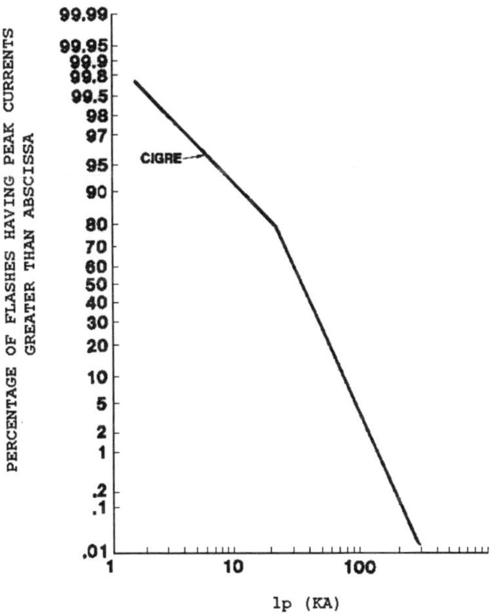

Fig. 4.10 Cumulative frequency distribution of peak return-stroke current of negative polarity [47].

To determine the cumulative probability, P_c, that the peak current in any lightning flash will exceed I_p kA, (4.8) needs to be integrated as follows:

$$\text{Putting,} \quad u = \frac{\ell n I_p - \ell n I_{pm}}{\sqrt{2}\sigma_{\ell n I_p}} ,$$

$$P_c(I_p) = \frac{1}{\sqrt{\pi}} \int_u^{\infty} e^{-u^2} du = 0.5\,\mathrm{erfc}(u) . \tag{4.9}$$

A much simpler form of an equation for P_c was given by Anderson, which fits the curve of Fig. 4.10 very well [48]. This is given by

$$P_c = \frac{1}{1 + \left(\frac{I_p}{31}\right)^{2.6}} , \tag{4.10}$$

where I_{pm} was assumed to be 31 kA, and $\sigma_{\ell nIp} = 0.691$. A comparison between P_c (rigorous) in (4.9) and P_c (Anderson) in (4.10) is shown in Table 4.2.

Table 4.2: Comparison between Rigorous and Anderson P_c

I_p, kA	P_c (Rigorous)	P_c (Anderson)
1.0	0.9990	0.9999
5.0	0.9701	0.9914
10.0	0.9132	0.9499
15.0	0.8545	0.8685
20.0	0.7995	0.7576
25.0	0.6822	0.6363
30.0	0.5685	0.5213
35.0	0.4672	0.4218
40.0	0.3809	0.3401

Generally, (4.10) is adequate for probability computation in backflashover. For low shielding currents in shielding designs, (4.9) would give more accurate results.

4.3.3 Distribution of return-stroke current waveshape

After analyzing Berger's extensive data on the return-stroke current, Anderson and Eriksson suggested the 'typical' waveshape of a return-stroke current as shown in Fig. 4.11 [47]. The equivalent front time, t_f, can be computed in two different ways in Fig. 4.11. $t_f=T_{10}/0.8$ or $t_f=T_{30}/0.6$, where T_{10} is the time interval between the 10 percent and the 90 percent points of the current wave, and T_{30} is the time interval between the 30 percent and the 90 percent points of the current wave.

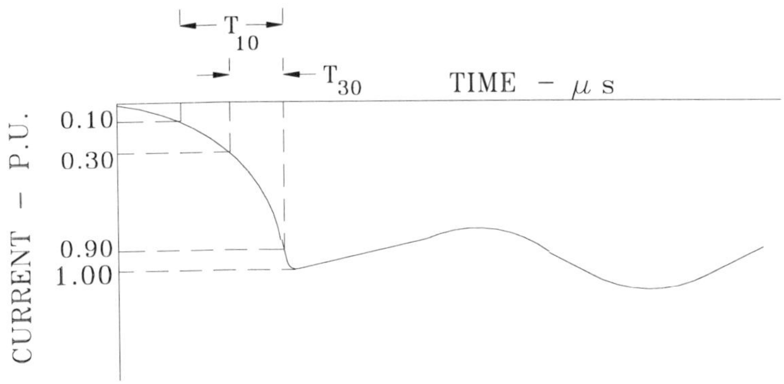

Fig. 4.11 Typical return-stroke current [47].

Similar to the current amplitude, the probability density function of the current front time can be expressed as:

$$p(t_f)=\frac{1}{\sqrt{2\pi}t_f\sigma_{\ell nt_f}}\exp[-0.5(\frac{\ell nt_f-\ell nt_{fm}}{\sigma_{\ell nt_f}})^2] \ . \tag{4.11}$$

The joint probability density function of the amplitude and the front time of the current can be expressed as:

$$p(I_p,t_f) = \frac{\exp[\frac{-0.5}{1-\rho^2}(f_1-f_2+f_3)]}{2\pi I_p t_f \sigma_{\ell nI_p}\sigma_{\ell nt_f}\sqrt{1-\rho^2}} \ , \tag{4.12}$$

$$\text{where} \quad f_1 = (\frac{\ell nI_p - \ell nI_{pm}}{\sigma_{\ell nI_p}})^2 \ , \tag{4.13}$$

$$f_2 = 2\rho\left(\frac{\ell nI_p - \ell nI_{pm}}{\sigma_{\ell nI_p}}\right)\left(\frac{\ell nt_f - \ell nt_{fm}}{\sigma_{\ell nt_f}}\right) , \tag{4.14}$$

$$f_3 = \left(\frac{\ell nt_f - \ell nt_{fm}}{\sigma_{\ell nt_f}}\right)^2 \text{ ,and} \tag{4.15}$$

ρ = coefficient of correlation.
If I_p and t_f are independently distributed, then $\rho = 0$, and

$$p(I_p,t_f) = p(I_f)p(t_f) . \tag{4.16}$$

The statistical parameters of the return-stroke current are as follows [47,49]:

Median time to crest, $t_{fm1} = 3.83$ μs; Natural log of standard deviation, $\sigma_{\ell ntf1} = 0.553$

For $I_p \leq 20$ kA: Median peak current, $I_{pm1} = 61.1$ kA (Fig. 4.10)
Log (to base e) of standard deviation, $\sigma_{\ell nIp1} = 1.33$

For $I_p > 20$ kA: Median peak current, $I_{pm2} = 33.3$ kA (Fig. 4.10)
Log (to base e) of standard deviation, $\sigma_{\ell nIp2} = 0.605$

Correlation coefficient (between time to crest and peak current), $\rho = 0.47$

4.4 Protective Shadow of a Structure

It is known that taller structures are hit more frequently by lightning than shorter structures. The rule of thumb has been that a vertical structure will attract lightning from a distance of twice its height (Fig. 4.12). In fact, the design of a lightning rod is based on this principle.

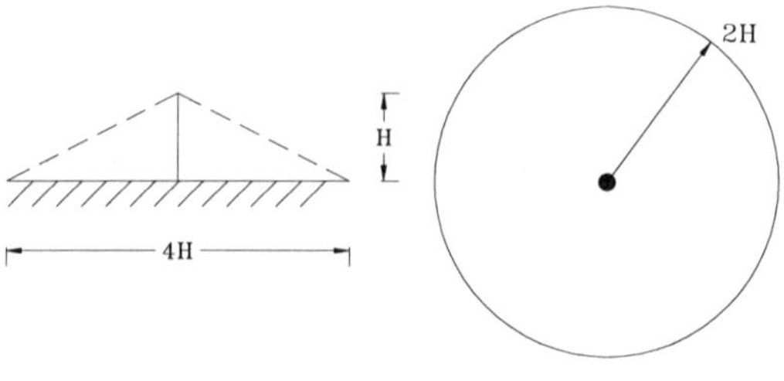

Fig. 4.12 Protective shadow of a lightning rod.

The principle of the protective shadow of a lightning rod has been extended to the attractive swath of an overhead line as shown in Fig. 4.13. It is as if an overhead line throws an electrical shadow on the ground below it, so that all lightning strokes falling within the width of this electrical shadow will hit the line instead of the ground. This width W is given by

$$W = b + 4H \quad (m) \tag{4.17}$$

where b is the horizontal separation distance between the two shield wires, and H is the average height of the shield wires. H is generally assumed to be

$$H = H_g - \frac{2}{3}(H_g - H_{gm}) \quad (m) \tag{4.18}$$

where H_g is the height of the shield wires at the tower, and H_{gm} is the shield-wire height at midspan. For a line with one shield wire, b is zero. For a line with no shield wires, b is the horizontal separation distance between the two outer phase conductors, and H pertains to the phase conductor height.

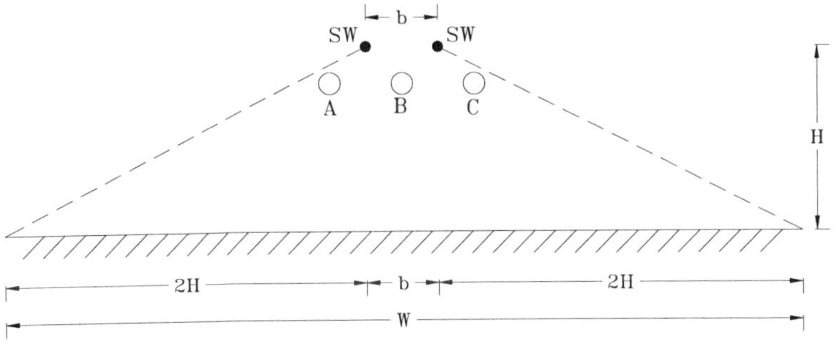

Fig. 4.13 Electrical shadow width of an overhead power line.
SW = shield wires; A,B,C = phase conductors.

4.5 Striking Distance

The lightning discharge starts at the cloud end for strokes to level ground or to low objects. In the initial stage, the leader stroke proceeds downwards without being influenced by grounded objects. As the charge of the cloud is lowered along the leader stroke, the electric field on the surface of grounded objects increases. Finally, at a certain distance from the tip of the leader stroke to the grounded object, the critical electric field for breakdown of air at the surface of the grounded object is reached, and an upward streamer starts from the object to meet the leader stroke. This distance from the leader tip to the grounded object which produces the upward streamer is called the striking distance. The striking distance is a very significant parameter in estimating the lightning performance of overhead power lines.

Based upon laboratory data on the breakdown of long air gaps, Golde assumed that the critical breakdown electric field between the leader tip and either a ground plane or a grounded vertical conductor or a grounded horizontal conductor will be about the same. According to Golde, this critical electric field will be 500 kV/m for a negative stroke and 300 kV/m for a positive stroke [50,51].

Golde also suggested that for vertical structures of low height (20 m or less) the striking distance is not affected by the structure height. For tall structures, a positively charged 'plume' is produced on the tip of the structure as a negatively charged leader approaches. The attraction of the negative leader stroke to the positively charged 'plume' would result in unusually long striking distances [52].

Golde's most significant contribution to the striking distance is the hypothesis that the striking distance r_s is a function of the return-stroke current. Whitehead formalized this relation by a series of analytical expressions, the one most widely used being given by [53]

$$r_s = 8\,(I_p)^{0.65} \quad (\mathrm{m}) \tag{4.19}$$

where I_p = peak return-stroke current in kA. An alternative equation, also given by Whitehead, has also been used [48,53]

$$r_s = 10\,(I_p)^{0.65} \quad (\mathrm{m}) \tag{4.20}$$

Two other alternative equations for the striking distance have also been suggested:

$$r_e = 6.4\,(I_p)^{0.75} \quad (\mathrm{m}) \tag{4.21}$$

$$r_c = 7.1\,(I_p)^{0.75} \quad (\mathrm{m}) \tag{4.22}$$

where r_e = the striking distance to earth, and
r_c = the striking distance to the line conductor.

It has, however, long been known that taller structures, e.g., chimneys and masts, attract more lightning strokes. Eriksson combined the effects of both the structure height H and the return-stroke current in kiloamperes on the attractive radius R_a of prominent structures, such as transmission-line towers, in the following equation [54, 55]:

$$R_a = 0.84\,H^{0.6}(I_p)^{0.74} \quad (m) \ . \tag{4.23}$$

Eriksson then assumed that the attractive radius R_a for horizontal conductors, such as the shield wires and the phase conductors of overhead power lines, will be reduced to about 80 percent, giving:

$$R_a = 0.67\,H^{0.6}(I_p)^{0.74} \quad (m) \ . \tag{4.24}$$

If the striking distance of a lightning stroke to an object is determined primarily by the critical electric field at the conductor surface or the average critical electric field between the leader tip and the grounded object, then several other parameters should also influence the striking distance, in addition to the structure height and the return-stroke peak current. Some of these parameters are the charge distribution along the leader stroke (return-stroke current waveshape), return-stroke velocity, the presence of other conductors, etc. Moreover, laboratory data on breakdown of long air gaps show that the breakdown depends upon the polarity of the high-voltage electrode (i.e., the leader stroke), as well as on the type of the air gap, for instance, rod-rod, vertical rod to horizontal conductor or rod-plane gaps. Therefore, the striking distances from the tip of the leader stroke to a transmission-line tower, to a horizontal phase conductor, or to flat ground will be different. It will also depend on the polarity of the leader stroke. These parameters were taken into account in a recent study which showed that some of these parameters are significant, and that intensive research is needed in this critical area [56].

In the interim period, however, it is suggested that (4.19) and (4.20) be used for overhead lines of short heights, e.g., power distribution lines, and (4.23) and (4.24) be used for overhead lines of tall structures, e.g., high-voltage power transmission lines.

4.6 Electrogeometric Model for Lightning Strike to an Overhead Line

According to the electrogeometric model, the striking distance, r_s, is a function of the peak return-stroke current; the higher the return-stroke current, the longer the striking distance r_s. If the tip of a leader stroke approaches within r_s, the leader stroke will perform the 'final jump' to the conductor. If the leader tip is beyond r_s from the conductor, then, by default, it will approach the earth, and perform a 'final jump' to the earth from a critical distance, r_e, from the earth's surface. Generally, r_e will be different from r_s, because the gap geometry between the leader and the earth is different from that between the leader and the conductor. This difference is evident from (4.21) and (4.22). r_e can generally be expressed as [48]

$$r_e = \beta \times r_s \; . \tag{4.25}$$

The electrogeometric model of lightning strike to a line is schematically shown in Fig. 4.14. The curve ($a_3a_2a_1vb_1b_2b_3$) enclosing the line conductor is the locus of stroke-tip positions which decides whether the leader stroke will strike the conductor or the earth. If the stroke tip is within this curved boundary, and its distance from the conductor is equal to or less than r_s, corresponding to the prospective return-stroke current of the leader stroke, then it will strike the conductor. However, even when the stroke tip is within this curved boundary, but during its descent its distance from the conductor is always greater than r_s, then it will pass through this boundary and eventually strike the earth.

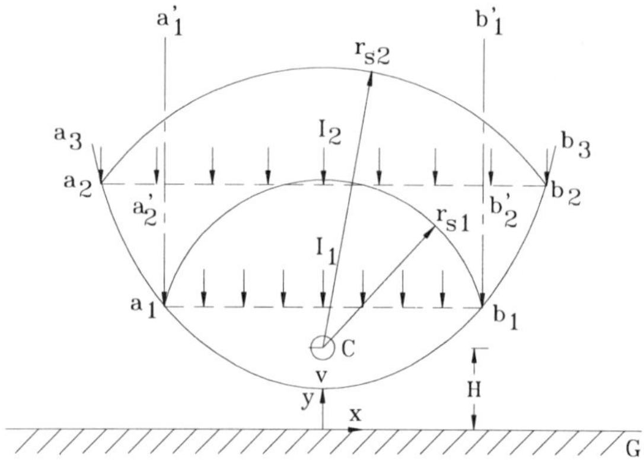

Fig. 4.14 Paraboloid of attraction for strike to overhead line.
$a_3a_2a_1vb_1b_2b_3$ = paraboloid; C = line conductor; G = ground.

It should be recognized that (4.25) is the equation of a conic section. For β less than, equal to or greater than one, the curve will be a hyperbola, parabola or an ellipse, respectively. Assuming the simple case when $\beta = 1.0$ for a parabolic curve, let us determine the attractive width through which a leader stroke of a given prospective return-stroke current must fall to be attracted to the line conductor. For a given return-stroke current I, the corresponding striking distance r_s can be determined from (4.19) or (4.24). Then, any point on the parabola is found by trigonometric manipulation as

$$r_s^2 = x^2 + (r_s - H)^2 \ . \tag{4.26}$$

For a parabola, $r_s = y$. Replacing r_s in (4.26),

$$y = \frac{x^2 + H^2}{2H} \ . \tag{4.27}$$

The parabola of attraction can now be drawn for a line of height H from (4.27). For a given return-stroke current I_1, a circle of radius r_{s1} is drawn with its center at the line conductor. This circle cuts the parabola at a_1 and b_1. Therefore, all return-stroke currents, $I \geq I_1$, falling within the width a_1b_1 will strike the line, C. The width a_1b_1 can be determined by rearranging (4.27) as follows:

$$x = \pm\sqrt{H(2y - H)} \ . \tag{4.28}$$

Putting, $y = r_{s1}$

$$a_1 b_1 = 2\sqrt{H(2r_{s1} - H)} \ . \tag{4.29}$$

If $r_{s1} \leq H$, then:

$$a_1 b_1 = 2 r_{s1} \ . \tag{4.30}$$

The attractive width a_2b_2 for a return-stroke current I_2, greater than I_1, can similarly be determined. If a stroke of current I_1 is falling through either a_2a_2' or b_2b_2', it will pass through the parabola and hit the earth.

4.7 Probability of Strikes to an Overhead Line

Once the attractive width of strike for a given current is known, the probability of strike to the line for the given current or higher can be calculated by determining:

(i) the exposed area of the line for the given current,
(ii) the ground flash density of the geographic area, and
(iii) the probability distribution of lightning current amplitudes [47,49].

For a line segment of 100 km, the exposed area of the 100-km segment of line to return-stroke currents, equal to or greater than I_1 is (Fig. 4.15)

$$A = 0.1\, a_1 b_1 \quad (\text{km}^2). \tag{4.31}$$

If the ground flash density of the geographic area is n_g strokes per km^2 per year, then the total number of strokes collected by 100-km segment of the line will be $n_g A$.

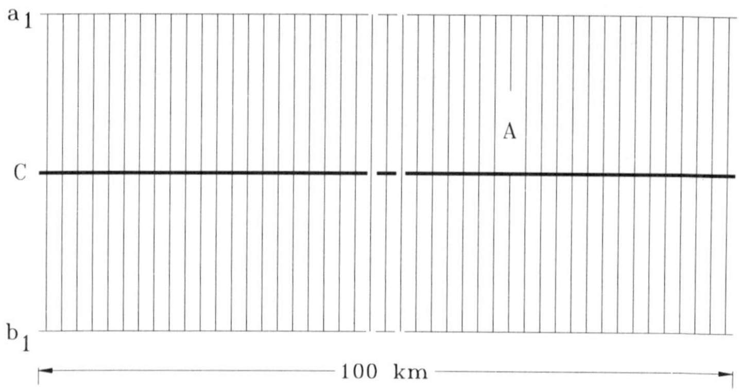

Fig. 4.15 Attractive area of a 100-km line segment.

If the probability of the return-stroke current exceeding I_1 is P_{c1}, then the number of strokes, exceeding the amplitude I_1, striking 100 km segment of the line per year will be given by

$$n_t = n_g A P_{c1}, \tag{4.32}$$

where n_g = ground flash density (number of ground flashes per km^2 per year), and P_{c1} can be computed from (4.9), (4.10), (4.12) or (4.16).

4.8 Concluding Remarks

The physical nature of lightning has not yet been completely unraveled. The ground flash density is one of the key parameters needed for the design of protective measures against lightning. The number of thunder days per year (keraunic level) used to be assumed to be directly proportional to the ground flash density, and many successful protective systems have been designed based on the keraunic level of the region. However, recent investigations [42,44,45] have shown the need for direct measurement of the regional ground flash density. As a result, lightning detection systems have been installed in many countries to measure the ground flash density directly.

The peak return-stroke current is another key parameter. Earlier data on peak current from magnetic links installed in power systems were later discarded as unreliable in preference to direct measurement. As tall structures attract more lightning, current transducers were installed on top of tall towers to collect data in a relatively short time. However, the return-stroke current is distorted by reflections from the ground as well as from the tower top [57]. In fact, the second peak of the current in Fig. 4.11 may be due to the reflection from the ground. Many researchers are endeavoring to estimate the return-stroke current by measuring its radiated electric and magnetic fields. A physical model of the return stroke is required for such estimation. Although several return-stroke models [58] are available, none can satisfactorily predict the radiated fields.

The return-stroke velocity significantly affects the response of an electrical system to lightning strokes to nearby ground. The return-stroke velocity is a function of the return-stroke current; it also varies along the lightning channel for the same stroke. Schonland [13] was perhaps the first to measure the return-stroke velocity. However, because of its complexity, the return-stroke velocity has remained one of the least known parameters of the lightning stroke.

Because of the uncertainties and the accompanying difficulties of taking measurements on natural lightning, researchers have turned to triggered lightning [59-62]. Lightning may be triggered by rapidly introducing a wire into the electric field of a thunderstorm by the use of a rocket. The rocket is fired upward at a velocity near 200 m/s and a grounded wire is unspooled from the ground. The leader is initiated from the tip of the rocket when it reaches a critical height. The lowest critical height was 50 m [61]. A good probability of triggering lightning occurs if the rocket is launched when the electrostatic field at the ground is between 5 to 10 kV/m [62]. The return stroke begins when the upward moving leader from the rocket tip reaches the cloud base. The rocket-triggered lightning emulates the subsequent strokes of a lightning flash. It cannot emulate the first stroke which generally causes the most damage. Nonetheless, the triggered lightning is a very powerful tool in calibrating measurement instruments and in evaluating theories of electromagnetic fields created by the lightning stroke.

In spite of many setbacks, exciting research on the lightning phenomena is continuing in many countries around the world. Such research will produce valuable data on the significant parameters of the lightning stroke, which will be applied in the design of more reliable and cost effective protective measures against lightning.

4.9 References to Chapter 4

1. R.A. Kerr, "Lightning Found on Venus at Last ?", ***Science***, Vol. 253, p. 1492, 27 September 1991.

2. D.A. Gurnett, et al., "Lightning and Plasma Wave Observations from the Galileo Flyby of Venus", ***Science***, Vol. 253, pp.1522-1525, 27 September 1991.

3. C.F. Wagner and J.M. Clayton, "Lightning Phenomena", Chapter 16, ***Electrical Transmission and Distribution Reference Book***. Westinghouse Electric Corp., East Pittsburgh, PA., Fourth Edition, 1964.

4. L.J. Battan, ***The Thunderstorm***. A Signet Science Library Book. Published by The American Library, New York, NY, 1964.

5. C.B. Moore and B. Vonnegut, "The Thundercloud", Chapter 3, ***Lightning***, Vol. 1: Physics of Lightning, edited by R.H. Golde. Academic Press, New York, NY, 1977.

6. P.H. Handel, "Polarization Catastrophe Theory of Cloud Electricity - Speculation on a New Mechanism for Thunderstorm Electrification", ***Jour. Geophysical Research***, Vol. 90, No. D3, pp. 5857-5863, 1985.

7. E.R. Williams, "The Electrification of Thunderstorms", ***Scientific American***, Vol. 259, No. 5, pp. 88-99, November 1988.

8. C.T.R. Wilson, "Investigations on Lightning Discharges and on the Electrical Field of Thunderstorms", ***Philosophical Trans. of the Royal Society***, Series A, Vol. 221, p. 73, 1920.

9. C.T.R. Wilson, "Some Thundercloud Problems", ***Jour. Franklin Institute***, Vol. 208, pp. 1-12, 1929.

10. G. Simpson, "The Mechanism of a Thunderstorm", ***Proc. Royal Society***, Series A, Vol. 114, p. 376, 1927.

11. G. Simpson and F.J. Scrase, "The Distribution of Electricity in Thunderclouds", ***Proc. Royal Society***, Series A, Vol. 161, p. 309, 1937.

12. G. Simpson and G.D. Robinson, "The Distribution of Electricity in Thunderclouds - II", ***Proc. Royal Society***, Series A, Vol. 177, p. 281, 1941.

13. B.F.J. Schonland and H. Collens, "Progressive Lightning", ***Proc. Royal Society***, Series A, Vol. 152, pp. 654-674, 1934.

14. B.F.J. Schonland, D.J. Malan and H. Collens, "Progressive Lightning II", ***Proc. Royal Society***, Series A, Vol. 152, pp. 595-625, 1935.

15. D.J. Malan and H. Collens, "Progressive Lightning III - The Fine Structure of Return Lightning Strokes", ***Proc. Royal Society***, Series A, Vol. 162, pp. 175-203, 1937.

16. B.F.J. Schonland, "Progressive Lightning IV", ***Proc. Royal Society***, Series A, Vol. 164, pp. 132-150, 1938.

17. B.F.J. Schonland, D.B. Hodges and H. Collens, "Progressive Lightning V - A Comparison of Photographic and Electrical Studies of the Discharge Process", ***Proc. Royal Society***, Series A, Vol. 166, pp. 56-75, 1938.

18. B.F.J. Schonland, D.J. Malan and H. Collens, "Progressive Lightning VI", ***Proc. Royal Society***, Series A, Vol. 168, pp. 455-469, 1938.

19. D.J. Malan and B.F.J. Schonland, "Progressive Lightning VII - Directly Correlated Photographic and Electrical Studies of Lightning from Near Thunderstorms", ***Proc. Royal Society***, Series A, Vol. 191, pp. 485-503, 1947.

20. B.F.J. Schonland, "The Pilot Streamer in Lightning and the Long Spark", ***Proc. Royal Society***, Series A, Vol. 220, pp. 25-38, 1953.

21. B.F.J. Schonland: "The Lightning Discharge", in ***Handbuch der Physik***. Springer-Verlag, Berlin, Vol. 22, pp. 576-628, 1956.

22. G.D. McCann, "The Measurement of Lightning Currents in Direct Strokes", ***AIEE Trans.***, Vol. 63, pp. 1157-1164, 1944.

23. A.M. Cravath and L.B. Loeb, "Mechanism of High Velocity of Propagation of Lightning Discharge", ***Physics***, Vol. 6, pp. 125-127, 1935.

24. J.M. Meek, "The Mechanism of Lightning Discharge", ***Phys. Rev.***, Vol. 55, pp. 972-977, 1939.

25. C.E.R. Bruce, "The Lightning and Spark Discharges", ***Nature***, Vol. 147, pp. 805-806, 1941.

26. C.E.R. Bruce, "The Initiation of Long Electrical Discharge", ***Proc. Royal Society***, Series A, Vol. 183, pp. 228-242, 1944.

27. C.F. Wagner and G.D. McCann, "Induced Voltage on Transmission Lines", ***AIEE Trans.***, Vol. 61, pp. 916-930, 1942.

28. C.E.R. Bruce and R.H. Golde, "The Lightning Discharge", ***Jour. IEE*** (G.B.), Part II, Vol. 88, pp. 487-505, 1941.

29. R. Rudenberg, "The Tip Velocity of Electrical Spark and Lightning", (in German), ***Wiss. Veröff. Siemens-Konzern***, Vol. 9, pp. 1-6, 1930.

30. R. Lundholm, "Induced Over-Voltage Surges on Transmission Lines and Their Bearing on the Lightning Performance at Medium Voltage Networks", ***Trans. Chalmers University of Technology***, Gothenburg, Sweden, No. 188, 1957.

31. S. Rusck, "Induced Lightning Over-Voltages on Power-Transmission Lines with Special Reference to the Over-Voltage Protection of Low-Voltage Networks", ***Trans. Royal Inst. of Tech.***, Stockholm, Sweden, No. 120, 1958.

32. C.F. Wagner, "Determination of the Wave Front of Lightning Stroke Currents from Field Measurements", ***AIEE Trans.***, Part III, Vol. 79, pp. 581-589, 1960.

33. K. Berger, "Results of the Lightning Measurements for the years 1947 to 1954 on the Mount San Salvatore", (in German), ***Bulletin SEV***, Vol. 46, pp. 405-424, 1955.

34. K. Berger, "Front Duration and Current Steepness of Lightning Strokes to the Earth", paper no.27, pp.63-73, ***Gas Discharges and the Electricity Supply Industry***. Edited by J.S. Forrest, P.R. Howard and D.J. Littler. Butterworths, London, U.K., 1962.

35. K. Berger, "The Earth Flash", Chapter 5, pp.119-190, ***Lightning***, Vol. 1 - Physics of Lightning. Edited by R.H. Golde. Academic Press, New York, 1977.

36. P. Chowdhuri, ***Voltage Surges Induced on Overhead Lines by Lightning Strokes.*** D.Eng. Dissertation, Rensselaer Polytechnic Institute, May 1966.

37. S. Yokoyama, et al., "Simultaneous Measurement of Lightning Induced Voltages with Associated Stroke Currents", ***IEEE Trans. on Power Apparatus and Systems***, Vol. PAS-102, pp. 2420-2429, 1983.

38. S. Yokoyama, et al., "Advanced Observations of Lightning Induced Voltage on Power Distribution Lines", ***IEEE Trans. on Power Delivery***, Vol. PWRD-1, pp. 129-139, 1986.

39. S. Yokoyama, et al., "Advanced Observations of Lightning Induced Voltage on Power Distribution Lines (II)", ***IEEE Trans. on Power Delivery***, Vol. 4, pp. 2196-2203, 1989.

40. R.H. Golde, ***Lightning Protection***, Chapter 2. Edward Arnold Ltd., London, 1973.

41. S.A. Prentice, "Frequency of lightning Discharges", Chapter 14 of ***Lightning***, Vol. 1, pp. 465-496. Edited by R.H. Golde. Academic Press, New York, 1977.

42. D.R. MacGorman, M.W. Maier and W.D. Rust, "Lightning Strike Density for the Contiguous United States from Thunderstorm Duration Records", Report prepared for the U.S. Nuclear Regulatory Commission, Washington, D.C., Report No. NUREG/CR-3759, May 1984.

43. R.B. Anderson, "Measuring Techniques", Chapter 13 of ***Lightning***, Vol. 1, pp. 437-463. Edited by R.H. Golde. Academic Press, New York, 1977.

44. K.L. Cummins, E.A. Bardo, W.I. Hiscox, R.B. Pyle and A.E. Pifer, "NLDN'95: A Combined TOA/MDF Technology Upgrade of the U.S. National Lightning Detection Network", International Aerospace & Ground Conference on Lightning and Static Electricity, Williamsburg, VA, 26-28 September 1995.

45. L.G. Byerley III, K.L. Cummnis, J. Tuel, D.J. Hagberg, Jr. and W. Bush, "The Measurement and Use of Lightning Ground Flash Density", International Aerospace & Ground Conference on Lightning and Static Electricity, Williamsburg, VA, 26-28 September 1995.

46. F. Popolansky, "Frequency Distribution of Amplitudes of Lightning Currents", ***Electra***, No. 22, pp. 139-147, 1972.

47. R.B. Anderson and A.J. Eriksson, "Lightning Parameters for Engineering Applications", ***Electra***, No. 69, pp. 65-102, 1980.

48. J.G. Anderson, "Lightning Performance of Transmission Lines", Chapter 12 of ***Transmission Line Reference Book 345 kV and Above***, Second Edition, Revised. Electric Power Research Institute, Palo Alto, CA., 1987.

49. A.J. Eriksson, "Notes on Lightning Parameters for System Performance Estimation", CIGRE Note 33-86 (WG33-01) IWD, July 15, 1986.

50. R.H. Golde, ***Lightning Protection***, Chapter 4. Edward Arnold Ltd., London, U.K., 1973.

51. R.H. Golde, "Lightning Conductor", Chapter 17 in ***Lightning***, Vol. 2. Edited by R.H. Golde. Academic Press, London, 1977.

52. R.H. Golde, "Lightning and Tall Structures", ***Proc. IEE*** (G.B.), Vol. 125, pp. 347-351, 1978.

53. IEEE Working Group Report, "Estimating Lightning Performance of Transmission Lines II - Updates to Analytical Models", ***IEEE Trans. on Power Delivery***, Vol. 8, pp. 1254-1267, 1993.

54. A.J. Eriksson, "The Incidence of Lightning Strokes to Power Lines", ***IEEE Trans. on Power Delivery***, Vol. PWRD-2, pp. 859-870, 1987.

55. A.J. Eriksson, "An Improved Electrogeometric Model for Transmission Line Shielding Analysis", ***IEEE Trans. on Power Delivery***, Vol. PWRD-2, pp.871-886, 1987.

56. P. Chowdhuri and A.K. Kotapalli, "Significant Parameters in Estimating the Striking Distance of Lightning Strokes to Overhead Lines", ***IEEE Trans. on Power Delivery***, Vol. 4, pp. 1970-1981, 1989.

57. A.M. Hussein, et al., "Simultaneous Measurement of Lightning Parameters for Strokes to the Toronto Canadian National Tower", ***Jour. Geophysical Research***, Vol. 100, N0. D5, pp. 8853-8861, 1995.

58. C.A. Nucci, "Lightning-Induced Voltages on Overhead Power Lines Part I: Return-Stroke Current Models with Specified Channel-Base Current for the Evaluation of the Return-Stroke Electromagnetic Fields", ***Electra***, No. 161, pp. 75-102,1995.

59. M. Brook, et al., "Artificial Initiation of Lightning Discharges", ***Jour. Geophysical Research***, Vol. 66, No. 11, pp. 3967-3969, 1961.

60. R.P. Fieux, et al., "Research on Artificially Triggered Lightning in France", ***IEEE Trans. on Power Apparatus and Systems***, Vol. PAS-97, pp. 725-733, 1978.

61. D.W. Clifford and H.W. Kasemir, "Triggered Lightning", ***IEEE Trans. on Electromagnetic Compatibility***, Vol. EMC-24, pp. 112-122, 1982.

62. M.A. Uman and E.P. Krider, "Natural and Artificially Initiated Lightning", ***Science***, Vol. 246, pp. 457-464, 1989.

Problems for Chapter 4

1. A horizontally configured unshielded three-phase overhead power line has the following dimensions:

   Line height, $h_p = 20$ m; Line length = 325 km
   Distance between adjacent phase conductors = 5 m.

   If the ground flash density, n_g, is 5 flashes per km^2 per year, then estimate the number of times per year the line will be hit by lightning strokes with return-stroke currents 20 kA and higher. Apply the electrical shadow method to determine the attractive area. Assume the probability, P(I), that the return-stroke current will exceed I kA as given by (4.9) and (4.10), and note the difference in the results, if any.

2. Repeat problem 1 by applying the electrogeometric model to determine the attractive area.

CHAPTER 5
GROUNDING FOR PROTECTION AGAINST LIGHTNING

The purpose of grounding for protection against lightning is to bypass the energy of the lightning discharge safely to the ground. In other words, most of the energy of the lightning discharge should be dissipated into the ground without raising the voltage of the protected system. Grounding is universally done by burying a metallic structure into the earth to bypass the discharge. Therefore, the rise in voltage of this grounding structure against a distant point on the earth should be minimal. This requires that the grounding resistance should be as small as possible.

5.1 Steady-State Tower-Footing Resistance

Power-line towers are erected on the ground surface, with their feet deeply buried in the earth. Precautions are taken so that the resistance of these tower feet against a distant point on the earth is low. This is called the tower-footing resistance. Ground wires, if used, are connected directly at the tower top. If lightning strikes the tower or the ground wire, its discharge should then be safely led to the earth and dissipated there. For wood poles, one or more ground rods are buried near the pole, and a copper down lead, running vertically from the top of the wood pole along its length, is connected to these rods. A low tower-footing resistance is essential for good lightning performance of an overhead power line. This will be discussed later.

The tower-footing resistance, R_{tf} is a significant parameter which determines the severity of the insulator-string stress caused by a direct stroke to the tower. The voltage wave traveling from the tower top is reflected back at the tower-footing resistance. If R_{tf} is smaller than the tower surge impedance, Z_t, then the reflected wave will be of opposite polarity which, on arrival at the tower top, will reduce the tower-top voltage. After a few successive reflections, the tower-top voltage will approach the voltage drop across R_{tf}, i.e., IR_{tf}. Therefore, R_{tf} should be as small as possible.

Several investigators have calculated the resistance of conductors of various shapes and sizes buried in the earth [1-6]. The equations for the ground resistance, derived by different investigators, do not always verify each other. Sunde's equations [3], which are widely used, are shown below. These are the static or steady-state ground resistances, which are valid for dc or power-frequency operations. The most common type of ground electrode is the driven ground rod. These are generally galvanized steel rods, 2.5 to 3 m in length and less than 2.5 cm in diameter; they are driven vertically down from the earth's surface. The resistance, R, of a driven ground rod is given by [3]:

$$R = \frac{\rho_g}{2\pi\ell}\left(\ell n\frac{4\ell}{a} - 1\right) \quad (\Omega), \ \ell \gg a \tag{5.1}$$

where ℓ=length of the buried rod (m), a=radius of rod (m), and ρ_g=soil resistivity (Ω-m). R decreases as the buried length of the rod increases. R also decreases as the rod radius, a, is increased; but the dependence on the radius is not that great. For practical applications, it is preferable to connect several ground rods in parallel. The resistance, R, will be inversely proportional to the number of parallel rods,n, provided the spacing between the rods, S, is large compared to their length. If the spacing is short and the ground rods are arranged on a circle of diameter D, then (5.1) still holds if the radius, a, is replaced by an equivalent radius, A, given by [3]:

$$R_n = \frac{\rho_g}{2\pi\ell}\left(\ell n\frac{4\ell}{A} - 1\right), \quad (\Omega), \ D \ll \ell \tag{5.2}$$

$$\text{where,} \quad A = \left[n\,a\left(\frac{D}{2}\right)^{n-1}\right]^{\frac{1}{n}}. \tag{5.3}$$

When the spacing between the adjacent n ground rods, arranged on a circle of diameter D, is equal to or larger than the length of a rod, the combined resistance of the n ground rods is given by [3]:

$$R_n = \frac{1}{n}\frac{\rho_g}{2\pi\ell}\left[\ell n\frac{4\ell}{a} - 1 + \frac{\ell}{D}\sum_{m=1}^{n-1}\frac{1}{\sin\frac{m\pi}{n}}\right], \ (\Omega), \ D \geq \ell. \tag{5.4}$$

It is apparent that the proximity effect between the ground rods tends to increase the combined resistance of the rods, thus diminishing the advantage of multiple ground rods.

Horizontal wires are sometimes buried in the soil to reduce the grounding impedance. When a horizontal wire is buried in the soil at a depth, d, its steady-state resistance is given by [3]:

$$R = \frac{\rho_g}{\pi \ell}\left[\ell n \frac{2\ell}{\sqrt{2ad}} - 1\right], \quad (\Omega), \ \ell \gg d . \tag{5.5}$$

Sometimes several horizontal wires are run radially from a common point which is connected to the foot of a tower. This arrangement lowers the initial surge impedance to lightning currents. The steady-state resistance of n radial horizontally laid wires is [3]:

$$R_n = \frac{\rho_g}{n \pi \ell}\left|\ell n \frac{2\ell}{\sqrt{2ad}} - 1 + N(n)\right|, \quad (\Omega), \tag{5.6}$$

$$\text{where} \quad N(n) = \sum_{m=1}^{n-1} \ell n \frac{1 + \sin\frac{\pi m}{n}}{\sin\frac{\pi m}{n}} . \tag{5.7}$$

The values of N(n) for various n are given below:

n =	2	3	4	6	8	12	100
N(n) =	0.7	1.53	2.45	4.42	6.5	11.0	116.0

As n becomes very large, N(n) approaches 1.22n, and in the limit

$$R_\infty = \frac{\rho_g}{2.57\ell} \quad (\Omega). \tag{5.8}$$

The counterpoise is a horizontal wire buried in the earth to a depth of about 1 m, and tied to the tower base. It may either be continuous, laid between towers, and parallel to the line conductors, or the radial type (crowfoot) consisting of several wires radiating from the tower base. Counterpoise has been used effectively to reduce the grounding impedance in power systems. The counterpoise resistance is in parallel with the tower-footing resistance. Equation (5.6) has been plotted in Fig. 5.1, which shows the advantage of short, multiple counterpoise over a long, single counterpoise.

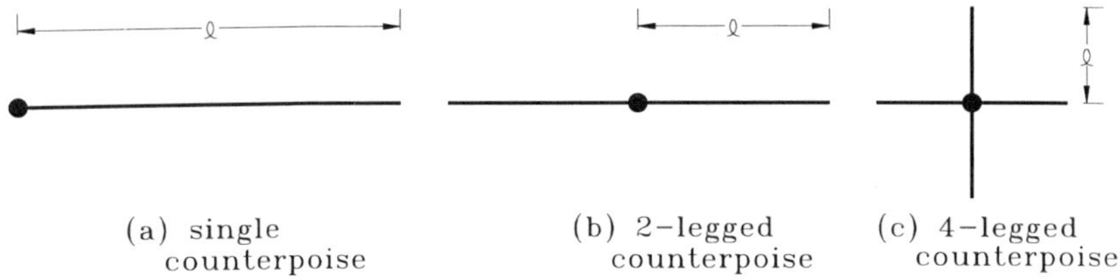

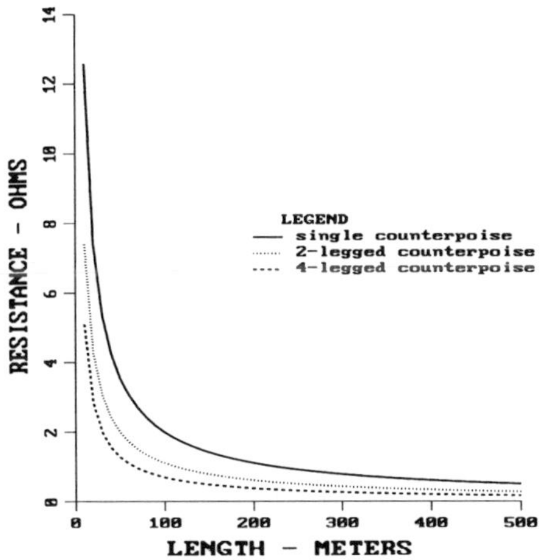

Fig. 5.1 Steady-state counterpoise resistance *vs.* its length.
Counterpoise radius, a=1 cm; depth of burial, d=1 m
Soil resistivity, ρ_g=100 Ω-m

5.2 Dynamic Tower-Footing Resistance

It has long been known that the tower-footing resistance is a function of the lightning return-stroke current flowing through the tower legs to the earth [7]. This dynamic resistance is time variant, and is significantly lower than its power-frequency value.

The grounding system of a tower can be broadly classified as (1) concentrated or (2) extended. The concentrated grounding can be represented by lumped circuit elements. An extended grounding system is defined as that system where the travel time of the electromagnetic field along the grounding electrode is comparable with that along the tower itself. This generally applies to grounding systems with physical dimensions exceeding 20 meters (*e.g.*, counterpoise).

5.2.1 Concentrated grounding system

The ground behaves essentially as a pure resistance at power frequencies. At high frequencies and/or high soil resistivities, electrolytic conduction mechanism sets in. The ground electrode-soil system then acts like an extremely lossy dielectric requiring a large number of frequency-dependent resistive and capacitive polarization components. At high impulse currents, the resulting electric field can ionize the soil structure, producing sparks and arcs. This is a highly nonlinear phenomenon. The condition of the soil during the discharge of a heavy transient current is shown in Fig. 5.2. Therefore, the factors which influence the dynamic impedance of ground electrodes are:

1. the impedance of the electrode itself,
2. the frequency-dependent electrical properties of the soil,
3. the electrical discharge process in the soil.

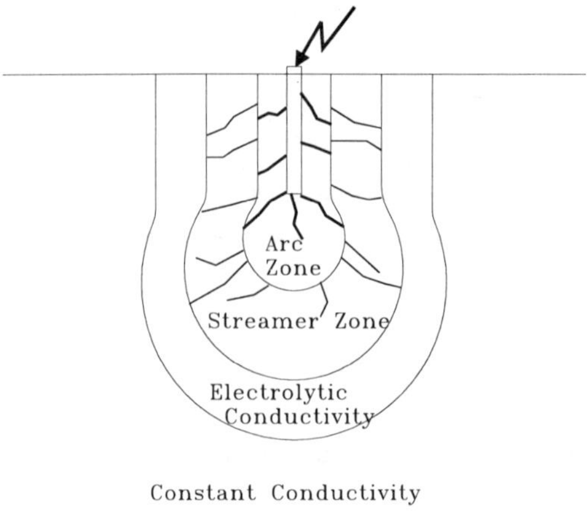

Fig. 5.2 Discharge of heavy impulse current through the grounding system.

For a concentrated grounding system, the impedance of the electrode itself is negligible. For the extended grounding system, the electrode is represented as a lossy transmission line. It has been shown by some investigators that the dynamic grounding impedance below the soil ionization level is practically the same as its commercial-frequency resistance. Therefore, the dependence of the soil parameters on frequency can be ignored.

The electrical conduction in soil takes place through electrolytic conduction and electrical discharge, both processes being nonlinear in character. Moreover, two types of electrical discharge take place: diffuse and streamer type. The streamer type discharge takes place near the electrode under heavy current. Both types of discharges have time delays. Experimental data from several researchers show that the dynamic grounding impedance is a function of both the magnitude and the waveshape of the discharge current.

The critical breakdown voltage gradient of soil is another important soil parameter. Many analyses are based on this parameter. However, this parameter is difficult to define and to measure. It has been measured by observing the breakdown of soil between two electrodes. But the critical breakdown voltage gradient should depend upon the size and shape of the electrodes as well as the composition of the soil. Moreover, Berger, amongst others, has shown that the nonlinearity of the grounding impedance becomes evident even before the soil actually breaks down [8]. Some observers have noted the looped volt-ampere characteristics of grounding electrodes at very small currents before any ionization could start (Fig. 5.3). It suggests then that the nonlinearity exists even when the conduction process is dominated by electrolytic conduction. Therefore, any analysis based solely on the critical breakdown voltage gradient will not be valid.

Experimental results have shown that the dynamic response of the grounding impedances resembles the nonlinear characteristics of a valve-type lightning arrester, such as the silicon carbide and the metal oxide varistor [8-13]. The nonlinear behavior of such valve-type arresters has been successfully characterized by a simple equation as follows:

$$R = kI^{\alpha} \quad (\Omega) , \tag{5.9}$$

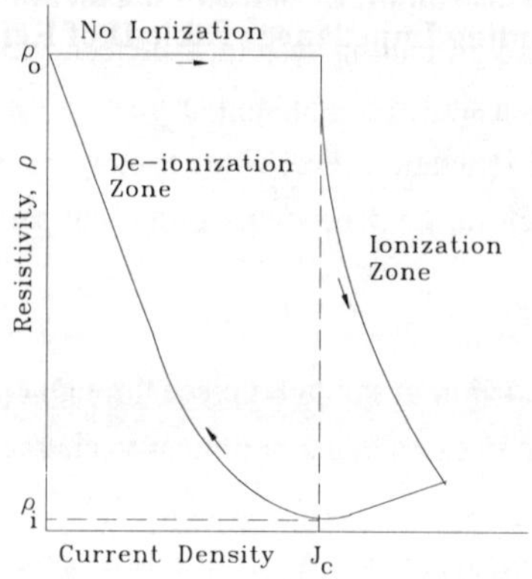

Fig. 5.3 Dynamic resistivity profile of soil during the flow of heavy impulse current [13].

where R is the nonlinear resistance of the valve block at a current level of I amperes. The constants k and α are either provided by the manufacturer or can be calculated from the R vs. I characteristic. The exponent α depends upon the internal property of the valve material and k depends upon the size of the block.

It seems appropriate to apply a similar relationship also to the grounding impedances. Tables 5.1-5.4 show the calculated values of the dynamic grounding impedances along with the measured values from four different experimental studies.

Table 5.1: Comparison of Measured vs. Calculated Dynamic Grounding Impedance (Fig. 3 of Ref. 13)

Current A	Measured Resistance Ω	Calculated Resistance Ω	k	α
13000	13	13		
9800	15	15	1738.25	-0.51682
3400	26	26		

Notes: Calculated resistance, $R = kI^{\alpha}$

0.61-m long 0.15-cm diameter rod driven into a soil of sand and gravel mix of 5000 Ω-cm resistivity.

Table 5.2: Comparison of Measured vs. Calculated Dynamic Grounding Impedance (Table II of Ref.10)

Current A	Measured Resistance Ω	Calculated Resistance Ω	k	α
7680	21.6	21.6		
5500	23.9	23.0	110.88	-0.18284
3500	26.0	24.9		
2050	27.5	27.5		
2240	26.5	26.5		
3330	27.9/28.5	24.8	95.31	-0.16593

Notes: Calculated resistance, $R = kI^{\alpha}$

Both sets of data were obtained with 6/13-μs positive-polarity current waves.

The ground electrodes were 10-ft. long, 1-inch diameter steel driven ground rods buried 90 inches into the ground.

Class of soil: shale and clay with mixture of gravel and sand.

Earth naturally moist.

Table 5.3: Comparison of Measured vs. Calculated Dynamic Grounding Impedance (Table I of Ref. 12)

Current A	Measured Resistance Ω	Calculated Resistance Ω	k	α
392	68.4	68.4		
632	51.6	54.2		
1090	34.9	41.5	1255.05	-0.48733
1133	35.1	40.8		
1210	35.9	39.5		
1679	35.7	33.7		

Notes: Calculated resistance, $R = kI^{\alpha}$

Hemispherical ground electrode of 2 cm radius embedded in earth of 2000 Ω-cm resistivity and impulsed with 1/40-μs positive-polarity voltage wave.

Table 5.4: Comparison of Measured vs. Calculated Dynamic Grounding Impedance (Table III of Ref. 11)

Current A	Measured Resistance Ω	Calculated Resistance Ω	k	α
1545	20.4	20.4		
3450	16.1	16.3	159.51	-0.28008
6240	13.6	13.8		
8880	12.5	12.5		
1550	20.8	20.8		
3360	17.6	16.5	191.07	-0.30189
6130	15.3	13.7		
9075	12.2	12.2		
1500	21.5	21.5		
3200	19.8	16.9	224.29	-0.32063
5550	16.5	14.1		
10850	11.4	11.4		
4320	12.5	12.5		
7250	11.5	11.2	72.81	-0.21051
11350	10.2	10.2		

Notes: Calculated resistance, $R = kI^{\alpha}$

First set of data: 20/49-μs positive-polarity current wave with driven ground rod F.

Second set of data: 20/49-μs negative-polarity current wave with driven rod F.

Third set of data: 20/50-μs negative-polarity current wave with driven ground rod G.

Fourth set of data: 20/46-μs negative-polarity current wave with driven ground rods F and G in parallel.

Both F and G ground rods are 1-inch diameter steel rods driven 10 ft. into naturally moist clay.

As can be seen from the tabulations, the agreement between the measured and the calculated values of the dynamic grounding resistances is fairly good, bearing in mind the inaccuracies in the measured values. The inaccuracies in the measured values should be evident by observing that sometimes the measured dynamic resistance increased with increased impulse currents.

Equation 5.9 does not show any explicit dependence on the critical breakdown voltage gradient nor on time delay of soil ionization. The dependence on the critical breakdown voltage gradient is implicit on the constants k and α. A simple equation like (5.9) is attractive. However, k and α for various combinations of electrodes and soil types must be known.

Korsuntsev's work [14], advanced by Popolansky [15], applied the similarity law to concentrated ground electrodes. A generalized curve, based on a large sample of experimental data, was proposed as a function of two dimensionless parameters, Π_1 and Π_2, given by:

$$\Pi_1 = \frac{Z_i\, s}{\rho_g} , \qquad (5.10)$$

$$\Pi_2 = \frac{I\, \rho_g}{s^2 E_c} , \qquad (5.11)$$

$$\Pi_1 = f(\Pi_2) , \qquad (5.12)$$

where ρ_g = soil resistivity, Ω-m,
s = characteristic dimension of electrode, m,
I = impulse current, kA,
E_c= critical soil ionization potential gradient, kV/m, and
Z_i= impulse impedance of ground electrode, Ω.

Korsuntsev [14] defined the characteristic dimension, s, as the distance from the geometric center of the electrode on the surface of the earth to the most distant point of the electrode. Oettle [17] suggested an alternative definition of s as follows:

$$s = \sqrt{h_1^2 + h_2^2 + d^2}\,, \tag{5.13}$$

where d is the depth of the electrode from the earth's surface, h_1 is the largest horizontal dimension of the electrode, and h_2 is the horizontal dimension which lies perpendicular to the largest dimension. This is illustrated in Fig. 5.4 [17].

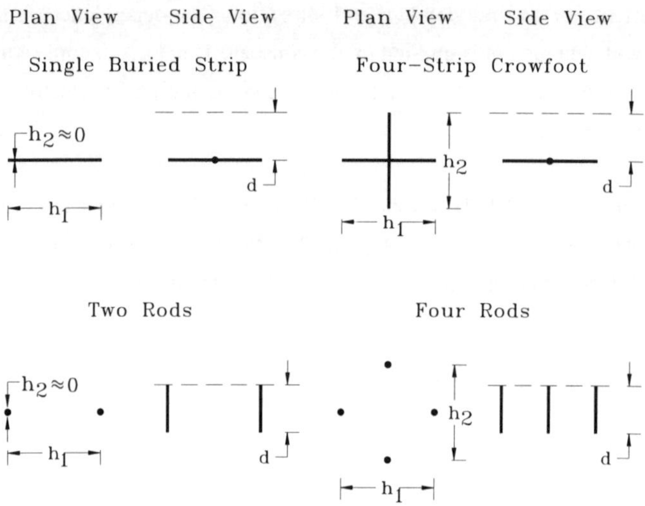

Fig. 5.4 Characteristic dimensions of ground electrodes according to Oettle [17].

The generalized curve can be divided into two regions: (1) geometry-dependent region and (2) geometry-independent region. The ground-electrode impedance in the geometry-dependent region is the power-frequency ground resistance which can be easily calculated from Section 5.1 [3]. Chisholm and Janischewskyj have proposed a simple equation for the power-frequency ground resistance of concentrated ground electrodes given by [16]:

$$\Pi_1^o = 0.4517 + \frac{1}{2\pi}\ell n\left(\frac{s^2}{A}\right), \tag{5.14}$$

where $\Pi_1{}^o$ is the geometry-dependent part of the Korsuntsev curve, s is the characteristic distance of the ground electrode, and A is its surface area. This equation was derived from Rudenberg's equation [2] for power-frequency resistance of concentrated ground electrodes, and, according to Chisholm and Janischewskyj, fits many types of concentrated ground electrodes.

In the geometry-independent region, soil ionization and the associated nonlinearity of the ground impedance set in. For this region, Oettle has proposed a simple relationship between Π_1 and Π_2 as [17]:

$$\Pi_1 = 0.24\,\Pi_2^{-0.3} \ . \tag{5.15}$$

Chisholm and Janischewskyj have proposed a similar equation based on a least-square fit of the Korsuntsev curve for Π_2, given by [16]:

$$\Pi_1 = 0.2631\,\Pi_2^{-0.3082} \ . \tag{5.16}$$

Both these equations are very similar. These equations cover a small region of the π_1 ***vs.*** π_2 plot. To cover the entire region, the log-log plot was piecewise linearized as follows:

$$\text{For } \Pi_2 \leq 5.0 \ : \ \Pi_1 = 0.2965\,\Pi_2^{-0.2867} \ , \tag{5.17a}$$

$$\text{For } 5.0 < \Pi_2 \leq 50.0 : \ \Pi_1 = 0.4602\,\Pi_2^{-0.6009} \ , \tag{5.17b}$$

$$\text{For } 50.0 < \Pi_2 \leq 500.0 \ : \ \Pi_1 = 0.9534\,\Pi_2^{-0.7536} , \tag{5.17c}$$

$$\text{For } \Pi_2 > 500.0 \ : \ \Pi_1 = 1.8862\,\Pi_2^{-0.8693} \ . \tag{5.17d}$$

The plot of Π_1 ***vs.*** Π_2 is shown in Fig. 5.5.

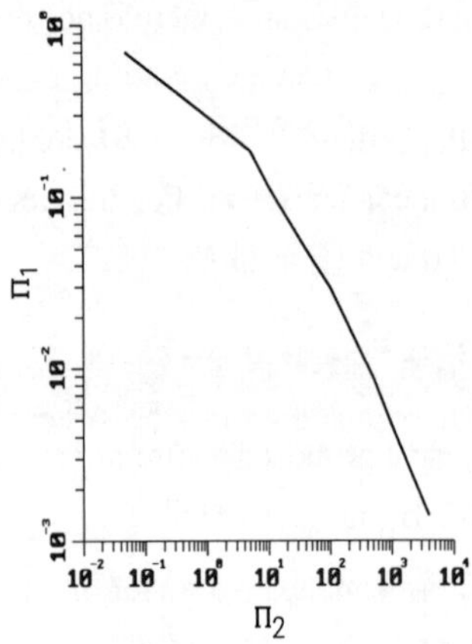

Fig. 5.5 Variation of geometry-independent impulse impedance of concentrated ground electrodes according to Korsuntsev [14].

The other important variable is the critical soil ionization potential gradient E_c. Oettle has proposed an equation relating the soil resistivity ρ_g to E_c, [17]:

$$E_c = 241\,\rho_g^{0.215} \quad (\text{kV/m}). \tag{5.18}$$

It appears that Korsuntsev's method, as expanded by others, is the simplest to use for calculating the impulse impedance of concentrated ground electrodes. However, several factors should be kept in mind. As pointed out by Weck [18], the Korsuntsev curve suggests that all ground electrodes act like hemispheres at high values of Π_2. According to Weck, this could be due to the fact that all results have been obtained for electrodes with limited dimensions, which may not represent practical ground electrodes. The second point to be considered is that the Korsuntsev curve does not show any dependence on the rate of rise of the impulse current. It is well known that the soil has an ionization time delay which can offset the results for steep-front lightning currents [13]. Insight into these two points can be obtained by well-designed laboratory experiments. In addition, the critical breakdown voltage gradient of soil, E_c, needs to be known accurately. E_c is difficult to define, and will be quite different in the field from that measured in the laboratory. E_c in the laboratory is calculated when the soil between the electrodes breaks down completely. In the field, the soil breakdown is incomplete. Even in the laboratory, E_c would be a function of the soil resistivity and the electrode configuration.

Although much work has been done on the dynamic behavior of the ground-electrode resistance under lightning discharge, there is no consensus yet on how to apply this knowledge to the design of the actual electrode system. The power-frequency grounding impedance, either measured or calculated, is being used at present to predict the lightning performance of power systems.

5.2.2 Extended grounding system

The counterpoise is an extended grounding system where voltage and current waves are dissipated while traveling along the length of the counterpoise. The counterpoise is simulated as a transmission line, as shown in Fig. 5.6, where L and Z_i are the inductance and internal impedance per meter of the counterpoise, and C and G are its capacitance and leakance per meter to ground.

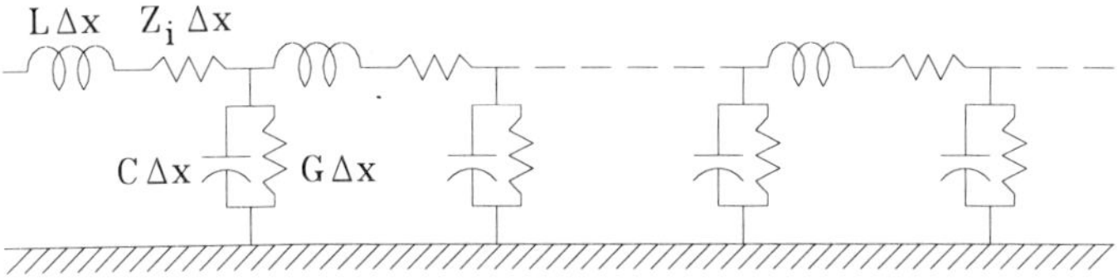

Fig. 5.6 Equivalent circuit of a counterpoise.

Following the same procedures as used in the derivation of the transmission line equations in Chapter 2, the voltage at a point x along the length of a counterpoise, in Laplace transform, can be derived as follows:

$$\frac{\partial V(x,s)}{\partial x} = -(Ls+Z_i(s))I(x,s) \text{ , and} \tag{5.19a}$$

$$\frac{\partial I(x,s)}{\partial x} = -(Cs+G)V(x,s) \text{ .} \tag{5.19b}$$

Differentiating (5.19a) with respect to x:

$$\frac{\partial^2 V(x,s)}{\partial x^2} = (Ls+Z_i(s))(Cs+G)V(x,s) \text{ .} \tag{5.20}$$

The solution of (5.20) is given by:

$$V(x,s) = V(0,s)e^{-\sqrt{(Ls+Z_i(s))(Cs+G)}\,x} . \tag{5.21}$$

V(0,s) is found by differentiating (5.21) with respect to x, and equating with (5.19a). This gives:

$$V(x,s) = Z_o\sqrt{\frac{s+b_o}{s+b_1}}\,I(x,s) , \tag{5.22}$$

where $Z_o=\sqrt{\frac{L}{C}}$, $b_o=\frac{Z_i(s)}{L}$, and $b_1=\frac{G}{C}$.

For a unit step-function input current,

$$V(0,s) = Z_o\frac{1}{s}\sqrt{\frac{s+b_o}{s+b_1}} . \tag{5.23}$$

V(x,s) in (5.21) can then be expressed as:

$$V(x,s) = Z_o\frac{1}{s}\sqrt{\frac{s+b_o}{s+b_1}}\,e^{-\frac{x}{v}\sqrt{(s+b_o)(s+b_1)}} , \tag{5.24}$$

where $v = \frac{1}{\sqrt{LC}}$.

Equation (5.24) is in Laplace transform. It can be inverted by computer programs such as the INLAP routine in the IMSL software package. However, it has been found that the effect of the counterpoise internal impedance, $Z_i(s)$, is not significant. Therefore, neglecting $Z_i(s)$, (5.24) can be rewritten as:

$$V(x,s) = Z_o\frac{e^{-\frac{x}{v}\sqrt{s(s+b_1)}}}{\sqrt{s(s+b_1)}} . \tag{5.25}$$

Inverting (5.25): $V(x,t) = 0$ when $0 < t \le \frac{x}{v}$

$$V(x,t) = Z_o e^{-0.5 b_1 t} I_o\left(0.5 b_1 \sqrt{t^2 - \frac{x^2}{v^2}}\right) \text{ when } t > \frac{x}{v}, \tag{5.26}$$

where I_o is the modified Bessel function of zero order, first kind. The earth leakage conductance (G), inductance (L), capacitance (C), and counterpoise internal impedance ($Z_i(s)$) in (5.22) were given by Sunde [3] as follows:

$$G = \frac{\pi}{\rho_g f_o}, \tag{5.27}$$

$$L = 2 \times 10^{-7} f_o; \; C = \frac{(k_g + 1) 10^{-9}}{36.0 f_o}, \text{ and} \tag{5.28}$$

$$Z_i(s) = \frac{0.5\sqrt{s}}{\pi a_c} \sqrt{\frac{\mu_o}{\sigma_c}}, \tag{5.29}$$

$$\text{where } f_o = \ell n \frac{2 \ell_c}{\sqrt{2 a_c d_c}} - 1, \tag{5.30}$$

k_g=soil dielectric constant, ℓ_c=counterpoise length, a_c=counterpoise radius, σ_c=counterpoise conductivity, d_c=depth of counterpoise beneath the earth's surface, and ρ_g=soil resistivity. The profiles of the counterpoise voltage at x=0 for unit step-function injected current are shown in Fig. 5.7. These profiles are also the counterpoise surge impedances for step-function input. The profile of the time-dependent surge impedance will be different for different waveshapes of the input current. The profound effects of the soil parameters should be noted in Fig. 5.7.

A comprehensive discussion on power-system grounding is given in [19].

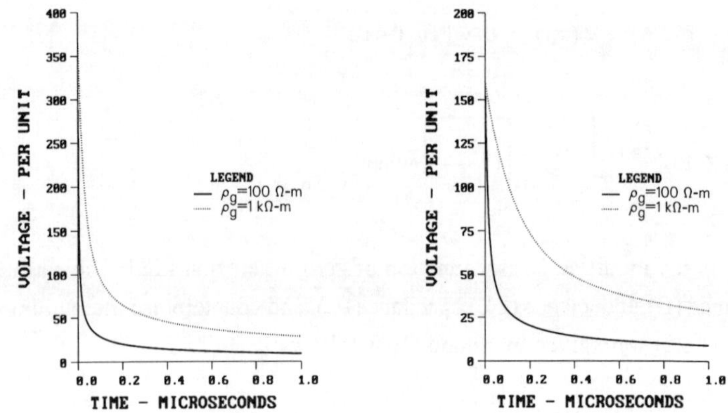

(a) (b)

Fig. 5.7 Transient voltage along a counterpoise.
Unit step-function input current
Counterpoise length=100 m; radius=1 cm
Soil dielectric constant: (a) k_g =1.0; (b) k_g=10.0
Depth of burial=1 m

5.3 Substation Grounding Grid

Tower-footing resistances are generally of the concentrated type; it is difficult to limit this resistance to below 10 ohms. However, the grounding system of a substation is the extended type, which covers the whole surface area of the substation. It is called the ground mat or grid. The substation grounding system has two main functions:

(i) human safety, and (ii) grounding of the neutrals of power apparatus.

Ground mats, buried in the ground to a depth of a few feet, are generally composed of heavy copper cables bonded at each crossover point; this is to reduce the potential gradient along the grounding system in case of a ground fault of any of the power apparatus, when a heavy current would flow through the grounding system. In addition, the ground mat is connected to driven ground rods at numerous points, preferably at each grid point to keep the potential gradient low. A good description of the substation grounding system is given in [20]. The resistance, R of a ground mat is approximately given by [21]:

$$R = \frac{\rho_g}{4r} + \frac{\rho_g}{\ell}, \ \Omega \tag{5.31}$$

where ρ_g = average soil resistivity, Ω-m
r = radius of a circle of the same area as that of the ground grid, m
ℓ = total length of buried conductor including the ground rods, m.

The resistance of a substation ground mat can be kept within 0.1 to 0.5 ohm. As the shielding structure is bonded to the grounding mat, the voltage drop across the grounding system will not be high enough to cause a backflashover for a direct lightning stroke to the substation shielding system. Figure 5.8 shows the schematic of a substation ground mat which is composed of many rectangular meshes.

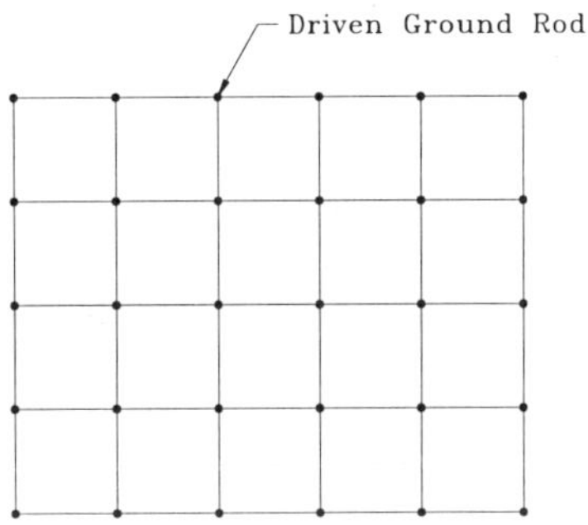

Fig. 5.8 Schematic of a substation ground mat.

The resistance of the substation ground mat (or grid) under impulse currents is different from that under power-frequency currents [22]. The impulse impedance of a grounding grid is defined as the ratio of the peak value of the voltage developed at the feeding point to the peak value of the injected current; the ratio of the impulse impedance to the power-frequency grounding resistance is defined as the impulse coefficient.

Gupta and Thapar showed [22] that the effect of the conductor radius on the impulse impedance of a grounding grid was not significant. It was also found that the impulse impedance decreases as the size of the grid increases, until a certain area is reached; increasing the grid area beyond this value does not appreciably change the impulse impedance. The area of the grid corresponding to the value of the impulse impedance within three percent of its final value is defined as the effective area. The effective areas of center- and corner-fed grids are shown in Fig. 5.9. The effective radius, r_e, is defined as the radius of a circle whose area is equal to the effective area. The authors showed that the effective radius and the impulse coefficient can be calculated using the following equations [22]:

$$r_e = K\sqrt{\rho_g t_f}\ , \ \mathrm{m} \tag{5.32}$$

where K = (1.45 - 0.05S) for center-fed grids,
= (0.6 - 0.025S) for corner-fed grids,
S = spacing between the grid conductors, m, and
t_f = front time of the lightning current, μs.

For r ≤ r_e, the impulse coefficient, A, is given by:

$$A = e^{0.333\left(\frac{r}{r_e}\right)^{2.3}}\ , \ \text{and} \tag{5.33}$$

$$\text{Impulse Impedance} = AR, \tag{5.34}$$

where R = power-frequency resistance of the grounding grid. **For r>r_e**, r = r_e.

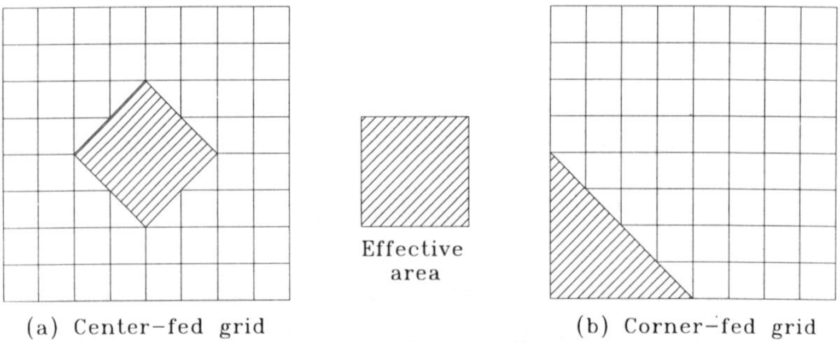

Fig. 5.9 Effective area of a grounding grid [22].
(a) center-fed grid; (b) corner-fed grid.

Figure 5.10 compares the impulse impedance of a grounding grid with its power-frequency value. It should be noticed that the impulse impedance is, in general, higher than the power-frequency resistance, and that the corner-fed grid has the highest value. This is contrary to what we have seen for the concentrated grounding system. The authors of [22] concluded that this happens because soil ionization is negligible for large-area grounding systems.

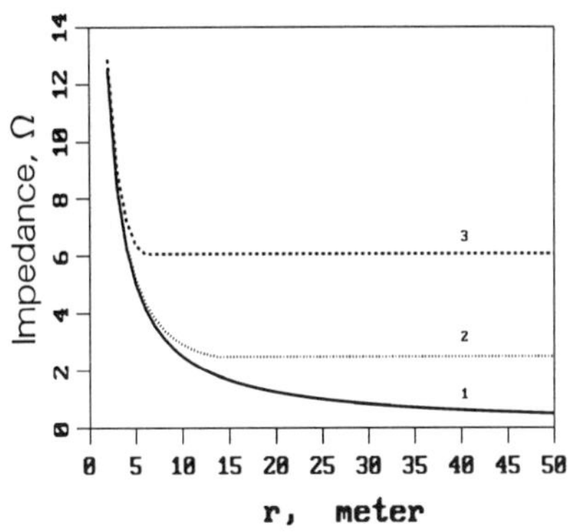

Fig. 5.10 Impulse impedances of a grounding grid [22].
ρ_g = 100 Ω-m; t_f = 1 μs ; S=1 m.
1: power-frequency grid resistance
2: impulse impedance of center-fed grid
3: impulse impedance of corner-fed grid

5.4 References to Chapter 5

1. H.B. Dwight,"Calculation of Resistances to Ground," ***Electrical Engineering***, Vol. 55, pp. 1319-1328, 1936.

2. R. Rudenberg,"Grounding Principles and Practice I - Fundamental Considerations on Ground Currents," ***Electrical Engineering***, Vol. 64, pp. 1-13, 1945.

3. E.O. Sunde: ***Earth Conduction Effects in Transmission Systems***. Dover Publications, New York, 1968.

4. E.K. Saraoja,"Lightning Earths," Chapter 18 of ***Lightning***, Vol. 2. Edited by R.H. Golde. Academic Press, New York, 1977.

5. J.G. Anderson,"Lightning Performance of Transmission Lines," Chapter 12 of ***Transmission Line Reference Book 345 kV and Above***, Second Edition, Revised. Electric Power Research Institute, Palo Alto, CA., 1987.

6. L.V. Bewley, "The Counterpoise," Chapter 10 of ***Traveling Waves on Transmission Systems***. John Wiley, New York, 1951.

7. H.W. Towne,"Impulse Characteristics of Driven Grounds," ***General Electric Review***, Vol. 31, pp. 605-609, 1928.

8. K. Berger,"The Behavior of Grounding under High Impulse Current," (in German), ***Bull. SEV***, Vol. 37,pp. 197-221, 1946.

9. R. Davis and J.E.M. Johnston,"The Surge Characteristics of Tower and Tower-Footing Impedances," ***Jour. IEE*** (G.B.), Vol. 88, Part II, pp. 453-465, 1941.

10. P.L. Bellaschi,"Impulse and 60-Cycle Characteristics of Driven Grounds," ***AIEE Trans.***, Vol. 60, pp. 123-128, 1941.

11. P.L. Bellaschi, R.E. Armington and A.E. Snowden,"Impulse and 60-Cycle Characteristics of Driven Grounds - II," ***AIEE Trans.***, Vol. 61, pp. 349-363, 1942.

12. T. Okubo and M. Hayashi,"Reducing Effect in Footing Resistance against Heavy Impulse Current," ***Memoirs of the Faculty of Engineering***, Kyoto University, Vol. 25, pp. 384-405, 1963.

13. A.C. Liew and M. Darveniza,"Dynamic Model of Impulse Characteristics of Concentrated Earths," ***Proc.IEE*** (G.B.), Vol. 121, pp.123-135, 1974.

14. A.V. Korsuntsev,"Application of Similarity Theory in the Calculation of the Impulse Characteristics of Concentrated Earth Electrodes," (in Russian), ***Elektrichestvo***, No. 5, pp. 31-35, 1958.

15. F. Popolansky,"Generalization of Model Measurement Results of Impulse Characteristics of Concentrated Earths," ***CIGRE Working Group Report***, No. SC 33-80(WG01) IWD, August 1980.

16. W.A. Chisholm and W. Janischewskyj,"Lightning Surge Response of Ground Electrodes," ***IEEE Trans. on Power Delivery***, Vol. 4, pp. 1329-1337, 1989.

17. E.E. Oettle,"A New General Estimation Curve for Predicting the Impulse Impedance of Concentrated Earth Electrodes," ***IEEE Trans. on Power Delivery***, Vol. 3, pp. 2020-2029, 1988.

18. K.H. Weck,"Remarks to the Current Dependence of Tower Footing Resistances," ***CIGRE Working Group Report***, No. 33-85(WG01)IWD.

19. E.R. Whitehead, "Fundamentals of Grounding for Lightning Protection of Transmission and Distribution Systems," A tutorial overview prepared for the 1979 Lightning Conference, University of Wisconsin - Extension, Madison, WI, 29-31 October, 1979.

20. ***Design Guide for Rural Substations***, REA Bulletin 65-1, June 1978, Rural Electrification Administration, U.S. Dept. Of Agriculture.

21. ***IEEE Guide for Safety in Substation Grounding***, IEEE Std. 80-1976.

22. B.R. Gupta and B. Thapar, "Impulse Impedance of Grounding Grids", ***IEEE Trans. on Power Apparatus and Systems***, Vol. PAS-99, pp. 2357-2362, 1980.

Problems for Chapter 5

1. Calculate the steady-state resistance and the impulse impedance of the following ground electrodes:
   (i) a 12.5-m long horizontal wire of 1-cm radius buried in the earth at a depth of 1 m.
   (ii) a four-legged crowfoot, each leg of which is a 12.5-m long horizontal wire of 1-cm radius, buried in the earth at a depth of 1 m.

   Assume: ρ_g=100 and 1000 Ω-m; I_p=10 and 100 kA.

2. Plot the impulse impedance of a grounding grid for the following conditions: (i) center-fed grid and (ii) corner-fed grid up to an equvalent radius of the grid of 50 m. The following parameters are given:
   ρ_g=500 Ω-m, t_f=5 μs, mesh size=1x1 m^2.

CHAPTER 6
DIRECT LIGHTNING STROKES TO OVERHEAD LINES WITHOUT SHIELD WIRES

6.1 Introduction

The benchmark of the survivability of a power apparatus against lightning-caused transient overvoltages is its BIL (basic impulse insulation level). This is the reference impulse insulation strength expressed as the peak of a 1.2/50-microsecond voltage wave. The BIL of a power apparatus is specified by the appropriate standard (ANSI, IEEE, IEC). The protection of a power apparatus against lightning is based on its BIL. For external insulation which is self restoring, e.g., line insulators, the critical flashover voltage (CFO) is used instead of BIL. CFO is defined as the impulse voltage level at which the probability of flashover of the insulation is 50 percent. CFO is also designated as V_{50}.

Very high voltages are generated if a lightning stroke hits the power line directly, whether it hits the phase conductors, the tower, or even the shield (or ground) wire. This is called the ***direct stroke***. The power-line tower which is at ground potential under the steady-state conditions rises in voltage transiently because of the flow of the lightning current through the surge impedance of the tower and also through the tower-footing resistance. An insulator flashover caused by a stroke to the tower is called the ***backflashover***. Even a stroke to the shield wire would raise the tower-top potential. Shield wires sometimes fail to shield the phase conductors; in that case, a ***shielding failure*** occurs. High transient voltages can also be generated on the power line for strokes to the nearby ground. These strokes are called the ***indirect strokes***, in order to distinguish them from the direct strokes. This is illustrated in Fig. 6.1.

The direct strokes are of concern for the protection of the high-voltage power transmission lines; the indirect strokes do not induce high enough voltages on a

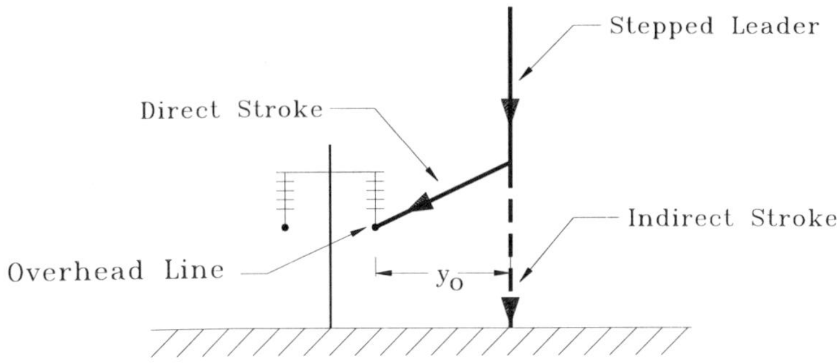

Fig. 6.1 Illustration of direct and indirect lightning strokes.

transmission line to cause outages. For power distribution lines, where the CFO is usually between 100 to 300 kV, the probability of survival from a direct stroke is minimal. However, the power distribution lines, because of their low height, do not attract many direct strokes. The indirect strokes, on the other hand, can induce voltages exceeding the CFO of the line; such strokes can be frequent.

The performance of an overhead power line under lightning depends upon many factors, some of which can only be estimated. The characteristics of lightning itself are certainly very important in estimating the lightning performance of overhead power lines. These characteristics, such as the magnitude and the time-to-crest of the return-stroke current, and the frequency of strike per square kilometer, are statistical quantities - both in time and in space. These, already discussed in Chapter 4, can be estimated, although they may vary widely from year to year. Another important factor is the distance through which a particular lightning stroke will be attracted to strike an object. The longer this attractive distance is, the higher will be the probability of a line outage; this distance is a function of the characteristics of both the object struck (e.g., its height and shape) and the lightning stroke (e.g., the return-stroke current). There are two methods to estimate this attractive distance: (1) the electrical shadow method, and (2) the electrogeometric model. The electrical shadow method is generally used to estimate the outage rates caused by direct strokes to power-line towers for lines shielded by shield wires, and for unshielded lines, for its simplicity; The electrogeometric method is applied for estimating the outage rates caused by shielding failure, and also for estimating the outage rates caused by strokes to nearby ground.

Most high-voltage power lines have one or more shield wires. However, some do not have any shield wires because either the ground flash density is low or the soil resistivity is very poor for the shield wires to be economically justifiable. Many power distribution lines are not equipped with shield wires, even though such lines have grounded neutral wires running below the phase conductors. Therefore, the neutral wires cannot shield the phase conductors from the direct strokes.

The response of an overhead line to a direct lightning stroke will depend upon the design of the line, i.e., whether the line has shield wires or not. The response will also depend upon the location of the strike on the line, i.e., strike to phase conductor, or to shield wire(s), or to the tower. All these cases will be treated separately.

Lightning may strike an unshielded line either somewhere along the span or at the tower. It is generally assumed that about 40 percent of the lightning strokes can be classified as midspan strikes, and the remaining 60 percent as striking near the towers [1].

6.2 Stroke to Line

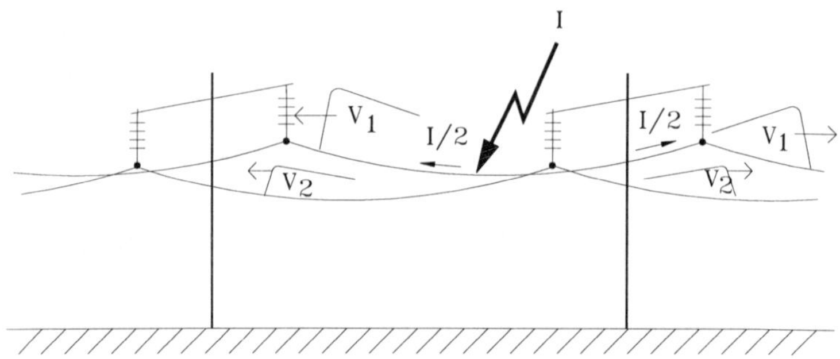

Fig. 6.2 Direct lightning stroke to line without shield wire.

As shown in Fig. 6.2, the stroke current will divide itself into two equal parts at the point of strike. These divided currents will generate voltages, V_1, which will travel in both directions along the line away from the point of strike. The voltage, V_1, is given by:

$$V_1 = \frac{I}{2} Z_p , \tag{6.1}$$

where Z_p is the surge impedance of the phase conductor. The voltage (to ground) induced on a second phase conductor will be:

$$V_2 = K_{12} V_1 , \tag{6.2}$$

where K_{12} is the coupling factor between phase conductors 1 and 2. The voltage between the conductors 1 and 2 is:

$$V_{12} = (1 - K_{12}) V_1 . \tag{6.3}$$

A flashover between two conductors may occur if the separation distance between them is not large enough to withstand V_{12}. Otherwise, V_1 and V_2 will travel along the phase conductors 1 and 2 towards the tower (or pole) where the phase conductors are supported by insulator strings. Similar situations will exist on the other phase conductors of the overhead line. As one end of each insulator string is connected to the ground (i.e., tower), the voltages $V_1, V_2, \ldots, V_n$ will appear across the insulator string corresponding to each of the phase conductors. Because V_1 is the highest, the insulator string connected to conductor 1 will be stressed the highest, and will flashover if V_1 exceeds the CFO of the insulator string. The critical lightning-stroke current, I_{cs}, to cause insulator flashover due to a midspan stroke is given by:

$$\mathrm{CFO} = V_1 = \frac{I_{cs}}{2} Z_p , \tag{6.4}$$

$$\text{or,} \quad I_{cs} = \frac{2\,\mathrm{CFO}}{Z_p} . \tag{6.5}$$

Any lightning stroke to midspan exceeding I_{cs} will cause an insulator-string flashover at the adjacent tower. The probability, $P(I_{cs})$, for a current to exceed I_{cs} can be calculated from Chapter 4. Applying the electrical shadow method, the overhead line will collect all lightning strokes within a width, W, given by (4.17). For a 100-km segment of the line, the area of collection (attractive area) of all lightning strokes is:

$$A = 0.1\,W \qquad (\mathrm{km}^2) . \tag{6.6}$$

If the ground flash density of the region is n_g flashes per square kilometer, then the outage rate of the line for midspan strokes is:

$$\begin{aligned} N_{ms} &= (0.4\,n_g)(0.1\,W)\,P(I_{cs}) \\ &= 0.04\,n_g W\,P(I_{cs}) , \end{aligned} \tag{6.7}$$

where the factor 0.4 denotes the 40 percent probability for a midspan stroke.

6.3 Stroke to Tower

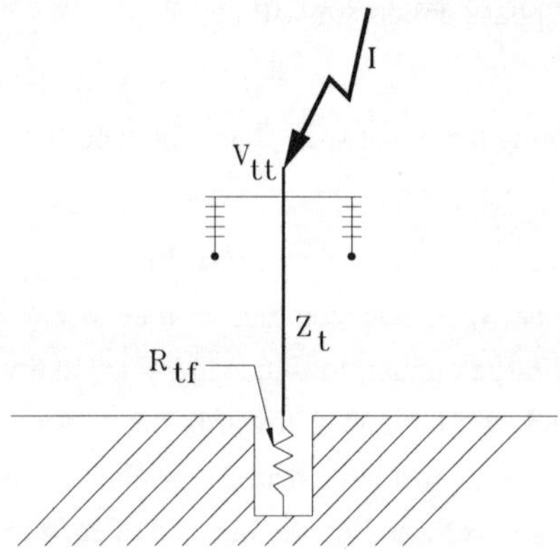

Fig. 6.3 Direct lightning stroke to tower without shield wire.

The tower is held at ground potential for steady-state operation. However, the tower-top potential immediately rises when it is struck by a lightning stroke. This tower-top potential is given by:

$$V_{tt} = IZ_t \,, \tag{6.8}$$

where I is the return-stroke current and Z_t is the tower surge impedance. V_{tt} travels down the tower until it hits the transition point at the foot of the tower, where the transition between the tower and the surrounding earth is represented by the tower-footing resistance, R_{tf}. The voltage is reflected at this transition point, and the reflected voltage travels back towards the tower top. Because the lightning stroke is considered to be a current source, the voltage reflected from the tower footing encounters a high impedance (ideally infinite) at the tower top. Therefore, a voltage of the same magnitude and polarity will start to travel back towards the tower footing; thus a lattice of successive reflections will take place. The tower-top voltage will then be a series of time-shifted voltage components, the time delays being measured by the up-and-down travel time of the wave along the tower length. The insulator-string voltage will be the difference between the tower-top voltage and the instantaneous power-frequency voltage of the phase conductor. For very high voltage lines where the tower is very tall and the distance between the tower top and the crossarm is considerable, the crossarm voltage should be calculated instead of the tower-top voltage. For short towers, such as for the power distribution line, where the travel time up and down the tower is short compared with the rise time of the return-stroke current, the traveling-wave theory and the accompanying lattice diagram can be waived. For such cases, the tower-top voltage is the voltage drop across the tower-footing resistance, i.e.,

$$V_{tt} = I R_{tf} \, . \tag{6.9}$$

If the insulator-string voltage is higher than its CFO, a backflashover will occur across the insulator string. The voltage across an insulator string is defined as the line-end terminal voltage with reference to the ground- (tower-) end terminal voltage. Therefore, for a stroke to tower, the insulator-string voltage has a polarity opposite to that of the return-stroke current. However, for a stroke to the line, the insulator-string voltage is of the same polarity as the return-stroke current. This is important to remember, because the withstand capability of an insulator can be sensitive to the polarity of the applied voltage.

Figure 6.4 shows the insulator-string voltage for a 30-kA stroke to the tower of an unshielded line for two different waveshapes of the return-stroke current, 5/50-μs and 1/50-μs respectively. In Fig. 6.4(a), the heights of the line and the tower were assumed to be 10 and 11 meters, respectively. The oscillations are actually the multiple reflections up and down the 11-m tower, which die out after a few reflections, and the tower-top voltage is eventually given by (6.9); in this case, it would be 300 kV for the assumed tower-footing resistance of 10 Ω. Figure 6.4(b) shows the case of a 33-m tower. The final value is the same as for the 10-m tower, but the voltage spikes are higher because of the longer travel time for the negative reflections from the tower footing to arrive at the tower top. These figures show the significance of the tower height and the rise time of the return-stroke current. For a rectangular current wave, the tower-top voltage would immediately rise to IZ_t. In our case, it would have been 3000 kV.

To estimate the outage rate for short towers, (6.9) is used to calculate the critical current, I_{ct}, for an insulator-string flashover for the given CFO. Then, similar to (6.7), the outage rate for strokes to tower is given by:

$$N_{tw} = (0.6\, n_g)(0.1\, W)\, P(I_{ct}) = 0.06\, n_g W\, P(I_{ct}) \, , \tag{6.10}$$

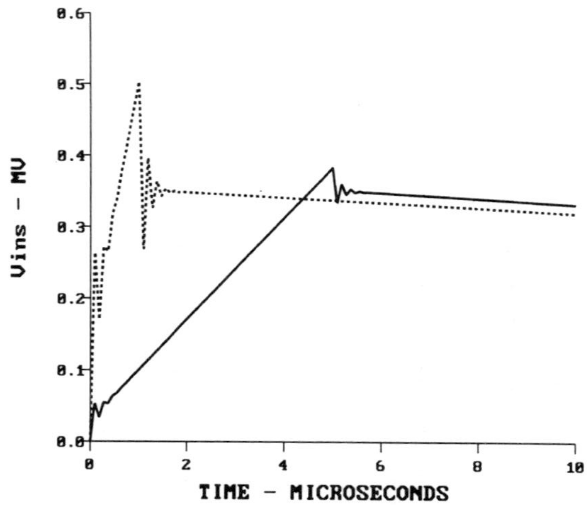

(a)

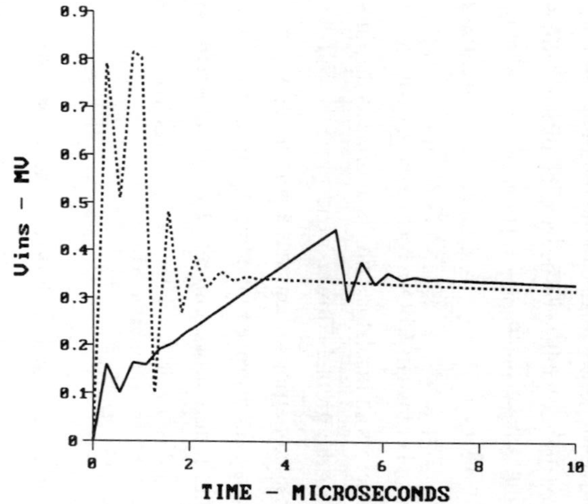

(b)

Fig. 6.4 Insulator-string voltages of an unshielded line for a stroke to tower.

I=30 kA; tower surge impedance=100 Ω; tower-footing resistance=10 Ω.
lightning-channel surge impedance=500 Ω.
solid line: 5/50-μs current; dotted line: 1/50-μs current.
(a) tower height=11 m; crossarm height=10.5 m; line height=10 m.
(b) tower height=33 m; crossarm height=31.5 m; line height=30 m.

where the factor 0.6 signifies the 60 percent probability of strike to tower. Insulator-string voltages for tall towers are sensitive to the waveshape of the lightning current, as was shown in Fig. 6.4. In that case, the probability of occurrence of the various wavefronts needs to be known, as was discussed in Chapter 4. However, it should be borne in mind that the magnitudes of the tower-footing resistances are not known precisely. The tower-footing resistance varies from tower to tower, from season to season and from year to year. Therefore, even for tall towers, the above procedure for calculating the outage rates can be calculated using the 'worst case' tower-footing resistance. The total outage rate, N_t, is then given, in terms of the number of outages per 100 km of line per year, by:

$$N_t = N_{ms} + N_{tw} \ . \tag{6.11}$$

6.4 Estimation of Attractive Area by Electrogeometric Model

The attractive area of strike to an overhead power line needs to be known in order to estimate the outage rate of the line caused by lightning strokes. As was described in Chapter 4, the electrical shadow method (ESM) is applied to estimate the attractive area to compute outage rates for direct strikes to the line or its towers. However, the electrogeometric model (EGM) is applied to compute the outage rates caused by shielding failure or by indirect lightning strokes. This is inconsistent.

According to ESM, the attractive width, W, is given by:

$$W = b + 4h \ , \tag{6.12}$$

where b is the separation distance between the two outermost phases, and h is the line height. According to EGM, the attractive width is given by:

$$W = b + 2\sqrt{r_s^2 - (\beta r_s - h)^2} \ , \tag{6.13}$$

where r_s is the striking distance to an overhead line, and βr_s is the striking distance to level ground for a particular magnitude of the return-stroke current, as discussed in Section 4.5. Therefore, it is expected that ESM and EGM will give two different results.

Grant and Anderson [2] applied the Monte Carlo technique to estimate the lightning incidence on a line by the EGM. With the help of random number generators, they randomized the occurrence of the peak return-stroke current and the point of incidence. If the striking distance corresponding to the return-stroke current produces an attractive width which encompasses the incident point, there will be a line strike, else it will be a ground flash. The line strikes were averaged over 200 years to arrive at the annual rate of line strike. Although correct, the Monte Carlo method is time consuming. An alternative method [3] is shown here which is significantly faster but as accurate as the Monte Carlo method.

A small return-stroke current, I_1, is chosen (e.g., I_1=1 kA) to initiate the computation. The corresponding attractive width, W_1, (Fig. 6.5) is calculated from (6.13). Any lightning flash with a return-stroke current equal to or greater than I_1 will hit the line if it descends within the width, W_1. If the ground flash density of the region is n_g (in number of lightning flashes per km^2 per year), then the number of stroke incidences (per 100 km of line per year) of the return-stroke currents exceeding I_1 will be given by:

$$N_{L1} = 0.1\, n_g W_1 P_1 \, , \tag{6.14}$$

where P_1 is the probability that the peak return-stroke current in a flash will exceed I_1. P_1 can be found from either (4.9) or from (4.10).

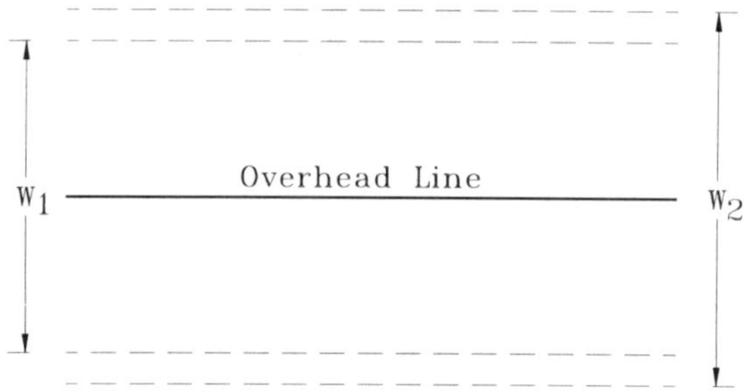

Fig. 6.5 Schematic of the attractive width of an overhead line.

The return-stroke current is then increased by a small step (e.g., ΔI=0.5 kA), and the attractive width, W_2, corresponding to the new current, I_2 (=I_1+ΔI), is calculated. The number of stroke incidences on the line for the return-stroke currents exceeding I_2 and falling within the strip, $\Delta W_{12}=W_2-W_1$ is given by:

$$\Delta N_{L1} = 0.1\, n_g \Delta W_{12} P_2 \, , \tag{6.15}$$

where P_2 is the probability that the peak return-stroke current will exceed I_2.

The incremental stroke incidence (ΔN_{L1}) is added to N_{L1}, the return-stroke current is increased by another step and the procedure repeated until the final current is reached (e.g., I_n=200 kA). The total number of stroke incidences per 100 km of line per year is then:

$$N_L = N_{L1} + \sum_{i=1}^{n-1} \Delta N_{Li} \,. \qquad (6.16)$$

If the line outage rate is required, the initiating current, I_1, will be the critical current of the line. Table 6.1 shows the results from three methods of computation. It is advisable to apply the EGM to estimate the outage rates of overhead power lines.

Table 6.1: Strike Incidence by Three Methods of Analysis

Line Height, m	Strike Incidence per 100 km per Year						
	EGM/Monte Carlo			EGM/Alternative			ESM
	β=0.6	β=0.8	β=1.0	β=0.6	β=0.8	β=1.0	
10	175.6	142.3	85.1	176.4	145	85.7	49.2
20	184.4	159.4	117.6	185.8	161.2	117.7	104.8
30	189.8	171.5	139	192.5	173.4	139.6	163
40	193.4	179.7	155.3	196.9	182.5	155.8	223

Note: For EGM, Striking Distance=$10I^{0.65}$

6.5 Tower Surge Impedance

The tower acts as a vertical transmission line for a direct stroke to the tower or to the shield wire. As such, it offers a surge impedance to the flow of the lightning current. Extensive research has been performed [4-15] on the response of vertical towers to lightning strokes, and research is still continuing. Sensitivity analysis shows that the tower surge impedance and the travel time along the tower are two very important parameters for the estimation of outage rates caused by direct lightning strokes to the tower [9,10]. Strictly speaking, this surge impedance is not constant, nor is the velocity of travel of a traveling wave along it. However, they are assumed to be constant for estimating the lightning performance of a line. Both analytical and experimental studies have been conducted to determine the tower surge impedance and the velocity of travel along the tower. The values of the tower surge impedance usually vary between 100 to 300 ohms. The velocity of travel along the tower varies between 0.7 to 0.9 of the velocity of light in free space. The equation for the surge impedance of three main types of tower are shown in Fig. 6.6 [1]. The value 2r for the Class 2 tower is the diameter of the down lead for the case of a wood pole or the diameter of the metal leg if it is a metal structure. The Class 3 tower is an approximation of a right cylinder.

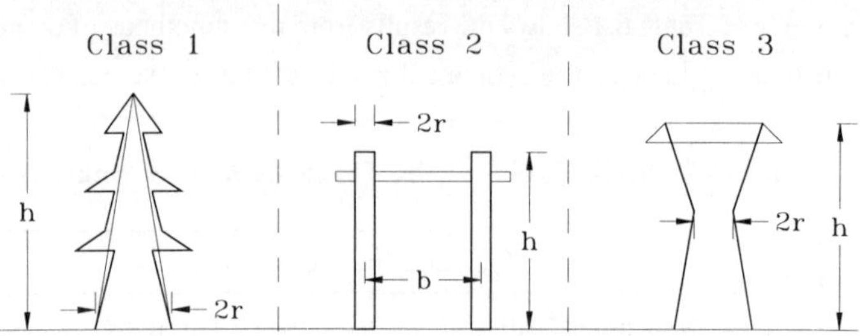

$$Z_t = 30\,\ln\left[\frac{2(h^2+r^2)}{r^2}\right] \qquad Z_t = 0.5(Z_s+Z_m) \qquad Z_t = \left[\ln\left(\sqrt{2}\,\frac{2h}{r}\right)-1\right]$$

$$Z_s = 60\,\ln(h/r)+90(r/h)-60\ ;\ Z_m = 60\,\ln(h/b)+90(b/h)-60$$

Fig. 6.6 Surge impedance for three types of tower.

6.6 References to Chapter 6

1. J.G. Anderson,"Lightning Performance of Transmission Lines," Chapter 12 of ***Transmission Line Reference Book 345 kV and Above***, Second Edition, Revised. Electric Power Research Institute, Palo Alto, CA., 1987.

2. I.S. Grant and J.G. Anderson, "A Comparison of the Electrogeometric Theory and the Shadow Method for Determination of the Incidence of Lightning to a Transmission Line Conductor", ***CIGRE 33-85 (WG 01) IWD 21.***

3. S. Mehairjan, "Analysis of Outages on Overhead Power Lines Caused by Direct Lightning Strokes", M.S. Thesis, Tennessee Technological University, Cookeville, Tennessee, May 1991.

4. C.A. Jordan, "Lightning Computation for Transmission Lines with Overhead Ground Wires Part II", ***General Electric Review***, Vol. 37, pp. 180-185,1934.

5. G.D. Breuer, A.J. Schultz, R.H. Schlomann and W.S. Price,"Field Studies of the Surge Response of a 345-kV Transmission Tower and Ground Wire," ***AIEE Trans. on Power Apparatus and Systems***, Vol. 76, pp. 1392-1396, 1958.

6. C.F. Wagner and A.R. Hileman,"A New Approach to the Calculation of the Lightning Performance of Transmission Lines III - A Simplified Method: Stroke to Tower," ***AIEE Trans. on Power Apparatus and Systems***, Vol. 79, pp. 589-603, 1960.

7. M. Kawai, "Studies of the Surge Response on a Transmission Line Tower", ***IEEE Trans. on Power Apparatus and Systems***, Vol. PAS-83, pp. 30-34, 1964.

8. M.A. Sargent and M. Darveniza,"Tower Surge Impedance," ***IEEE Trans. on Power Apparatus and Systems***, Vol. PAS-88, pp. 680-687, 1969.

9. W.A. Chisholm, Y.L. Chow and K.D. Srivastava,"Lightning Surge Response of Transmission Towers," ***IEEE Trans. on Power Apparatus and Systems***, Vol. PAS-102, pp. 3232-3242, 1983.

10. W.A. Chisholm, Y.L. Chow and K.D. Srivastava,"Travel Time of Transmission Towers," ***IEEE Trans. on Power Apparatus and Systems***, Vol. PAS-104, pp. 2922-2928, 1985.

11. K. Okumura and A. Kijima,"A Method for Computing the Surge Impedance of Transmission Line Tower by Electromagnetic Field Theory," ***Electrical Engineering in Japan***, Vol. 105, pp. 56-64, 1985.

12. K. Okumura and A. Kijima,"Analysis of Characteristics of Ground Wire and Tower System by Electromagnetic Field Theory," ***Electrical Engineering in Japan***, Vol. 106, pp. 81-89, 1986.

13. M. Ishii, et al., "Multistory Transmission Tower Model for Lightning Surge Analysis", ***IEEE Trans. on Power Delivery***, Vol. 6, pp. 1327-1335, 1991.

14. M.E. Almeida and M.T. Correia de Barros, "Tower Modelling for Lightning Surge Analysis Using Electro-Magnetic Transients Program", ***Proc. IEE - Generation, Transmission and Distribution*** (G.B.), Vol. 141, pp. 637-639, 1994.

15. T. Hara and O. Yamamoto, "Modelling of a Transmission Tower for Lightning-Surge Analysis", ***Proc. IEE - Generation, Transmission and Distribution*** (G.B.), Vol. 143, pp. 283-289, 1996.

Problems for Chapter 6

1. For a stroke to tower, plot the profiles of the insulator-string voltage up to 10 μs for the following parameters:

   (a) tower height=11 m; crossarm height=10.5 m; line height=10 m; tower surge impedance = 100 Ω; tower-footing resistance=10 Ω; return-stroke current=20 kA with linearly rising front of i. t_f=5 μs, and ii. t_f=1 μs.
   Assume the velocity of traveling wave along tower = 0.8 p.u.

   (b) tower height=33 m; crossarm height=31.5 m; line height=30 m; tower surge impedance=100 Ω; tower-footing resistance=10 Ω; return-stroke current=20 kA with linearly rising front of i. t_f=5 μs, and ii. t_f=1 μs.
   Assume the velocity of traveling wave along tower = 0.8 p.u.

2. If the CFOs are 300 kV and 1000 kV of system 1a and system 1b, respectively, what are the outage rates of the two systems for a ground flash density, n_g=10 km²/year? Assume the phase-conductor radius to be equal to 1.5 cm. Compute the outage rates by both electrical shadow method and electrogeometric model. Include outages caused by strokes to midspan and to tower.

3.

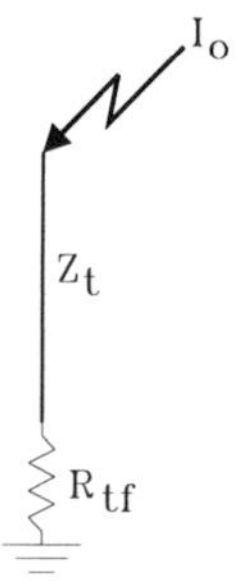

A tower of an unshielded overhead line is struck by lightning, the return-stroke current of which is I_o. If the surge impedance of the tower is Z_t, and the tower-footing resistance is R_{tf}, find an expression for the tower-top voltage, V_{tt}. Show that V_{tt} eventually approaches the value I_oR_{tf}.

CHAPTER 7
DIRECT LIGHTNING STROKES TO OVERHEAD LINES WITH SHIELD WIRES

7.1 Introduction

Originally, ground wires were installed to reduce the voltages induced on phase conductors by lightning strokes to nearby ground. Later research showed that induced voltages by indirect strokes were no threat to high-voltage power lines. The duty of the ground wires was then redefined. Since then, the ground wires have been used to shield phase conductors from direct lightning strokes. Hence, ground wires are also called shield wires. Ground (or shield) wires are usually strung above the phase conductors, and supported by the same towers as the phase conductors. The phase conductors are electrically isolated from the towers by means of insulator strings to withstand the steady-state power-frequency voltage as well as the transient voltages created by switching and lightning. The shield wires are directly connected to the towers, and through the towers to the ground below. Occasionally, the shield wires are supported on a separate structure. It is important, however, that the shield wires be connected to the ground as frequently as possible, e.g., at each tower. For the shield wires to be effective, the grounding impedance should be as low as possible. Besides shielding the phase conductors against direct lightning strokes, the shield wires also provide the following beneficial effects:

1. For strokes to tower, they divert a part of the stroke current from the tower so that the tower-top voltage is reduced; in other words, the shield wires reduce the impedance through which the stroke current flows.

2. Currents flowing through the shield wires induce voltages on the phase conductors by electromagnetic coupling, thus reducing the voltages across the insulator strings.

3. The presence of the shield wires reduces the effective surge impedance of the phase conductor. Therefore, in case a stroke hits the phase conductor directly, the voltage developed on the struck phase conductor will be lower than the phase-conductor voltage without the shield wire.

4. Reflections at the top of the struck tower will reduce the tower-top voltage.

5. Reflections from adjacent towers will further lower the tower-top voltage.

6. Shield wires reduce the voltage induced by lightning strokes to nearby ground (indirect stroke).

7. Shield wires reduce the magnetic fields created by the load current of an overhead power line (non-lightning phenomenon).

The principal function of the shield wire is to shield the phase conductors against direct lightning strokes. To perform this function well, the shield wire must be positioned properly so that all lightning strokes of return-stroke current exceeding a critical level strike the shield wire instead of the phase conductors. It will be a ***shielding failure*** if lightning strikes a phase conductor instead of the shield wire. All strokes, of return-stroke current less than the critical level, to the phase conductors will be withstood by the line, causing no outages. If a shield wire is properly placed, then either the shield wire or the tower will be struck by lightning of return-stroke current greater than the critical current level. Before discussing the shielding theory, let us discuss the response of the line for a strike to either the shield wire or the tower.

7.2 Stroke to Tower

The schematic for stroke to tower is shown in Fig. 7.1a. The stroke current, I, will divide into three parts; one part, I_t, will flow through the tower, and the rest will divide equally and flow in opposite directions over the shield wire,

$$I = I_t + I_s \ . \tag{7.1}$$

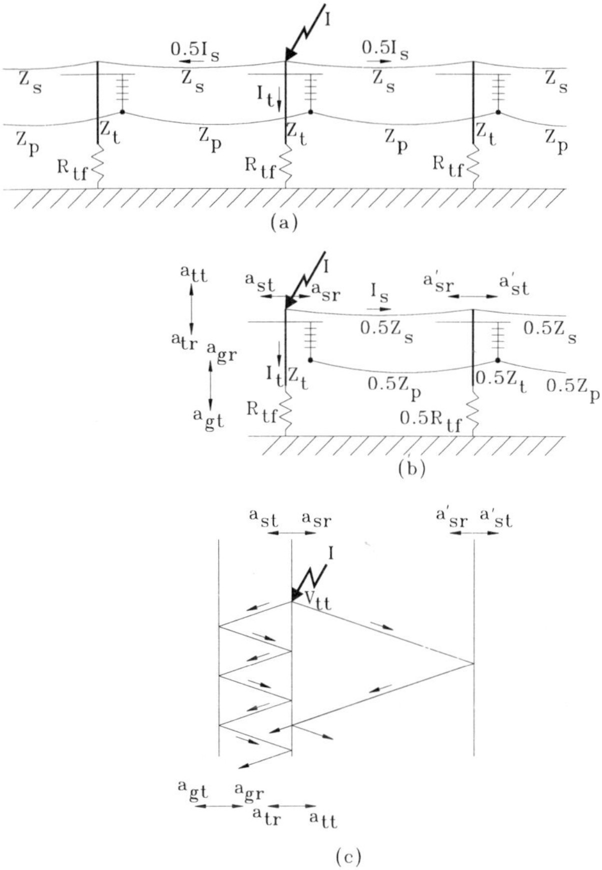

Fig. 7.1 Stroke to tower for a shielded line.
(a) schematic of line; (b) line folded back on itself; (c) voltage lattice diagram.

Three voltage waves, all equal to the ***initial*** tower-top voltage, V_o, will travel away from the struck point; one, along the tower towards the ground with a velocity somewhat less than that of light in free space, and the other two in opposite directions from the struck point along the shield wire with a velocity equal to that of light in free space. Each of the three waves will be reflected and transmitted at the nearest transition points. The tower voltage will be repeatedly reflected between the tower-footing resistance and the tower top; the shield-wire voltages will be reflected and further transmitted at the adjacent towers. The transmitted voltage waves at the adjacent towers will travel further towards the remote towers, and also up and down the adjacent towers. This process will repeat itself as the transmitted voltage waves travel along the shield wire.

Because of the symmetry around the struck tower, the power line along with the shield wires can be folded on itself at the struck tower as shown in Fig. 7.1b. Consequently, all the impedances will be halved, except the Z_t and R_{tf} of the struck tower. The reflected voltage from the adjacent towers will be of a polarity opposite to that of the incident voltage wave, because of the parallel combination of Z_s and Z_t at the adjacent towers. Therefore, the tower-top voltage of the struck tower will be reduced once the reflections come back from the adjacent towers. In the meantime, the voltage along the struck tower will also be reflected from its tower-footing resistance. The polarity of this reflected voltage wave will depend upon the magnitude of R_{tf} relative to Z_t. If $R_{tf} < Z_t$, the reflected voltage wave will be of opposite polarity, and will reduce the tower-top voltage, V_{tt}, on its arrival at the tower top. However, if $R_{tf} > Z_t$, then V_{tt} will be increased. Because of the shorter length of the tower compared to the span length, reflections from the tower-footing resistance will arrive at the tower top sooner. Therefore, it is of utmost importance that R_{tf} be made as low as possible.

The voltage at any point in the overhead-line system can be tracked by means of the lattice diagrams. Such an endeavor can be tedious. The insulator-string voltages determine the outage rate of a line. Moreover, the insulator-string voltages of the struck tower experience the highest voltages. These, then, are the voltages which need to be calculated for an actual system. The successive reflections up and down the struck tower will be the determining factors for the outage probability of the insulator strings. The addition of the first reflections from the adjacent towers will add to the accuracy. Further reflections can be ignored without much loss in accuracy. Figure 7.1c shows the lattice diagram of the voltage waves along the struck tower as well as along the shield wire.

The equivalent surge impedance seen by the return-stroke current as it strikes the tower top (Fig. 7.1b) is given by:

$$Z_{eq} = \frac{Z_s Z_t}{Z_s + 2Z_t} . \tag{7.2}$$

The initial tower-top voltage of the struck tower, V_o, is

$$V_o = IZ_{eq} . \tag{7.3}$$

The voltage reflection and the transmission coefficients at the tower-footing resistance, R_{tf}, are:

$$a_{gr} = \frac{R_{tf} - Z_t}{R_{tf} + Z_t} , \quad \text{and} \tag{7.4}$$

$$a_{gt} = 1 + a_{gr} . \tag{7.5}$$

As the voltage, reflected at R_{tf}, travels back up the tower, a part of it is reflected back down the tower, and the rest is transmitted at the tower top to the shield wire. These voltage coefficients are given by:

$$a_{tr} = \frac{Z_s - 2Z_t}{Z_s + 2Z_t} , \quad \text{and} \tag{7.6}$$

$$a_{tt} = 1 + a_{tr} . \tag{7.7}$$

Simultaneous with these multiple reflections along the struck tower, multiple reflections also take place along the shield wire between the struck tower and its adjacent towers. The reflection and the transmission coefficients at the adjacent tower are given by:

$$a_{sr}' = - \frac{Z_s}{Z_s + 2Z_t} , \quad \text{and} \tag{7.8}$$

$$a_{st}' = 1 + a_{sr}' = \frac{2Z_t}{Z_s + 2Z_t} . \tag{7.9}$$

The corresponding coefficients at the struck tower are:

$$a_{sr} = \frac{2Z_t - Z_s}{2Z_t + Z_s}, \quad \text{and} \tag{7.10}$$

$$a_{st} = 1 + a_{sr} . \tag{7.11}$$

The tower-top voltage, V_{tt}, is then given by:

$$V_{tt} = V_o(t)u(t) + a_{tt}a_{gr}\Sigma\{(a_{tr}a_{gr})^{n-1}V_o(t-2n\tau_t)u(t-2n\tau_t)\} +$$

$$a_{st}a'_{sr}V_o(t-2\tau_s)u(t-2\tau_s) , \tag{7.12}$$

where τ_s and τ_t are the travel times along the span length and the tower length, respectively; n varies between 1 and integer value of $t/2\tau_t$. If the tower crossarm is very near the tower top, then the insulator-string voltage is given by:

$$V_{ins} = (1 - K_{sp})V_{tt} . \tag{7.13}$$

If the crossarm is not very near the tower top, then the crossarm voltage, V_{ca}, can be determined from the tower lattice diagram. Then, the insulator-string voltage will be:

$$V_{ins} = V_{ca} - K_{sp}V_{tt} . \tag{7.14}$$

In fact, the crossarm voltages will be different from the tower-top voltage, especially for vertically configured lines. In that case, the voltage of the topmost crossarm will have the highest voltage, because the reflection from the tower footing will arrive there the latest. Similarly, the voltage of the lowest crossarm will be the lowest. However, the electromagnetic coupling of the top-most phase conductor to the shield wire will be the highest, and that for the lowest phase conductor will be the smallest. Therefore, it is not possible to state which insulator string will have the highest voltage without computing each of them. If V_{ins} is greater than the insulator-string CFO, then a ***backflashover*** will occur across that insulator string. Therefore, it should be obvious that a critical magnitude of the return-stroke current exists, above which a stroke to the tower will cause a backflashover.

7.3 Stroke to Shield Wire

Figure 7.2a is the schematic for the case of a strike to the shield wire at midspan. The return-stroke current, I, will be divided into two equal halves which travel in opposite directions along the shield wire towards the nearest towers. The corresponding traveling voltage wave, V_s, on the shield wire will be

$$V_s = Z_s \frac{I}{2} , \tag{7.15}$$

where Z_s is the surge impedance of the shield wire. The voltage induced on a phase conductor, because of electromagnetic coupling, will be given by

$$V_p = K_{sp} V_s , \tag{7.16}$$

where $K_{sp} = Z_{sp}/Z_s$, and
Z_{sp} = mutual surge impedance between the shield wire and the phase conductor.

The voltage difference between the shield wire and the phase conductor is given by

$$V_{sp} = (1 - K_{sp}) V_s . \tag{7.17}$$

These voltages will persist until reduced by the reflections from the adjacent towers. This reduction (i.e. negative reflection coefficient) is possible, because V_s encounters Z_s and Z_t in parallel on arrival at the adjacent towers. If the "length" of the span in μs is τ_s, then that time must elapse before relief arrives. In the meantime, flashover between the shield wire and a phase conductor must not occur. This will not occur if the separation distance between the shield wire and the phase conductor is such that the sparkover voltage across this air gap is not reached before the voltage waves of reduction return from the adjacent towers.

The profiles of the various voltages, caused by multiple reflections between the towers and the stroke point, can be traced with the help of the lattice diagram. Because of the symmetry around the stroke point, the line can be folded back on itself and the impedances halved, as shown in Fig. 7.2b. The lattice diagram is shown in Fig. 7.2c.

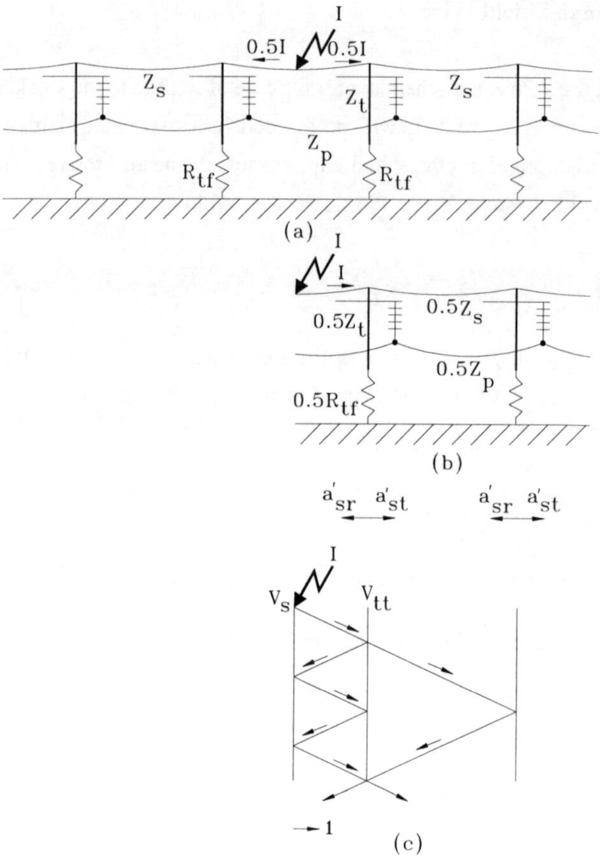

Fig. 7.2. Stroke to shield wire at midspan for a shielded line.
(a) schematic of line; (b) line folded back on itself; (c) voltage lattice diagram.

At the adjacent tower,

$$Z_{eq} = \frac{0.5\, Z_s Z_t}{Z_s + Z_t}\,, \tag{7.18}$$

$$a'_{sr} = -\frac{Z_s}{Z_s + 2 Z_t}\,, \tag{7.19}$$

$$a'_{st} = 1 + a'_{sr} = \frac{2Z_t}{Z_s + 2Z_t}, \tag{7.20}$$

$$V_{tt} = a'_{st}V_s = \frac{Z_s Z_t}{Z_s + 2Z_t} I, \tag{7.21}$$

where a'_{sr} = coefficient of voltage reflection,
a'_{st} = coefficient of voltage transmission, and
V_{tt} = tower-top voltage.

The reflection coefficient at the point of strike is 1. It should be noticed that (7.19) and (7.20) are the same as (7.8) and (7.9). Also, (7.21) is the same as the initial tower-top voltage, V_o, in (7.3) for a stroke to the tower. From now on, the phenomenon of multiple reflections along the tower, as well as along the shield wire, is the same as for a stroke to the tower.

If the tower height is short and Z_t is neglected, then Z_t can be replaced by the tower-footing resistance, R_{tf}. In that case, no reflections along the length of the tower will be considered. For the case where Z_t can be neglected, the various voltages as functions of the return-stroke current are tabulated in Table 7.1. If R_{tf} in Table 7.1 is replaced by Z_t, then the voltages at these locations are the initial voltages before the arrival of the reflected voltage waves.

The effects of shield wire on the insulator-string voltage are shown from Figs. 7.3 and 7.4. In Fig. 7.3, only the reflections along the struck tower were considered, (i.e., the reflections from the adjacent towers were neglected). The reflections from the adjacent towers were included in Fig. 7.4.

Comparing Fig. 7.3 with Fig. 6.4, it should be noted that the shield wires reduce the insulator-string voltage significantly by electromagnetic coupling. The negative reflections from the adjacent towers further reduce the insulator-string voltage (Fig. 7.4). The effects of the negative reflections are particularly significant when the front time of the stroke current is longer than the travel time of the negative reflections to reach the tower crossarm.

Table 7.1: Various Voltages as a Function of Return-Stroke Current Tower Surge Impedance, Z_t, Neglected

Initial Voltage	For Stroke to Tower	For Stroke to Midspan
V_s (tower)	$\frac{R_{tf} Z_s I}{2R_{tf} + Z_s}$	$\frac{R_{tf} Z_s I}{2R_{tf} + Z_s}$
V_s (midspan)	$\frac{R_{tf} Z_s I}{2R_{tf} + Z_s}$	$Z_s \frac{I}{2}$
V_p (tower)	$\frac{R_{tf} Z_{sp} I}{2R_{tf} + Z_s}$	$\frac{R_{tf} Z_{sp} I}{2R_{tf} + Z_s}$
V_p (midspan)	$\frac{R_{tf} Z_{sp} I}{2R_{tf} + Z_s}$	$Z_{sp} \frac{I}{2}$
V_{ins} (tower)	$\frac{R_{tf}(Z_s - Z_{sp}) I}{2R_{tf} + Z_s}$	$\frac{R_{tf}(Z_s - Z_{sp}) I}{2R_{tf} + Z_s}$
V_{ins} (midspan)	$\frac{R_{tf}(Z_s - Z_{sp}) I}{2R_{tf} + Z_s}$	$\frac{(Z_s - Z_{sp}) I}{2}$

I = return-stroke current, R_{tf} = tower-footing resistance
V_s = shield-wire voltage, V_p = phase-conductor voltage
V_{ins} = insulator-string voltage, Z_s = shield-wire surge impedance
Z_{sp} = mutual surge impedance between shield wire and phase conductor

7.4 Shielding of Overhead Lines against Direct Strokes

As was mentioned in Chapter 4, the protective shadow width of a structure of height H is 4H (Fig. 4.12). Therefore, the half angle, θ, of the protective cone should be arctan2 (=63.4°). All other structures within this cone should then be protected by the structure of height, H. By the same reasoning, a phase conductor will be protected by a shield wire if a line joining it to the shield wire makes an angle of 63.4° or less to the vertical line passing through the shield-wire cross section. However, previous experience with various power lines indicated that although 20° was giving satisfactory

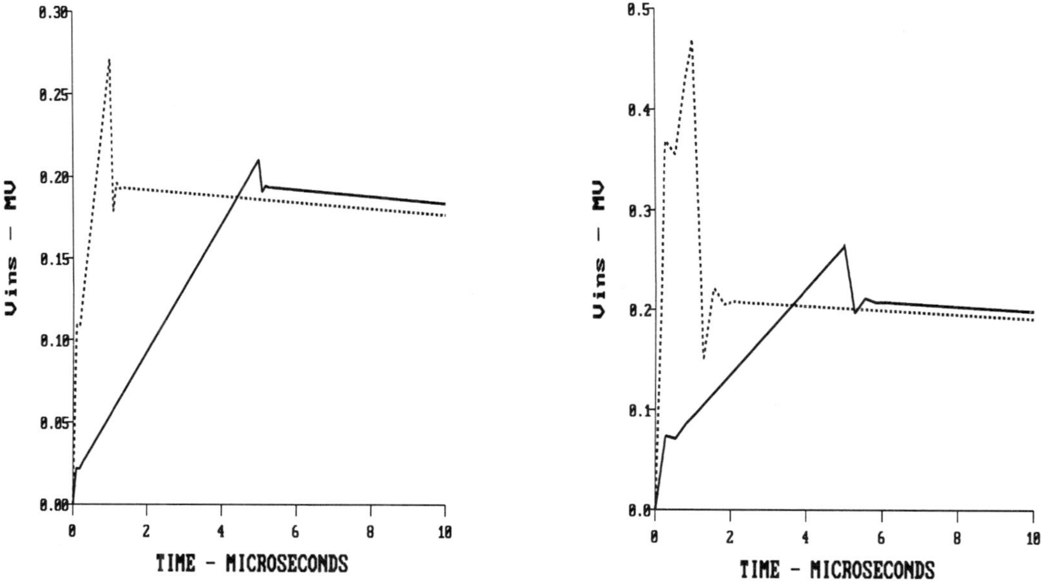

(a) (b)

Fig. 7.3 Insulator-string voltages of a shielded line for a stroke to tower.
Negative reflections from adjacent towers neglected.
All parameters same as in Fig. 6.4.
solid line: 5/50-μs current; dotted line: 1/50-μs current.
(a) tower height=11 m; crossarm height=10.5 m; line height=10 m.
(b) tower height=33 m; crossarm height=31.5 m; line height=30 m.

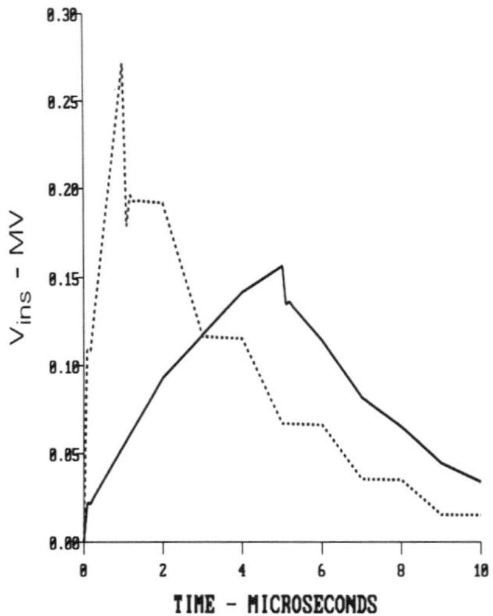

(a)

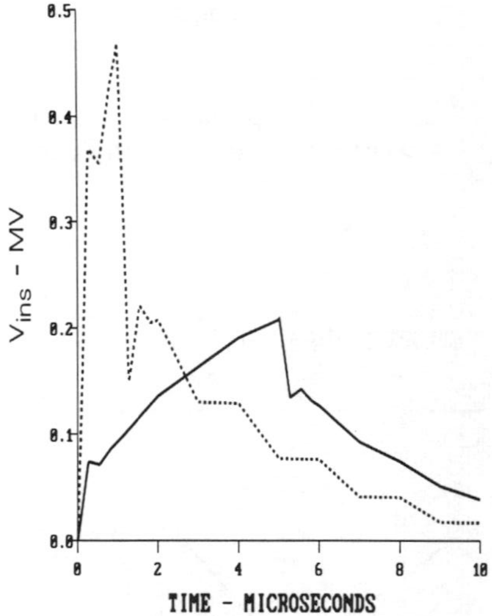

(b)

Fig. 7.4 Insulator-string voltages of a shielded line for a stroke to tower.
Negative reflections from adjacent towers **included**.
Span length=300 m; all other parameters same as in Fig. 6.4.
solid line: 5/50-μs current; dotted line: 1/50-μs current.
(a) tower height=11 m; crossarm height=10.5 m; line height=10 m.
(b) tower height=33 m; crossarm height=31.5 m; line height=30 m.

operation, some lines with 45° were operating poorly.It was concluded from simulated laboratory tests that a shielding angle of 45° should be satisfactory when the line was erected on level ground, and that when the tower was erected on a hillside the angle should be decreased by the angle of the slope of the hill [1].

The above rule of thumb was successful in the design of high-voltage power lines. However, the anomalous lightning performances of the first 345-kV line in the U.S.A. and the 275-kV line on the British Grid gave a new impetus to re-examine the established methods of predicting the lightning performance of overhead lines, and subsequently suggested the development of new theories. As a result, much fruitful research was initiated during the late 1950's. This research showed that the concept of the tower surge impedance was very important in estimating the lightning performance of high-voltage power lines. The prevailing shielding practices were also thoroughly scrutinized. The assumption that the striking distance of a lightning stroke to a structure is solely a height-dependent parameter was seriously questioned, and Golde's original proposition that the striking distance is a function of the return-stroke current appeared to be more convincing [2]. In other words, it is now accepted that the striking distance is an electrical parameter, rather than a simple geometric one. The phase conductor can withstand a direct stroke without a flashover provided the stroke current does not exceed a critical current, I_c, because of the line's specified CFO (I_c=2CFO/Z_p). The shield wire must be geometrically placed in relation to the phase conductor so that for stroke currents exceeding I_c, the stroke will be 'trapped' to strike either the shield wire or the earth surface. This model, which combines the striking distance with the geometrical dimensions of the shield wire and the phase conductor for power-line shielding, is called the electrogeometric model. This model was also developed at that time.

7.5 Electrogeometric Model of Shielding

Several electrogeometric models have been proposed [3-7]. Of these, the electrogeometric model proposed by Whitehead and colleagues is now being almost universally used. Figure 7.5a should help in understanding the logic behind the electrogeometric model of shielding. In this Figure, S is the shield wire and P is a phase conductor. It is assumed that the striking distances to the shield wire and to the phase

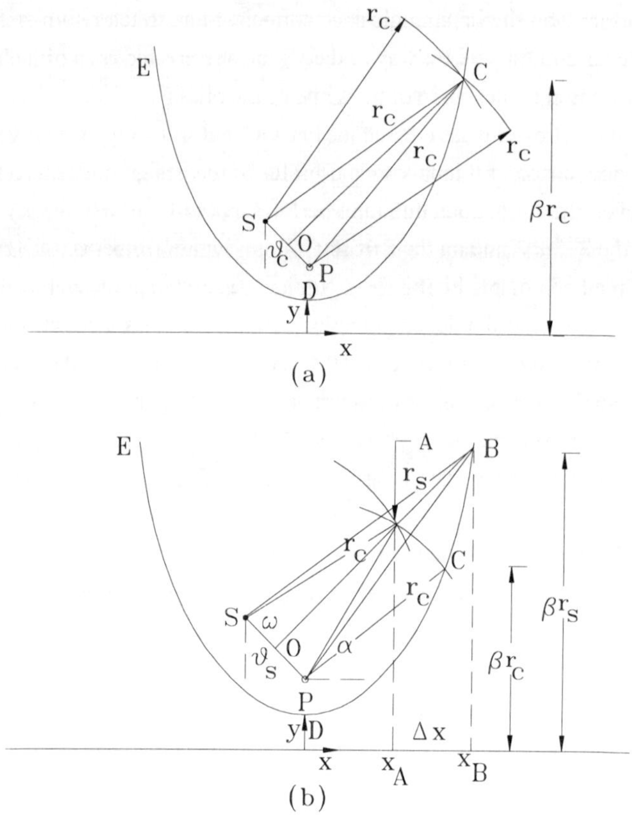

Fig. 7.5 Electrogeometric model for shielding of an overhead line.
(a) perfect shielding; (b) imperfect shielding.

conductor are equal (=r), and that to the surface of the earth is βxr where β is a constant. The locus of all points whose heights are β times their distances from the phase conductor will fall on a curve EDC. If β=1, this curve will be a parabola; if β>1, it will be an ellipse, and if β<1, it will be a hyperbola. In general, β is assumed to be less than one. For simplicity and ease in understanding, let us assume that β=1, which makes the curve EDC a parabola. Similarly, all points equidistant from the shield wire and the phase conductor will lie on the perpendicular bisector OC of the line joining S and P. All lightning strokes descending outside the curve EDC will eventually strike the earth's surface; all those falling within the boundary of EDC will either hit the shield wire or the phase conductor, provided the distance of the descending stroke tip to either S or P becomes shorter than the striking distance corresponding to the return-stroke current. If the distance of the stroke tip always exceeds the striking distance of the prospective return-stroke current, then the stroke will bypass both S and P, and hit the earth's surface. Once the stroke tip has crossed the line OC, but still within the curve DC, it will strike the phase conductor if its distance from P ever becomes shorter than the striking distance corresponding to its return-stroke current; otherwise, it will pass through the curve and hit the earth's surface. The remotest point within the sector OCD is the point C, whose distance from P is r_c. The return-stroke current, I_c, corresponding to r_c is given by:

$$r_c = 8(I_c)^{0.65} , \tag{7.22}$$

where I_c is in kA. For no outage to occur if I_c strikes the phase conductor, the following relation must hold:

$$I_c \leq \frac{2\,CFO}{Z_p} , \tag{7.23}$$

where Z_p is the surge impedance of the phase conductor. I_c is called the critical return-stroke current. The shielding angle, θ_c, is called the critical shielding angle (Fig. 7.5a), and is given by:

$$\theta_c = \cos^{-1}\left(\frac{h_s - h_p}{SP}\right) , \tag{7.24}$$

where h_s and h_p are the heights of the shield wire and the phase conductor, respectively.

In many cases, a few outages caused by shielding failure are allowed. The shielding angle is designed such that the outage rate for shielding failure is within the specified level. Figure 7.5b shows the construction for an imperfect shielding design with a given shield-wire position $(-x_s,h_s)$.

It should be clear from Fig. 7.5a that any return-stroke current, I, which has a corresponding striking distance greater than r_c, will either hit the earth's surface or the shield wire. In other words, the phase conductor is protected against any stroke current greater than I_c. Let us call this the shielding current, I_s, with the corresponding r_s. In Fig. 7.5a, I_s=I_c, where I_c =2CFO/Z_p. If the shielding angle, θ_s, is increased (Fig. 7.5b) then the perpendicular bisector will rotate counterclockwise, increasing its length OB in the process. I_c and r_c, however, remain fixed because of the relations, I_c=2CFO/Z_p, and $r_c=8I_c^{0.65}$. The point B (Fig. 7.5b), where the perpendicular bisector, OB, crosses the parabolic curve EDB, is a limiting point; the striking distance, PB (=r_s), created by the point corresponds to the stroke current, I_s. Any current greater than I_s will hit either the shield wire or the earth's surface. Any current, I ($I_c \leq I \leq I_s$), descending through the wing-like sector, BAC, will strike the phase conductor and cause a line flashover. Any current less than I_c will generate a line voltage ($<I_cZ_p/2$) which is less than the critical flashover voltage of the line.

The horizontal projection, Δx, of the line AB is then the shadow width of concern. Δx can be found as follows:

$$\Delta x = x_B - x_A = r_s\cos\alpha - r_c\cos(90^\circ - \omega + \theta_s)$$

$$= r_s\cos\alpha + r_c\sin(\theta_s - \omega), \qquad (7.25)$$

where α = angle between the line BP and the horizontal line,

$$= \sin^{-1}\left(\frac{\beta r_s - h_p}{r_s}\right)$$

θ_s = angle between the line SP and the vertical line,

$$= \tan^{-1}\left(\frac{-x_s}{h_s - h_p}\right), \text{ and}$$

$$\omega = \angle ASP = \cos^{-1}\left(\frac{SP}{2r_c}\right).$$

All the parameters in the above equation are known except r_s. To determine r_s, the coordinates of the point, B, need to be found first. The equation for the line OB is

$$y = mx + b, \qquad (7.26)$$

where $m=\tan\theta_s$, and $b= h_p + \frac{SP}{2\cos\theta_s}$.

Assuming the curve EDB to be a parabola, the equation for any point (x,y) on this curve is:

$$r^2 = x^2 + (y - h_p)^2 , \tag{7.27}$$

where r is the distance of this point from P. As r=y for a parabola, the coordinates of any point on the parabola are given by:

$$y = \frac{x^2 + h_p^2}{2h_p} . \tag{7.28}$$

Equations 7.26 and 7.28 are equated to find the coordinates of the point B, i.e.,

$$mx + b = \frac{x^2 + h_p^2}{2h_p} . \tag{7.29}$$

Equation 7.29 gives a quadratic equation:

$$x^2 - 2h_p mx - 2h_p b + h_p^2 = 0 . \tag{7.30}$$

Once the x-coordinate of point B is found by solving (7.30), its y-coordinate can be found from (7.26). However, $y=r_s$. Now, the shadow width can be found from (7.25), as well as the attractive area, A, per 100 km length of the line. The probabilities of return-stroke currents exceeding I_c and I_s can be determined from Chapter 4. The number of outages caused by shielding failure per year per 100 km of the line is then:

$$N_{sf} = n_g A[P(I_c) - P(I_s)] , \tag{7.31}$$

where n_g = ground flash density per year per km^2,
A = attractive area per 100 km length of line = 0.2Δx,
$P(I_c)$ = probability that the stroke current will exceed I_c,
$P(I_s)$ = probability that the stroke current will exceed I_s.

The factor of 0.2 in the attractive area (A) takes into account both sides of the line. If N_{sf} is not the specified shielding failure rate, then the shield-wire position $(-x_s,h_s)$ is adjusted, and N_{sf} is recalculated until it is within the specified value.

This, in essence, is the Whitehead electrogeometric model for shielding of overhead lines. It should be borne in mind that the striking distance is a significant parameter in this model. The factor β for the striking distance to the earth's surface is not known very well. Although a parabolic curve was assumed in this analysis, other curves, e.g, ellipse (β>1) and hyperbola (β<1) can also be incorporated in this analysis. Whitehead and Brown also considered the effects of non-vertical strokes on shielding [7]. As knowledge about the non-vertical strokes is sparse, consideration of these types of strokes is not of any practical value. Improvements to the Whitehead model have been proposed by several researchers [8-11]. However, research is continuing and is expected to continue in the future before a definitive theory is established.

7.6 Placement of Shield Wires

We discussed the theory of the shield wire in the previous section, assuming that the position of the shield wire is known. Actually, the designer needs to determine the coordinates of the position of the shield wire from the known coordinates of the phase conductors. The designer also needs to decide how many shield wires are needed to protect all the phase conductors. We will discuss the placement of the shield wires in this section.

Let us consider a horizontally configured three-phase overhead line, and focus on the right outer phase conductor, P, of Fig. 7.5a. Assuming its coordinates to be (0, h_p), its parabola can be drawn from (7.28). Knowing the critical current of the line (I_c=2CFO/Z_p), the critical striking distance, r_c, is determined. Then, an arc is drawn with P as the center and r_c as the radius. This arc will intersect the parabola at the point C, as shown in Fig. 7.5a. An arc is drawn with C as the center and r_c as the radius. This arc will go through P and should also contain the shield wire for perfect shielding. This is shown in Fig. 7.6a by simplifying Fig. 7.5a. A vertical line is drawn from the middle conductor, P_2 , to intersect the arc drawn with C as the center. A shield wire placed at the point of intersection of the vertical line and the arc will protect both P_3 and P_2. P_1, being symmetrically placed, will also be protected by this shield wire. However, two other points need to be considered before a final choice is made.

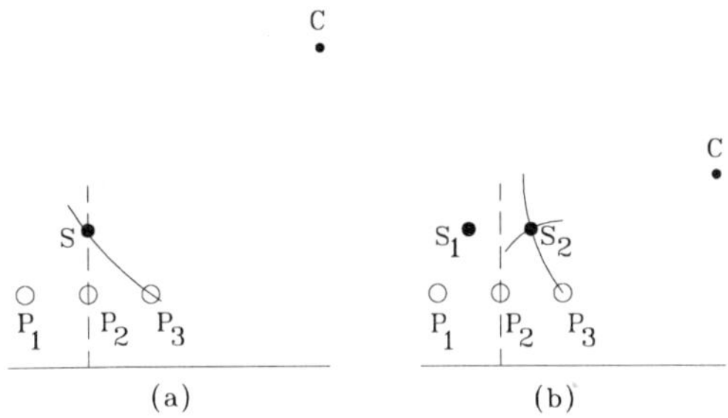

Fig. 7.6 Placement of shield wires.
(a) Protection by single shield wire; (b) Protection by two shield wires.

The shield wire being at ground potential, there must be a minimum clearance, s_{ps}, between the shield wire and the phase conductor. If the vertical distance between the shield wire, S, and P_2 in Fig. 7.6a is less than s_{ps}, then the height of the shield wire must be increased until this separation distance is equal to s_{ps}. Of course, this will increase the tower height.

The phase separation distance may be such that the vertical line from P_2 cannot intercept the arc drawn from the point, C, as shown in Fig. 7.6b. In that case, an arc should be drawn with P_3 as the center and s_{ps} as the radius. This arc will intersect the arc drawn from the point C. The shield wire, S_2, should be placed at this intersection point. The coordinates of this position can be easily determined from the geometrical configuration. In this case, a second shield wire, S_1, must be placed on the other side of P_2 to protect the left outer phase conductor, P_1. The position of S_1 will be the mirror image of S_2 with respect to P_2.

In the final phase of the shield-wire design, it should be ascertained that these two shield wires will also protect P_2. Or, will lightning hurtle through the space between the two shield wires and strike P_2? This is shown in Fig. 7.7. Circles of radius, r_c, are drawn with centers at S_1 and S_2. These two circles intersect on the y-axis of P_2. If the height of this point of intersection is less than (h_p+r_c), then an arc of unprotected zone remains through which lightning may strike P_2 (Fig. 7.7a). If the height is at least equal to (h_p+r_c), then the two shield wires will protect P_2, in addition to protecting P_1 and P_3 (Fig. 7.7b).

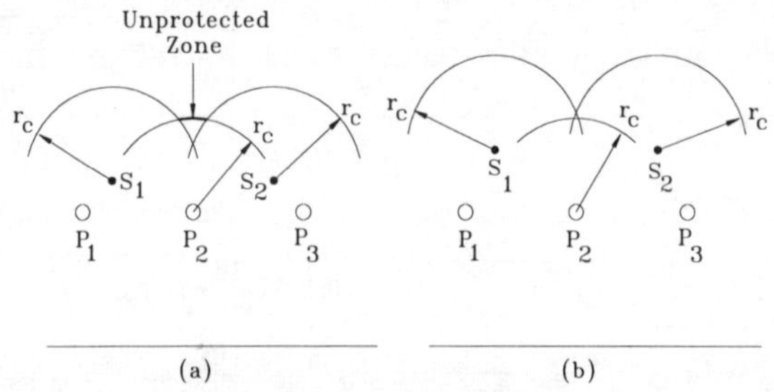

Fig. 7.7 Protection of middle phase conductor by shield wires.

The geometric construction of shielding, as discussed above, is very effective in understanding the shielding principles. However, it is convenient to have a set of algebraic equations for the actual design, especially if a digital computer is used for the design. This is discussed next.

Let us assume that the coordinates of P_2 and P_3 are $(0,h_p)$ and (d_p,h_p) in Fig. 7.6. The ordinate of the point C, in Figs. 7.5 and 7.6, is $y_c=\beta r_c$. The abscissa, x_c, of C is:

$$x_c = d_p + \sqrt{r_c^2 - (\beta r_c - h_p)^2}, \quad \text{for } \beta r_c \geq h_p \ . \tag{7.32}$$

The shield wire is situated on the circle whose center is at C and radius is r_c. The equation of this circle is:

$$(x_c - x_s)^2 + (y_c - y_s)^2 = r_c^2 . \tag{7.33}$$

If $x_c \leq r_c$, then this circle will intersect the vertical line through P_2, and only one shield wire will be able to protect all three phase conductors. The height of the shield wire, h_s, is determined by putting $x_s=0$, and $y_s=h_s$ in (7.33):

$$h_s = y_c - \sqrt{r_c^2 - x_c^2} . \tag{7.34}$$

To meet the requirement for minimum clearance between a live part and a grounded object (s_{ps}), $h_s \geq (h_p+s_{ps})$. If this condition is not met, then the height of the shield wire is increased to (h_p+s_{ps}). This will obviously overprotect P_1 and P_3 to some extent. If h_s in

(7.34) is very high compared with h_p, the cost of towers will be very high. The designer will have to decide whether one shield wire with high towers will be more cost effective than two shield wires with shorter towers. If $x_c>r_c$ or the design with two shield wires and shorter towers is more cost effective, then the right shield wire (S_2 in Fig. 7.6b) will be situated at the point of intersection between the circle around C with radius, r_c, and the circle with its center at P_3 and radius, s_{ps}, whose equation is given by:

$$(x_s - d_p)^2 + (y_s - h_p)^2 = s_{ps}^2. \tag{7.35}$$

The coordinates of S_2 (x_s, y_s) are determined from (7.33) and (7.35). The coordinates of S_1 will be ($-x_s$, h_s). Alternatively, by a simpler trigonometric manipulation of Fig. 7.5a, it can be shown that for $\beta r_c>h_p$:

$$\theta_c = \cos^{-1}\left(\frac{S_{ps}}{2r_c}\right) + \cos^{-1}\left(\frac{x_c - d_p}{r_c}\right) - 90^\circ , \tag{7.36}$$

$$x_s = d_p + S_{ps}\sin\theta_c,\ h_s = h_p + S_{ps}\cos\theta_c , \tag{7.37}$$

where S_{ps}=PS in Fig. 7.5a. If $\beta r_c=h_p$, the point C in Fig. 7.5a will still be on the curve EDC, but will be horizontally displaced from the point P. If $\beta r_c<h_p$, the point C will be inside the curve. In either case, the coordinates of the point C will be:

$$x_c = d_p + r_c,\ y_c = h_p \quad \text{for } \beta r_c \le h_p. \tag{7.38}$$

From (7.38), it should be obvious that at least two shield wires will be required, because $x_c<r_c$. The coordinates of the shield wires can be determined from (7.33) and (7.35). Alternatively,

$$\theta_c = 90^\circ - \cos^{-1}\left(\frac{S_{ps}}{2r_c}\right) \quad \text{for } \beta r_c \le h_p. \tag{7.39}$$

The coordinates of the shield wire (x_s,h_s) are still given by (7.37).

If perfect shielding is not desired, then the shielding current, I_s, needs to be decided upon. The striking distance, r_s, is determined from I_s, and r_c in all the equations are replaced by r_s.

The shielding design of outdoor substations against direct lightning strokes follows the same general principles of the electrogeometric model. However, because of the much smaller attractive area of the substation compared with that of an overhead power line, vertical masts and combinations of vertical masts and shield wires, in addition to the shield wires alone, can be deployed in the shielding design. Moreover, because of the destructive nature of failure of the substation equipment and the associated cost, the design philosophy is somewhat different for substation shielding. This will be discussed in Chapter 14.

7.7 References to Chapter 7

1. A.C. Monteith, E.L. Harder and J.M. Clayton,"Line Design Based upon Direct Strokes," Chapter 17 of ***Electrical Transmission and Distribution Reference Book***, 4th. Edition, Westinghouse Electric Corp., 1964.

2. R.H. Golde,"The Frequency of Occurrence and the Distribution of Lightning Flashes to Transmission Lines, ***AIEE Trans.***, Vol. 64, Part III, pp. 902-910, 1945.

3. F.S. Young, J.M. Clayton and A.R. Hileman,"Shielding of Transmission Lines," ***IEEE Trans.***, Vol. S82, pp. 132-154, 1963.

4. F. Schwab,"Lightning-Proof Overhead Lines," ***Bull. SEV***, Vol. 55, pp. 87-90, 1964.

5. P. Chowdhuri, E.T.B. Gross and S.P. Negreira,"Shielding of Transmission Lines against Lightning," ***Jour. Franklin Institute***, Vol. 286, pp. 269-279, 1968.

6. H.R. Armstrong and E.R. Whitehead,"Field and Analytical Studies of Transmission Line Shielding," ***IEEE Trans. on Power Apparatus and Systems***, Vol. PAS-87, pp. 270-281, 1968.

7. G.W. Brown and E.R. Whitehead,"Field and Analytical Studies of Transmission Line Shielding: Part II," ***IEEE Trans. on Power Apparatus and Systems***, Vol. PAS-88, pp. 617-626, 1969.

8. B.P. McRae, J.L. Diesendorf and G.T. Glasson,"A New Theory of Shielding against Direct Lightning Strokes Based on Recent Lightning Discharge Data," ***Proc. Fifth International Conference on Gas Discharges*** (G.B.), IEE Conf. Publ. No. 165, pp. 108-111, 1978.

9. A.J. Eriksson,"An Improved Electrogeometric Model for Transmission Line Shielding Analysis," ***IEEE Trans. on Power Delivery***, Vol. PWRD-2, pp. 871-886, 1987.

10. A.M. Mousa and K.D. Srivastava,"Effect of Shielding by Trees on the Frequency of Lightning Strokes to Power Lines," ***IEEE Trans. on Power Delivery***, Vol. 3, pp. 724-732, 1988.

11. A.M. Mousa and K.D. Srivastava,"A Revised Electrogeometric Model for the Termination of Lightning Strokes on Ground Objects," ***Proc. 1988 International Aerospace and Ground Conf. on Lightning and Static Electricity***, Paper No. 10B.4, pp. 342-352, 1988.

Problems for Chapter 7

1a. A single-conductor overhead line with one shield wire has the following dimensions:

Tower height, h_{gw}=11 m; crossarm height, h_{ca}=10.5 m; line height, h_l=10 m;
Conductor radius, r_l=1.5 cm; shield wire radius, r_{gw}=1.0 cm; span length=300 m;
Tower surge impedance, Z_t=100 Ω; tower-footing resistance, R_{tf}=10 Ω;
Return-stroke current, I_o=20 kA; traveling-wave velocity along tower=0.8 p.u.

The shield wire is situated on the same vertical plane as the phase conductor. For a stroke to tower and for a linearly-rising and linearly-falling return-stroke current 5x50-μs, plot the voltage profile across the insulator string at the stricken tower for a period of 10 μs. Include the reflections from the adjacent towers; however, neglect the reflections from the tower-footing resistance of the adjacent towers, and the reflections from the subsequent towers.

1b. Repeat (a) with R_{tf}=50 Ω.

1c. Repeat (a) with h_{gw}=33 m; h_{ca}=31.5 m; h_l=30 m.

1d. Repeat (c) with a 0.5x50-μs return-stroke current.

2.

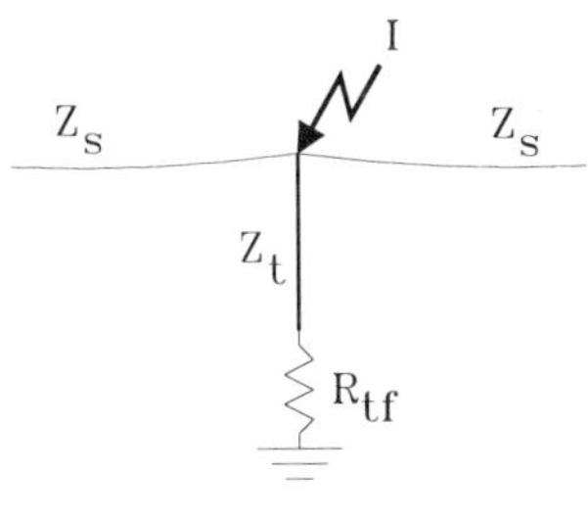

For a direct lightning stroke to tower with a shield wire, as shown above, show that the final tower-top voltage is independent of the tower surge impedance. Neglect reflections from the adjacent towers.

3.

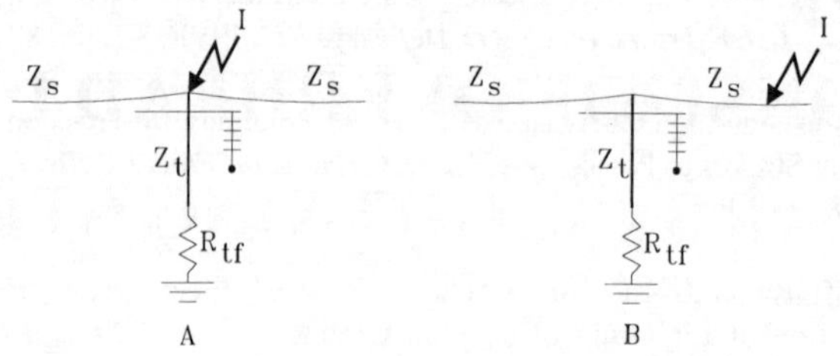

Find the following:

a. For a direct lightning stroke to tower (Fig.A):
(i) insulator-string voltage at the tower, and
(ii) voltage between shield wire and phase conductor at midspan.

b. For a direct lightning stroke to midspan (Fig.B):
(i) insulator-string voltage at the adjacent tower, and
(ii) voltage between shield wire and phase conductor at midspan.

These are the **initial** voltages.

4. A single-conductor overhead line has the following data:

Height above ground, h_p = 10 m ; radius of conductor, r_p = 1.5 cm
Minimum clearance from grounded object, S_b = 1 m ; CFO = 500 kV

Find the height and the lateral distance of the shield wire from the phase conductor for perfect shielding against lightning strokes.

5. For the conductor of problem 4, the shield wire is now placed vertically above it at a distance 1 m from the conductor. If the radius of the shield wire is 1 cm, then, in case of a direct hit on the line (shielding failure), find the magnitude of the return-stroke lightning current to cause flashover. Assume the shield wire to be at ground potential.

CHAPTER 8
RESPONSE OF OVERHEAD LINES TO NEARBY LIGHTNING STROKES

8.1 Review of Research on Lightning-Induced Voltages

During the early years of power transmission, protection against a direct lightning stroke was thought to be impossible. Therefore, power lines were designed with overhead ground wires to protect against transient voltages induced from nearby lightning strokes to ground. As the transmission voltages became higher and higher with the consequent increase in the insulation levels of overhead lines, it was realized that the transient voltages induced from nearby lightning strokes were not the chief source of disturbance, and that protection against direct strokes is indeed technically feasible and economically justifiable [1].

Much work has been done to date on the theories of protection of transmission lines against direct strokes. In contrast, the study of the mechanism of induced voltages by nearby strokes has generally been neglected, although it is considered that such voltages could be dangerous to transmission systems of 50 kV and below. Field studies in many countries have generated renewed interest in the induction effects of nearby lightning strokes [2-8].

K.W. Wagner [9] made the first attempt to study induced voltages on overhead lines caused by nearby lightning strokes. According to Wagner, a thundercloud above an overhead line induces a charge of opposite polarity on the line. The distribution of charge on the line is determined by the fact that the line is held electrostatically at ground potential by its leakage path to ground over the insulator strings, and also by connection to ground through neutrals of transformers. The bound charge on the line is released when the cloud discharges, giving rise to traveling waves which move in either direction of the line. According to Wagner's theory, the traveling waves will be rectangular in form (Fig. 8.1), and the induced voltage is given by the product of the line height and the inducing electrostatic field caused by the charged cloud prior to the discharge.

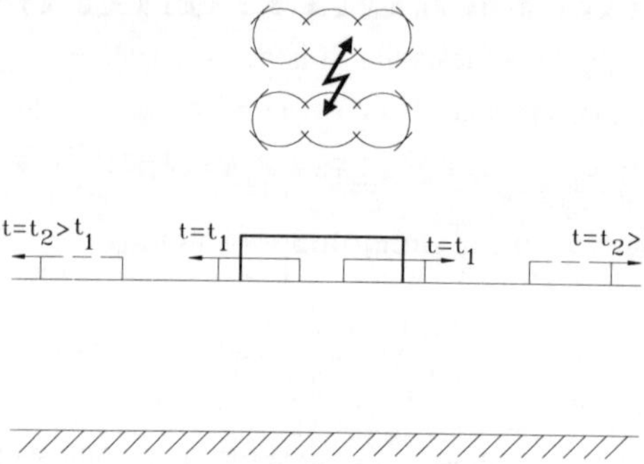

Fig. 8.1 Induced waves from rectangular bound charges according to K.W. Wagner.

Wagner's approximation of instantaneous cloud discharge was modified by Bewley [10]. Bewley assumed that the cloud is discharged gradually, the law of cloud discharge being a function of time. As the first step, the induced traveling voltage wave can be calculated on the basis of instantaneous cloud discharge; then the effect of the law of cloud discharge can be accounted for either by applying Duhamel's theorem or by step-by-step graphical method.

Neither Wagner nor Bewley considered the effect of the lightning stroke itself on the induced voltage. Aigner [11] recognized this problem and calculated the magnetically induced voltage on a transmission line caused by a lightning stroke. In his analysis, Aigner neglected the electrostatic component of the induced voltage. Moreover, he assumed the stroke current to be sinusoidally varying with a single frequency. No attempt was made to modify the analysis for the aperiodically varying lightning currents which are encountered in practice.

Norinder [12] studied the induced voltage both analytically and experimentally. In his analysis, he followed Bewley with the further assumption of a half-sine wave for current variation. He calculated the rate of change of the cloud charge from this current waveshape. His experimental study was in no way correlated with his analytical study.

C.F. Wagner and G.D. McCann [13] did a comprehensive analytical study of the mechanism of induced voltages. They assumed uniform charge density along the stepped leader and a rectangular current wave in the return stroke propagating with a constant velocity. They considered both the electrostatic as well as the magnetic field components in the analysis of the induced voltage. But later they neglected the magnetic field component on the hypothesis that it is negligible. Their results show that for a stroke very near the line, an induced voltage of infinite amplitude will arise, contrary to experience.

Szpor [14] calculated the induced voltage caused by lightning discharge in the vicinity of a line by considering both the electric and the magnetic fields of the discharge. However, as he treated the problem as quasi-stationary, he estimated that his results were applicable to points within 100 meters of the lightning discharge.

Golde [15] investigated the induced voltages on power transmission lines by considering only the change in the electrostatic field by neutralization of the charge in the stepped leader.

Lundholm [16] laid the ground work for later research with an extensive and rigorous analysis of the induced voltages on power transmission lines. He assumed the waveshape of the current to be rectangular and later calculated the induced voltages for a linearly-rising front of the current wave. Although he considered both the electric and the magnetic fields of the return stroke, he ignored the initial conditions imposed by the stepped leader. Moreover, his analysis was based on a relationship between the return-stroke current and its velocity, which could not be completely justified.

Rusck [17], likewise, studied the induced voltages on power lines, modifying Lundholm's expression for the relationship between the current and the velocity of the return stroke. This modified relationship is empirical in nature, and variation in the constant of that relationship can modify the results considerably. Rusck found that the induced voltage on a conductor was not affected by the presence of other conductors. This cannot be justified from the physical principles.

Pepet-Lepine [18] analyzed induced voltages by assuming both the charge and the current distributions along the return stroke to be varying with height. His analysis did not consider the traveling-wave nature of the induced voltage. As a consequence, the magnitude of the induced voltage becomes very small at some distance along the line.

Owa [19] has studied induced voltages on power lines caused by lightning with Griscom's model [20] of the stepped leader. According to this model, the distribution of charge along the stepped leader is fairly uniform except at the tip where the charge concentration is unusually high. Owa calculated the field variation as well as the induced voltage caused by each step of the leader stroke. He concluded that the induced voltage can be high under certain conditions of charge in the stepped leader. This high induced voltage seems to come mainly from the concentrated charge at the tip of the stepped leader. Actual measurements [21,22] show, however, that the variation of the field caused by the individual steps of the leader stroke is very small compared to that caused by the return stroke. Owa's analysis of the component of the voltage induced by the return stroke is similar to that of Rusck.

In spite of the divergence in assumptions, several common conclusions can be drawn from these studies:

(i) the magnetic effect of the return-stroke current is considerably less than the electrostatic effect of the charge in the return stroke, and therefore, the net effect of the return stroke is still electrostatic;

(ii) because of the net electrostatic effect, the induced voltage on a line will be unipolar in nature, and of a polarity opposite to that of the return-stroke current;

(iii) the voltage on an overhead line induced by lightning is entirely a traveling wave in character.

Records of induced voltages on a 500-kV experimental line in the Pennsylvania Electric Company showed that many of them were bipolar. Some of the induced-voltage oscillograms were reproduced in [27]. Records from several other field studies, where the records were taken on shorter time scales, also showed the bipolar nature of the induced voltages [23-25]. Additional field data [26] also cast doubt on the conclusion that the induced voltage is entirely a traveling-wave phenomenon.

Chowdhuri and Gross [27,28] studied the induced-voltage phenomena with the same basic assumptions as Lundholm's, but with different initial conditions and different analytical techniques. Their main conclusions were:

(i) the magnetic effects of the return-stroke current can be significant, if the rise time of this current is short;

(ii) the induced voltage can be bipolar in character, depending upon the waveshape of the return-stroke current;

(iii) the induced voltage is not entirely a traveling-wave phenomenon;

(iv) the presence of other conductors would, in general, affect the induced voltage of a conductor.

The Chowdhuri-Gross analysis was later improved upon by several researchers [29-33].

Singarajah [34] presented an elegant analysis on the induced voltage, in which he considered the upward streamer from the grounded object before the beginning of the return stroke. He showed that this upward streamer is responsible for the bipolar characteristic of the induced voltage. Although upward streamers have been frequently observed for strokes to tall objects, such streamers have very seldom been observed for strokes to flat ground or to short structures. Even if they exist for strokes to low objects, their length would be too small to be of much significance to the induction process.

The electric field was assumed to be vertical in all the previous studies. A horizontal component of the electric field will exist in the direction of propagation if the ground is imperfect, i.e., the soil resistivity is finite. A horizontal electric field along the length of an overhead line will then exist if the electromagnetic field of a lightning stroke is radiating in the axial direction of the line. Wait [35] found that for a flat and homogeneous ground, the wavetilt function relating the Fourier transform of the vertical electric field to that of the horizontal electric field at ground may be approximated by:

$$W(j\omega) = \frac{E_H(j\omega)}{E_V(j\omega)} = \frac{1}{\sqrt{\epsilon_r + \dfrac{\sigma}{j\omega\epsilon_o}}}, \qquad (8.1)$$

where W = wavetilt,
E_H = Fourier transform of horizontal electric field,
E_V = Fourier transform of vertical electric field,
ε_r = relative permittivity of soil,
σ = soil conductivity, and
ε_o = permittivity of free space.

Koga, Motomitsu and Taguchi [36] used this horizontal electric field to calculate the voltage induced on an overhead line by a nearby lightning stroke. Master [37] and also de la Rosa [38] have used this wavetilt principle to calculate the horizontal electric field component of the lightning stroke, and also to compute the voltage induced on the line, utilizing the coupling model proposed by Agrawal, Price and Gurbaxani [39,40].

The wavetilt principle has long been used in high-frequency communication technology. It should be evident from (8.1) that the wavetilt is a frequency-dependent function. Its use in time-domain phenomena, such as lightning, can hardly be justified. Moreover, wavetilt is applicable to radiated fields which originate at a long distance from the receiver. Voltages induced on an overhead line from such a source would be too small to be of any concern to power system engineers. It should also be borne in mind that a horizontal electric field perpendicular to the line axis will have no effect.

No electric field exists along the surface of the perfect earth. However, above the earth's surface, the residual charge in the lightning channel will produce a component of the electric field in the horizontal plane even when the lightning channel is vertical. This horizontal electric field can be decomposed into two component fields: (i) along the axis of the overhead line, and (ii) perpendicular to the line axis. The axial component will be zero at the point directly across the lightning channel. In addition, a non-vertical lightning channel can produce an axial component of the electric field. Branching and tortuosity of the lightning channel can also produce the axial electric field component. The axial component will affect the voltage induced on the line. Several researchers [41-44] have investigated the effect of the axial electric field produced by a vertical lightning channel on the induced voltage. The results are, however, significantly different from one another. For instance, Diendorfer [41] concluded that the maximum line voltage does not necessarily occur at the point closest to the return stroke channel. On the other hand, Nucci, Rachidi, Ianoz and Mazzetti [44] concluded that the line voltage at the point closest to the lightning channel will be significantly higher than the voltages induced at points farther from the strike point. Nucci et al. [44] also concluded that for a strike point parallel to a line of finite length, the voltage at the far end of the line will be smaller and of slower front time than that at the near end of the line, contrary to field experience [38].

The intensities of the axial electric field from the other sources, namely non-vertical, branched and tortuous strokes, are unknown. The intensity of the axial electric field produced by non-vertical strokes can be high relative to that from a vertical stroke, depending upon its angle of inclination to the vertical plane. In view of these uncertainties, the axial electric field is neglected in the following discussion on lightning-induced voltages.

8.2 Mechanism of Voltage Induction [27,47]

From the discussion of the structure of a lightning stroke, it should be evident that the voltage induced on an overhead line would have ***four*** components:

1. The charged cloud above the line induces bound charges on the line while the line itself is held electrostatically at ground potential by the neutrals of connected transformers and by leakage over the insulators. When the cloud is partially or fully discharged, these bound charges are released and travel in both directions on the line giving rise to the traveling voltage and current waves.

2. The charges lowered by the stepped leader further induce charges on the line. When the stepped leader is neutralized by the return stroke, the bound charges on the line are released and thus produce traveling waves similar to that caused by the cloud discharge.

3. The residual charges in the return stroke induce an electrostatic field in the vicinity of the line and hence an induced voltage on it.

4. The rate of change of current in the return stroke produces a magnetically induced voltage on the line.

If the lightning has subsequent strokes, then the subsequent components of the induced voltage will be similar to one or the other of the four components discussed above.

It has been shown by several investigators that the magnitudes of the voltages induced by the release of the charges bound either by the cloud or by the stepped leader are small compared with the voltages induced by the return stroke [13,16-18].

8.3 Basic Assumptions

a. Only the electrostatic and the magnetic components induced by the return stroke are considered, i.e., the electrostatic effects of the cloud and the stepped leader are neglected.

b. The charge distribution along the leader stroke is uniform.

c. The shape of the return-stroke current is rectangular. The result with the rectangular current wave can be transformed to that with currents of any other waveshape by the convolution integral (Duhamel's theorem). The computations are extended to current waves having a linearly rising front, in order to study the effect of the waveshape of the current on the induced voltage wave.

d. The velocity of the return stroke is constant. It was thought to be of advantage to be able to take the return-stroke velocity as a parameter and to study its effects on the induced voltage. Later, the analysis was extended to include a return-stroke velocity which is a function of the return-stroke current.

e. The lightning stroke is vertical.

f. The overhead line is loss-free and the earth is perfectly conducting.

A rectangular system of coordinates in space is chosen so that the origin of the system is the point where lightning strikes the surface of the earth. The line conductor is located at a distance y_o meters from the origin, having a mean height above ground of h meters and running along the x direction (Fig. 8.2). The origin of time (t=0) is assumed to be the instant when the return stroke starts at the earth level.

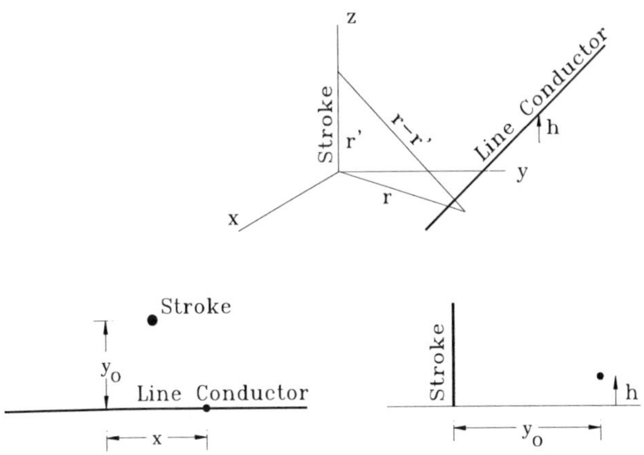

Fig. 8.2 Coordinate system of line conductor and lightning stroke.

The total electric field associated with the charge and the current in the lightning stroke at any point in space is

$$\mathbf{E_i} = \mathbf{E_{ei}} + \mathbf{E_{mi}} = -\nabla\phi - \frac{\partial \mathbf{A}}{\partial t}, \tag{8.2}$$

where ϕ is the ***inducing*** scalar potential created by the residual charge in the return stroke, and **A** is the ***inducing*** vector potential created by the return-stroke current. ϕ and **A** can be easily computed by assuming the lightning channel to be a vertical line of charge and current (Fig. 8.3).

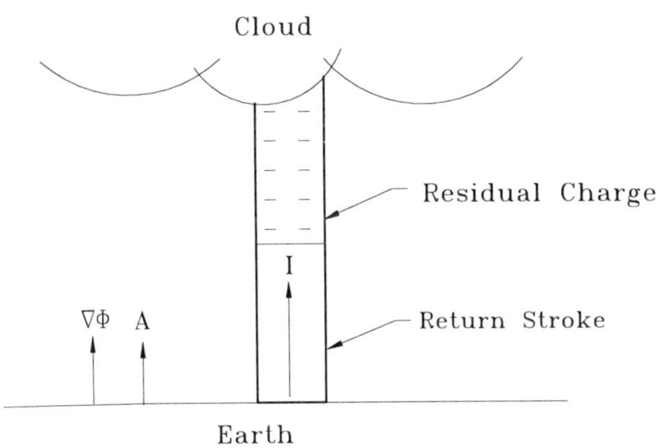

Fig. 8.3 Return stroke with the residual charge column.

The inducing scalar potential and the inducing vector potential are given by:

$$\phi(\mathbf{r},t) = \frac{1}{4\pi\epsilon}\int \frac{\rho(\mathbf{r}',t - \sqrt{\epsilon\mu}|\mathbf{r} - \mathbf{r}',|)}{|\mathbf{r} - \mathbf{r}',|}dv', , \tag{8.3}$$

$$\mathbf{A}(\mathbf{r},t) = \frac{\mu}{4\pi}\int \frac{\mathbf{J}(\mathbf{r}',t - \sqrt{\epsilon\mu}|\mathbf{r} - \mathbf{r}',|)}{|\mathbf{r} - \mathbf{r}',|}dv', , \tag{8.4}$$

where $\mathbf{r},\mathbf{r}'$ = field and source points, respectively,
ρ = charge density,
$\mathbf{J}$ = current density, and the integrations are taken over a volume enclosing the sources.

The equations for the electromagnetic potentials (known as the retarded potentials) indicate that at a given field point $\mathbf{r}$ and a given time t, the potentials are determined by the charge and current which existed at the source point $\mathbf{r}'$ at an earlier time; the difference in time is the time required to travel the distance between the source point and the field point with a velocity $\upsilon=1/\sqrt{(\varepsilon\mu)}$. In our problem, the source is the lightning channel, and the field point is any point along the overhead line. Because the overhead line is suspended in the air, the effect of the lightning stroke travels towards the line with a velocity $c=1/\sqrt{(\varepsilon_o\mu_o)}=3\text{x}10^8$ m/s. From the distribution of charge and current in the lightning channel, the retarded electromagnetic potentials ϕ and $\mathbf{A}$ can be deduced. Knowing ϕ and $\mathbf{A}$, in turn, the inducing electric field $\mathbf{E}_i$ can be computed from (8.2). The line integral of $\mathbf{E}_i$ will then provide the ***inducing*** voltage at the overhead line:

$$V_i=-\int_0^{h_p}\mathbf{E}_i\cdot\mathbf{dz}=-\int_0^{h_p}\mathbf{E}_{ei}\cdot\mathbf{dz}-\int_0^{h_p}\mathbf{E}_{mi}\cdot\mathbf{dz}=V_{ei}+V_{mi} \ . \tag{8.5}$$

As the height, h_p, of the line conductor is small compared with the height of the cloud, i.e., the length of the lightning channel, the inducing electric field below the line conductor can be assumed to be constant, and equal to that on the ground surface,

$$V_i = \left(\nabla\phi + \frac{\partial\mathbf{A}}{\partial t}\right)\cdot\mathbf{h}_p \ . \tag{8.6}$$

The inducing voltage will act on each point along the length of the overhead line.

One would indeed measure the inducing voltage only in the absence of any conductor. The inducing voltage at different distances along the length of the overhead line will be different. The overhead line being a good conductor of electricity, these differences of voltage will tend to be equalized by the flow of currents. Therefore, the actual voltage measured between a point on an overhead line and the point on the ground vertically below it will be different from the inducing voltage. This voltage, which can actually be measured on the overhead line, is defined as the ***induced*** voltage. The calculation of this voltage is the primary objective.

Neglecting losses, an overhead line may be represented as consisting of distributed series inductance L (H/m), and distributed shunt capacitance C (F/m). The effect of the inducing voltage will then be equivalent to connecting a voltage source along each point of the line (Fig. 8.4). The partial differential equation for such a configuration will be:

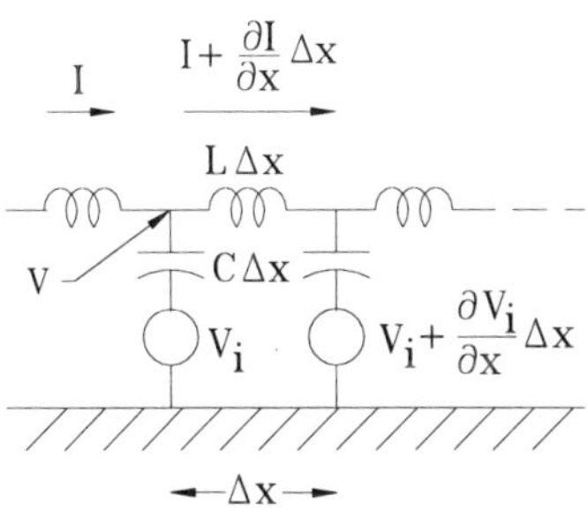

Fig. 8.4 Equivalent circuit of transmission line with inducing voltage.

$$-\frac{\partial V}{\partial x}\Delta x = L\frac{\partial I}{\partial t}\Delta x \text{ , and} \tag{8.7}$$

$$-\frac{\partial I}{\partial x}\Delta x = C\frac{\partial}{\partial t}(V - V_i)\Delta x \text{ .} \tag{8.8}$$

Differentiating (8.7) and eliminating I, the voltage equation can be written as:

$$\frac{\partial^2 V}{\partial x^2} - \frac{1}{c^2}\frac{\partial^2 V}{\partial t^2} = -\frac{1}{c^2}\frac{\partial^2 V_i}{\partial t^2} = F(x,t) \text{ ,} \tag{8.9}$$

where $c = \frac{1}{\sqrt{LC}} = \frac{1}{\sqrt{\epsilon_o \mu_o}} = 3\times10^8 m/s$, and $F(x,t) = -\frac{1}{c^2}\frac{\partial^2 V_i}{\partial t^2}$.

Equation 8.9 is an inhomogeneous wave equation for the induced voltage along the overhead line. Its solution can be obtained by assuming F(x,t) to be a superposition of impulses which involves the definition of Green's function [45,46].

To obtain the voltage caused by a distributed source F(x), we calculate the effects of each elementary portion of source and add them all. If G(x;x′) is the voltage at a point x along the overhead line caused by a unit impulse source at the source point x′, the voltage at x caused by a source distribution F(x′) is the integral of G(x;x′)F(x′) over the whole domain of x' occupied by the source, provided that F(x′) is a piecewise continuous function in the domain $a \le x \le b$,

$$V(x) = \int_a^b G(x;x',)F(x',)dx', \ . \tag{8.10}$$

The function G(x;x′), called the Green's function, is therefore a solution for a case which is homogeneous everywhere except at one point. Green’s function, G(x;x′), has the following properties:

(a) $G(x;x'+0) - G(x;x'-0) = 0$, (8.10a)

(b) $\left(\frac{dG}{dx}\right)_{x'+0} - \left(\frac{dG}{dx}\right)_{x'-0} = 1$, (8.10b)

(c) G(x;x′) satisfies the homogeneous equation everywhere in the domain, except at the point x=x′, and

(d) G(x;x′) satisfies the prescribed homogeneous boundary conditions.

8.4 Doubly Infinite Single-Conductor Line

A doubly infinite line can be represented by a line whose terminals are sufficiently distant from the point of strike of lightning that reflections of the induced voltage at the terminals can be neglected. A doubly infinite single-conductor line is relatively simple to analyze; furthermore, it gives an insight into the physical nature of the phenomena.

8.4.1 Inducing voltage: rectangular return-stroke current

The assumption of a rectangular waveshape for the return-stroke current keeps the analysis general, so that for any particular current waveshape the analysis can be transformed to the desired case by Duhamel's theorem as discussed in Chapter 1. Under the assumption of rectangular waveshape, the charge along the leader stroke is uniformly distributed as is also the current in the return stroke.

The expression for the inducing voltage was given in (8.6). Because of the retardation effect, ϕ and $\mathbf{A}$ at the field point $\mathbf{r}$ and at time t must be originated at the source point $\mathbf{r}'$ at an earlier time:

$$t' = t - \frac{|r - r'|}{c} . \qquad (8.11)$$

It is convenient to choose the instant at which the return stroke starts from the ground as the origin of time, i.e., t=0. If the velocity of the return stroke is a fraction of the velocity of light in free space ($v=\beta c$), the disturbance of an element of charge or current at a height z' above the ground and along the return stroke will be felt at a field point $\mathbf{r}$ on the ground at time

$$t = \frac{z'}{v} + \frac{\sqrt{z'^2 + x^2 + y_o^2}}{c} . \qquad (8.12)$$

Therefore, the earliest time at which the disturbance from the return stroke will reach a field point would be

$$t_o = \frac{\sqrt{x^2 + y_o^2}}{c} . \qquad (8.13)$$

In other words, the inducing voltage at a field point remains zero until $t=t_o$. Hence,

$$V_i = \psi(x,t)\,u(t - t_o)\ , \tag{8.14}$$

where $u(t\text{-}t_o)$ is the shifted unit step function. The continuous function, $\psi(x,t)$ can be evaluated from the assumed structure of the lightning stroke [17], and is given by:

$$\psi(x,t) = -\frac{60 I_o h_p}{\beta}\left[\frac{1-\beta^2}{\sqrt{\beta^2 c^2 t^2 + (1-\beta^2) r^2}} - \frac{1}{\sqrt{h_c^2 + r^2}}\right], \tag{8.15}$$

where I_0=return-stroke current, A
h_p=height of line above ground, m
β=v/c,
r=distance of point x on line from point of strike, m, and
h_c=height of cloud charge center above ground.

8.4.2 Induced voltage: rectangular return-stroke current

The initial and the boundary conditions of the partial differential equation (8.9) for the induced voltage on the line conductor are: (a) V is bounded for $x \rightarrow \pm\infty$, and (b) $V(x,0)=V'(x,0)=0$. From (8.14):

$$\frac{\partial^2 V_i}{\partial t^2} = \psi''(x,t)u(t-t_o)$$

$$+2\psi'(x,t)\delta(t-t_o)+\psi(x,t)\delta'(t-t_o), \tag{8.16}$$

where the primes represent differentiation with respect to t, and $\delta(t\text{-}t_o)$ is the shifted Dirac delta function. Taking the Laplace transform of (8.9), and considering the initial conditions, we get:

$$\frac{\partial^2 V(x,s)}{\partial x^2} - \frac{s^2}{c^2}V(x,s) = F(x,s)\ , \tag{8.17}$$

$$\text{where } F(x,s) = -\frac{s^2}{c^2}e^{-st_o}\ \mathcal{L}\{\psi(x,t+t_o)\}\ . \tag{8.18}$$

To solve this equation, we find Green's function, G(x;x',s) of (8.17):

$$\frac{\partial^2 G(x;x',s)}{\partial x^2} - \frac{s^2}{c^2} G(x;x',s) = 0 . \tag{8.19}$$

The general solution of (8.19) is given by:

$$G(x;x',s) = Ae^{-\frac{sx}{c}} + Be^{\frac{sx}{c}} .$$

The constants, A and B, are found from the boundary conditions and from the properties of Green's function. As G(x;x',s) is finite for $x \to -\infty$ and $x \to +\infty$,

$$G_1 = Ae^{\frac{sx}{c}} \text{ for } x<x'; \; G_2 = Be^{-\frac{sx}{c}} \text{ for } x>x'.$$

$$\text{From (8.10a): } Ae^{\frac{sx}{c}} = Be^{-\frac{sx}{c}} \text{ ,i.e., } B = Ae^{\frac{2sx'}{c}} .$$

$$\text{From (8.10b): } A = -\frac{ce^{-\frac{sx'}{c}}}{2s}; \text{ hence, } B = -\frac{ce^{\frac{sx'}{c}}}{2s}.$$

$$G_1(x;x',s) = -\frac{c}{2s}\exp\left\{-\frac{s(x'-x)}{c}\right\} \quad \text{for } x<x' \tag{8.20a}$$

$$G_2(x;x',s) = -\frac{c}{2s}\exp\left\{\frac{s(x'-x)}{c}\right\} \quad \text{for } x>x' . \tag{8.20b}$$

By applying (8.10):

$$\begin{aligned} V(x,s) &= -\frac{c}{2s}\int_{-\infty}^{x} G_1(x;x',s)F(x',s)dx' \\ &\quad -\frac{c}{2s}\int_{x}^{\infty} G_2(x;x',s)F(x',s)dx' = V_1(x,s) + V_2(x,s). \end{aligned} \tag{8.21}$$

Putting the value of F(x,s) from (8.18), taking the inverse Laplace transform and considering the properties of the unit step function as well as the Dirac delta function, the solution for (8.9) is obtained for a rectangular return-stroke current. The results are given below:

$$V(x,t) = (V_{11} + V_{12} + V_{21} + V_{22})u(t - t_o) , \tag{8.22}$$

where

$$V_{11} = \frac{30I_o h(1-\beta^2)}{\beta^2(ct-x)^2+y_o^2}\left[\beta(ct-x)+\frac{(ct-x)x-y_o^2}{\sqrt{(ct)^2+(1-\beta^2)\dfrac{x^2+y_o^2}{\beta^2}}}\right] \tag{8.23a}$$

$$V_{12} = -\frac{30I_o h}{\beta}\left[1-\frac{1}{\sqrt{K_1^2+1}}-\beta^2\right]\frac{1}{ct-x}, \tag{8.23b}$$

$$V_{21} = \frac{30I_o h(1-\beta^2)}{\beta^2(ct+x)^2+y_o^2}\left[\beta(ct+x)-\frac{(ct+x)x+y_o^2}{\sqrt{(ct)^2+(1-\beta^2)\dfrac{x^2+y_o^2}{\beta^2}}}\right] \tag{8.23c}$$

$$V_{22} = -\frac{30I_o h}{\beta}\left[1-\frac{1}{\sqrt{K_2^2+1}}-\beta^2\right]\frac{1}{ct+x}, \tag{8.23d}$$

$$K_1 = \frac{2h_c(ct-x)}{y_o^2+(ct-x)^2}, \quad K_2 = \frac{2h_c(ct+x)}{y_o^2+(ct+x)^2}, \tag{8.23e}$$

$$t_o = \frac{\sqrt{x^2+y_o^2}}{c}, \tag{8.23f}$$

$u(t-t_o)$ = shifted unit step function,
I_o = peak return-stroke current,
h_c = height of cloud, and
$\boldsymbol{\beta}$ = ratio of return-stroke velocity to velocity of electromagnetic wave in free space.

Actual return-stroke currents do not have simple waveshapes. However, most waveforms can be represented as piecewise linear. Therefore, it would be advantageous to determine the induced voltage for a linearly-rising and linearly-falling return-stroke current. Such a current wave can be analytically constructed by algebraically adding two linearly-rising functions, one being shifted by a time equal to the front time of the desired current wave (Fig. 1.1b). The two functions are analyzed separately, and the solutions are superposed to get the analytical expression for the induced voltage. In other words,

$$I(t)=\alpha_1 tu(t)-\alpha_2(t-t_f)u(t-t_f)=I_1(t)+I_2(t)\ . \tag{8.24}$$

The induced voltage caused by I(t) is

$$V(t) = V_1(t) + V_2(t)\ , \tag{8.25}$$

where $V_1(t)$ and $V_2(t)$ are the induced voltages caused by $I_1(t)$ and $I_2(t)$, respectively. $V_1(t)$ and $V_2(t)$ are found by applying Duhamel's theorem to the expression for the induced voltage caused by a unit step-function return-stroke current.

Haldar and Liew [33] have presented an alternative solution of the Chowdhuri-Gross model. This solution gives the induced voltage directly in terms of the inducing voltage due to a step current. This method may greatly reduce the algebraic manipulations of the analysis.

8.4.3 Parametric effects on the induced voltage [48]

The magnitude of the induced voltage varies directly as the height of the overhead line above ground, h_p, and also as the peak amplitude of the return-stroke current, I_p. The variation of the induced voltage with several other parameters was found to be very complex. These parameters are: distance along the line from the struck point on the ground (x), perpendicular distance of the struck point from the line (y_o), waveshape of return-stroke current (t_f, t_h), return-stroke velocity (β) and the cloud height (h_c). This parametric dependence is shown in Figs. 8.5 through 8.8. The following parameters were assumed in the plots unless otherwise specified in the figures:

h_c=3 km; h_p=10 m; x=0 m; y_o=100 m

β=0.3; I_p=30 kA; t_f=5 μs; t_h=5 ms

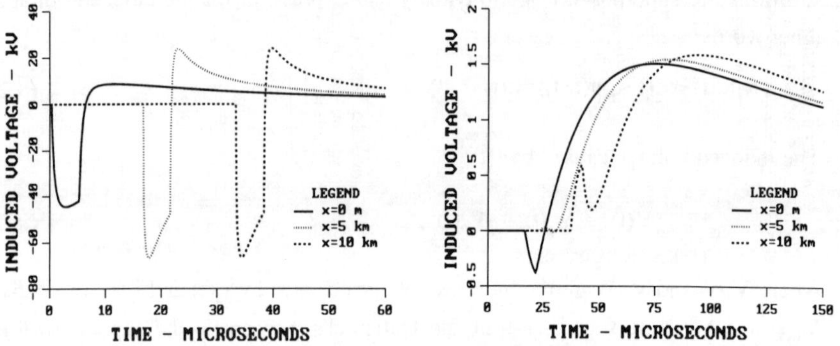

(a) (b)

Fig. 8.5 Voltage profiles along an overhead line induced by nearby lightning stroke. Effect of strike distance from line, y_o.
(a) y_o=100 m; (b) y_o=5 km.

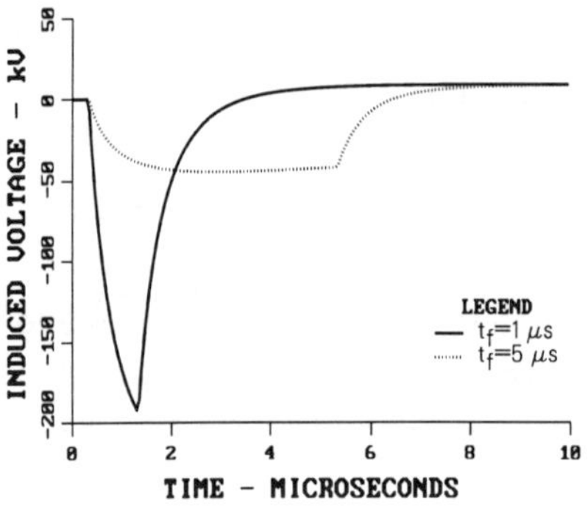

Fig. 8.6 Line voltage profiles: Effect of t_f. t_h=5 ms.

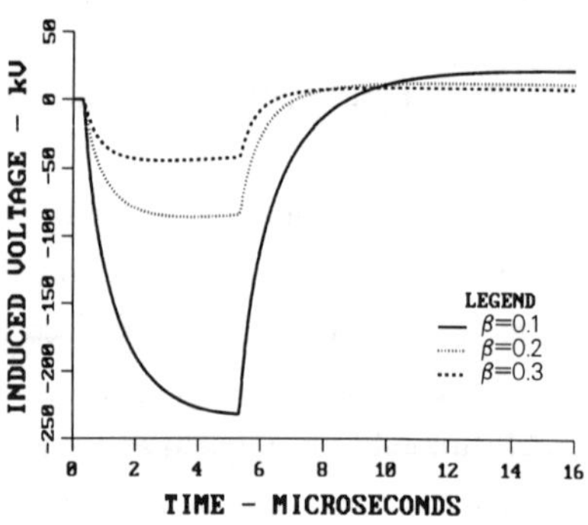

Fig. 8.7 Line voltage profiles: Effect of stroke velocity, β.

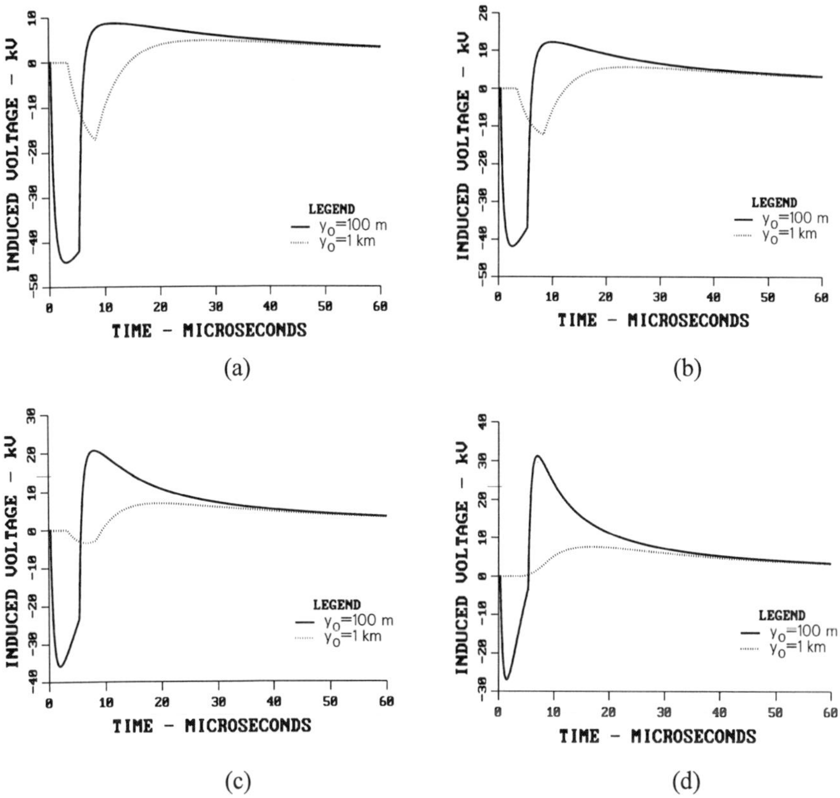

Fig. 8.8 Line voltage profiles: Effect of cloud height, h_c.
(a) h_c=3 km; (b) h_c=2 km; (c) h_c=1 km; (d) h_c=0.5 km.

8.5 Multiconductor Overhead Lines [28,31,49]

Overhead power lines are usually three-phase lines with or without shield wires. Sometimes several three-phase circuits are strung from the same tower. Even low-voltage, single-phase lines have their neutral conductor running nearby. How do the neighboring conductors affect the induction process? Do the shield wires and the neutral conductors affect the induced voltage? These questions will be explored in this section. The equivalent circuit of a two-conductor line is shown in Fig. 8.9.

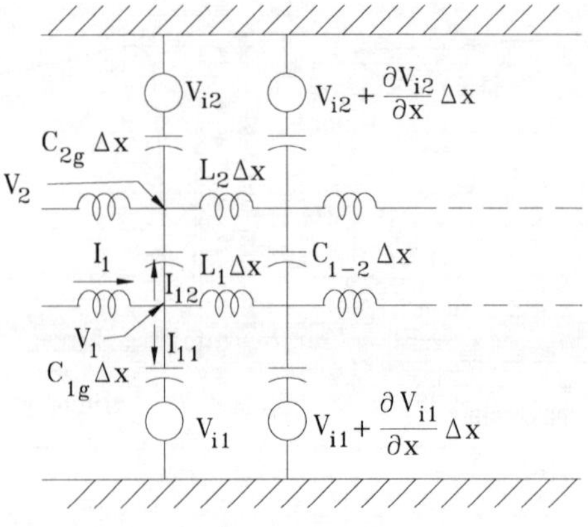

Fig. 8.9 Equivalent circuit of a two-conductor line.

According to Fig. 8.9:

$$I_1 = I_{11} + I_{12} \, , \tag{8.26}$$

$$-\frac{\partial I_{11}}{\partial x} = C_{1g}\frac{\partial}{\partial t}(V_1 - V_{i1}) \, , \text{ and} \tag{8.27}$$

$$-\frac{\partial I_{12}}{\partial x} = C_{1\ 2}\frac{\partial}{\partial t}(V_1 - V_2) \, . \tag{8.28}$$

$$\text{Then, } -\frac{\partial I_1}{\partial X} = -\frac{\partial}{\partial x}(I_{11} + I_{12})$$

$$= (C_{1g}+C_{1-2})\frac{\partial V_1}{\partial t} - C_{1g}\frac{\partial V_{i1}}{\partial t} - C_{1-2}\frac{\partial V_2}{\partial t}. \tag{8.29}$$

However [50],

$$C_{11}=C_{1g}+C_{1-2}; \; C_{22} = C_{2g}+C_{1-2}; \; C_{12} = -C_{1-2} \, , \tag{8.30}$$

where C_{11}, C_{12} and C_{22} are the elements of the matrix:

$$[C] = [p]^{-1} , \tag{8.31}$$

and [p] is the matrix of potential coefficients, discussed in Chapter 3. Then,

$$-\frac{\partial I_1}{\partial x} = C_{11}\frac{\partial V_1}{\partial t} + C_{12}\frac{\partial V_2}{\partial t} - (C_{11}+C_{12})\frac{\partial V_{i1}}{\partial t} . \tag{8.32}$$

Extending to an n-conductor system and putting in matrix form:

$$-\frac{\partial [I]}{\partial x} = [C]\frac{\partial [V]}{\partial t} - [I'] , \tag{8.33}$$

where

$$[I'] = \begin{bmatrix} C_{1g}\dfrac{\partial V_{i1}}{\partial t} \\ . \\ . \\ C_{jg}\dfrac{\partial V_{ij}}{\partial t} \\ . \\ . \\ C_{ng}\dfrac{\partial V_{in}}{\partial t} \end{bmatrix} = [C_g]\frac{\partial [V_i]}{\partial t} , \tag{8.34}$$

$C_{jg} = C_{j1}+C_{j2}+...+C_{jn}$, and $[C_g]$ = a diagonal matrix of elements C_{jg} (j=1,...,n).

The other equation for Fig. 8.9 is:

$$-\frac{\partial [V]}{\partial x} = [L]\frac{\partial [I]}{\partial t} . \tag{8.35}$$

Combining (8.33) and (8.35):

$$\frac{\partial^2 [V]}{\partial x^2} - [L][C]\frac{\partial^2 [V]}{\partial t^2} = -[L]\frac{\partial [I']}{\partial t} . \tag{8.36a}$$

Replacing [I′] by (8.34):

$$\frac{\partial^2[V]}{\partial x^2}-\frac{1}{c^2}\frac{\partial^2[V]}{\partial t^2}=-[L][C_g]\frac{\partial^2[V_i]}{\partial t^2}=-[M]\frac{\partial^2[V_i]}{\partial t^2} \quad . \tag{8.36b}$$

For the j-th conductor, the partial differential equation for its induced voltage will be:

$$\frac{\partial^2 V_j}{\partial x^2}-\frac{1}{c^2}\frac{\partial^2 V_j}{\partial t^2}=-\left(M_{j1}\frac{\partial^2 V_{i1}}{\partial t^2}+.+M_{jj}\frac{\partial^2 V_{ij}}{\partial t^2}+.+M_{jn}\frac{\partial^2 V_{in}}{\partial t^2}\right) \quad . \tag{8.37a}$$

If the ratio of the inducing voltage of the m-th conductor to that of the j-th conductor is k_{mj} (m=1,2,...,n), then

$$\frac{\partial^2 V_j}{\partial x^2}-\frac{1}{c^2}\frac{\partial^2 V_j}{\partial t^2}=-c^2(M_{j1}k_{1j}+..+M_{jj}+..+M_{jn}k_{nj})\frac{1}{c^2}\frac{\partial^2 V_{ij}}{\partial t^2}$$

$$= (M_j c^2)F_j(x,t) \quad , \tag{8.37b}$$

where $M_j=M_{j1}k_{1j}+..+M_{jj}+..+M_{jn}k_{nj}$, and $F_j(x,t)=\frac{1}{c^2}\frac{\partial^2 V_{ij}}{\partial t^2}$.

If the j-th conductor in its present position existed alone, the partial differential equation of its induced voltage V_{js} would have been the same as (8.9):

$$\frac{\partial^2 V_{js}}{\partial x^2} - \frac{1}{c^2}\frac{\partial^2 V_{js}}{\partial t^2} = F_j(x,t) \quad . \tag{8.38}$$

Therefore, the ratio of the induced voltage of the j-th conductor in an n-conductor system to that of a single conductor at the same position (i.e., in the absence of the other conductors) would be:

$$\frac{V_j}{V_{js}} = M_j c^2 \quad . \tag{8.39}$$

The inducing voltage of a conductor is directly proportional to its height above ground. However, the dependence of the inducing voltage on the lateral distance of the strike point is complex. Generally, this lateral distance is large compared to the separation distance between the phase conductors. Therefore, the various conductors can be assumed to be equidistant from the strike point without any appreciable error. In that case, the ratio between the inducing voltages is equal to the ratio between the respective conductor heights. For a horizontally configured system, the inducing voltages of all the conductors are approximately equal under this condition. M_j in (8.37b) then becomes:

$$M_j = M_{j1} + M_{j2} + \dots + M_{jn} = \frac{1}{c^2} . \tag{8.40}$$

Equation (8.37b) is then identical to the equation for a single-conductor system. In other words, the induced voltages of all the conductors in a horizontally configured n-conductor system are approximately equal to each other and to that of a single-conductor system of the same height when the distance of the strike point is large compared to the separation distances between the conductors. However, if the conductors are vertically configured, the induced voltages will be different from each other, and also different from that of a single-conductor line of the same height.

8.5.1 Induced voltage with shield wires

For a (n+r)-conductor line of which r conductors are grounded, the voltage equation is

$$-\frac{\partial[V_{n+r}]}{\partial x} = [L_{n+r}]\frac{\partial[I_{n+r}]}{\partial t} , \tag{8.41a}$$

where the matrices encompass all (n+r) conductors. By partitioning:

$$-\frac{\partial}{\partial x}\begin{bmatrix} V_n \\ V_r \end{bmatrix} = \begin{bmatrix} L_{nn} & L_{nr} \\ L_{rn} & L_{rr} \end{bmatrix}\frac{\partial}{\partial t}\begin{bmatrix} I_n \\ I_r \end{bmatrix} , \tag{8.41b}$$

where V_n, I_n = nx1 matrices of the induced voltages and currents of the ungrounded phase conductors, and

V_r, I_r = rx1 matrices of the induced voltages and currents of the grounded shield wires.

From (8.41b):

$$-\frac{\partial[V_n]}{\partial x} = [L_{nn}]\frac{\partial[I_n]}{\partial t}+[L_{nr}]\frac{\partial[I_r]}{\partial t}\ , \text{ and} \tag{8.42}$$

$$-\frac{\partial[V_r]}{\partial x} = [L_{rn}]\frac{\partial[I_n]}{\partial t} + [L_{rr}]\frac{\partial[I_r]}{\partial t}\ . \tag{8.43}$$

As $[V_r] = [0]$,

$$\frac{\partial[I_r]}{\partial t} = -[L_{rr}]^{-1}[L_{rn}]\frac{\partial[I_n]}{\partial t}, \text{ and} \tag{8.44}$$

$$-\frac{\partial[V_n]}{\partial x} = [L_{nn}]\frac{\partial[I_n]}{\partial t} - [L_{nr}][L_{rr}]^{-1}[L_{rn}]\frac{\partial[I_n]}{\partial t}$$

$$= [L']\frac{\partial[I_n]}{\partial t}\ , \tag{8.45}$$

where $[L'] = L_{nn} - [L_{nr}][L_{rr}]^{-1}[L_{rn}]$.

The current equation can be similarly rearranged from (8.33) and (8.34):

$$-\frac{\partial}{\partial x}\begin{bmatrix} I_n \\ I_r \end{bmatrix} = \begin{bmatrix} C_{nn} & C_{nr} \\ C_{rn} & C_{rr} \end{bmatrix}\frac{\partial}{\partial t}\begin{bmatrix} V_n \\ V_r \end{bmatrix} - \begin{bmatrix} C_{gn} & 0 \\ 0 & C_{gr} \end{bmatrix}\frac{\partial}{\partial t}\begin{bmatrix} V_{in} \\ V_{ir} \end{bmatrix}, \tag{8.46}$$

where the elments of the submatrices C_{nn}, C_{nr}, C_{rn} and C_{rr} are the elements of the (n+r)x(n+r) capacitance matrix, given by $[C]=[p]^{-1}$ and p is the matrix of potential coefficients of the (n+r) conductors, C_{gn} is an nxn diagonal matrix, whose j-th element is given by:

$$C_{jgn}(j \le n) = \sum_{k=1}^{n+r} C_{jk}\ , \text{ and} \tag{8.47}$$

C_{gr} is an rxr diagonal matrix, whose j-th element is given by:

$$C_{jgr}(n<j\leq n+r) = \sum_{k=n+1}^{n+r} C_{jk} \ . \tag{8.48}$$

As $[V_r]=[0]$,

$$-\frac{\partial[I_n]}{\partial x} = [C_{nn}]\frac{\partial[V_n]}{\partial t} - [C_{gn}]\frac{\partial[V_{in}]}{\partial t} \ . \tag{8.49}$$

Combining (8.45) and (8.49):

$$\frac{\partial^2[V_n]}{\partial x^2} - [L'][C_{nn}]\frac{\partial^2[V_n]}{\partial t^2} = -[L'][C_{gn}]\frac{\partial^2[V_{in}]}{\partial t^2} \ . \tag{8.50a}$$

As $[L'][C_{nn}] = \dfrac{[1]}{c^2}$,

$$\frac{\partial^2[V_n]}{\partial x^2} - \frac{1}{c^2}\frac{\partial^2[V_n]}{\partial t^2} = -[L'][C_{gn}]\frac{\partial^2[V_{in}]}{\partial t^2}$$

$$= -[M_g]\frac{\partial^2[V_{in}]}{\partial t^2} \ . \tag{8.50b}$$

The induced voltage, V_j, of the j-th conductor is:

$$\frac{\partial^2 V_j}{\partial x^2} - \frac{1}{c^2}\frac{\partial^2 V_j}{\partial t^2} = -\left(M_{gj1}\frac{\partial^2 V_{i1}}{\partial t^2} + .. + .. + M_{gjn}\frac{\partial^2 V_{in}}{\partial t^2}\right) \tag{8.51a}$$

$$= -c^2\left(M_{gj1}k_{ij} + .. + M_{gjj} + .. + M_{gjn}k_{nj}\right)\frac{1}{c^2}\frac{\partial^2 V_{ij}}{\partial t^2} \tag{8.51b}$$

$$= (M_{gj}c^2)F_j(x,t) \ , \tag{8.51c}$$

where k_{nj} is the ratio of the inducing voltage of the n-th conductor to that of the j-th conductor. The ratio of the induced voltage on the j-th conductor to that for a single-conductor line is then given by:

$$\frac{V_j}{V_{js}} = M_{gj} c^2 . \tag{8.52}$$

Defining the protective ratio as the ratio of the induced voltages on the j-th conductor with and without the shield wires in place:

$$\text{Protective Ratio} = \text{P.R.} = \frac{M_{gj}}{M_j} , \tag{8.53}$$

where M_j is given by (8.37b).

8.5.2 An example [49]

The ratios of the induced voltages of the three conductors of a three-phase line to that of a single-conductor line were calculated using (8.37b). Table 8.1 shows the dimensions of the lines, and Table 8.2 shows the ratios of the induced voltages. The voltages $V(h_1)$, $V(h_2)$ and $V(h_3)$ are the induced voltages of a single-conductor line of heights above ground h_1, h_2 and h_3 respectively.

According to Table 8.2, the phase 1 conductor of the vertical configuration (which has the lowest height) will experience a 14 percent increase of the induced voltage because of the presence of the other two phase conductors. On the other hand, the phase 3 conductor (which has the highest elevation) will experience a 9 percent decrease of the induced voltage. As the conductors 2 and 3 are at higher elevations than conductor 1, their induced voltages will be higher than that of conductor 1. Because of the electromagnetic coupling among the three conductors, the voltage of conductor 1 will be pulled up. Similarly, the induced voltages of conductors 1 and 2 will tend to pull down the voltage of conductor 3 by electromagnetic coupling. The conductor 2, situated in the middle, will experience the least deviation. A three-phase line with horizontal configuration will show practically no deviation, provided that the strike distance (y_o in Fig. 8.2) from the line is large compared to the horizontal distance of the phase separation. Table 8.2 shows this.

However, the strike point may occasionally be close to the overhead line. For instance, the strike point to the ground may be close to the line if the return-stroke current is small. A strike to the tower or pole is another example of a strike where the strike distance is comparable to the horizontal distance between the phase conductors. Under these circumstances, the presence of the other phase conductors will be manifested on the induced voltages, particularly for low return-stroke velocities.

Table 8.3 shows the effect of shield wires on the induced voltages. One shield wire was used for the vertically configured line, and two shield wires for the horizontally configured line. The heights of the shield wires were varied to observe the effect on the protective ratio.

Table 8.1: Dimensions of Multiconductor Lines

Config.	r_p, mm	r_s, mm	Height above Ground, m					Separation between Conductors, m			
			h_1	h_2	h_3	h_4	h_5	d_{12}	d_{23}	d_{31}	d_{45}
V	9.14	3.96	10.0	13.66	17.31	6.95	-	3.66	3.66	7.32	-
						8.47					
						18.84					
						20.36					
H	9.14	3.96	10.0	10.0	10.0	6.95	6.95	3.66	3.66	7.32	3.66
						8.47	8.47				
						11.52	11.52				
						13.05	13.05				

V: Vertical configuration; H: Horizontal configuration
r_p: radius of phase conductors; r_s: radius of shield wires

Table 8.2: Per-Unit Induced Voltages on 3-Conductor Lines without Shield Wires

Configuration	$V_1/V(h_1)$	$V_2/V(h_2)$	$V_3/V(h_3)$
Vertical	1.14	1.0	0.91
Horizontal	1.0	1.0	1.0

Table 8.3: Effect of Shield Wires on Induced Voltages

Config.	h_4 or h_5 m	Phase 1		Phase 2		Phase 3	
		$V_1/V(h_1)$	P.R.	$V_2/V(h_2)$	P.R.	$V_3/V(h_3)$	P.R.
V	6.95	0.90	0.79	0.87	0.87	0.83	0.91
	8.47	0.80	0.70	0.84	0.84	0.81	0.89
	18.84	0.96	0.84	0.79	0.79	0.60	0.65
	20.36	0.98	0.86	0.81	0.81	0.66	0.73
H	6.95	0.73	0.73	0.67	0.67	0.73	0.73
	8.47	0.67	0.67	0.57	0.57	0.67	0.67
	11.52	0.66	0.66	0.58	0.58	0.66	0.66
	13.05	0.70	0.70	0.65	0.65	0.70	0.70

V_1, V_2 and V_3 = induced voltage on conductor 1, 2 and 3 with shield wire(s) in place
$V(h_1)$, $V(h_2)$ and $V(h_3)$ = induced voltage on a single-conductor line of height h_1, h_2 and h_3 with no shield wires. P.R. =protective ratio

The effectiveness of the shield wires, as shown in Table 8.3, is optimistic, bearing in mind that the shield wires were assumed to be held at ground potential. Currents will flow along the shield wires over their surge impedances. These currents will eventually flow to the ground through the grounding impedances at discrete points along the shield wires. Hence, the shield wires will never be at the ground potential. Therefore, the effectiveness of the shield wires will be less than the idealized cases shown in Table 8.3.

It has been shown by Cinieri and Muzi [51] that the outage rate of a power distribution system can be significantly reduced by proper placement of shield wires. The outage rate of overhead power lines by lightning-induced voltages will be discussed in Section 8.7.

8.6 Lightning-Induced Voltages on Overhead Lines of Finite Length [52]

The analysis in the previous sections on the lightning-induced voltages was based on overhead lines of infinite length. However, actual overhead lines are of finite length. They are sometimes terminated by high impedances, such as transformers, and by low impedances, such as underground cables. The pole-mounted distribution apparatus are normally protected by lightning arresters; these constitute termination by nonlinear resistors. This section extends the previous analysis to overhead lines of finite length and terminated by resistors, including short- and open-circuit. This is done by imposing boundary conditions as required by the different line terminations. This analysis is based on vertical E- and A-fields. The overhead line is shown schematically in Fig. 8.10.

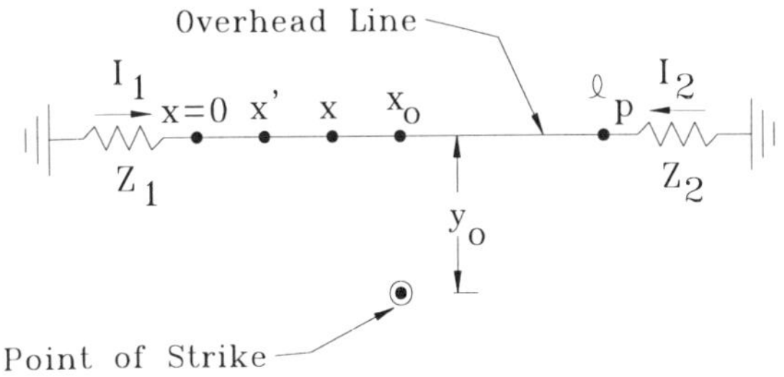

Fig. 8.10 Coordinate system for overhead line and lightning stroke point.
ℓ_p=line length; x=field point; x′=source point; x_o,y_o=coordinates of point of strike; Z_1, Z_2=terminal resistances.

Equation (8.17) is still applicable in the computation of induced voltages on overhead lines of finite length. The solution of (8.17) is given in terms of Green's function as:

$$V(x,s) = \int_0^{\ell_p} G(x;x',s)F(x',s)dx' . \tag{8.54}$$

The time-domain solution of the induced voltage, V(x,t), at a point x on the overhead line is then obtained by taking the inverse Laplace transform of (8.54).

8.6.1 Evaluation of Green's function for overhead lines of finite length

The Green's function in (8.19) for the line of finite length is similar to the previous solution:

$$G_1(x;x',s) = A_1 e^{\frac{sx}{c}} + B_1 e^{\frac{-sx}{c}} \text{ ,for } x<x', \text{ and} \tag{8.55a}$$

$$G_2(x;x',s) = A_2 e^{\frac{sx}{c}} + B_2 e^{\frac{-sx}{c}} \text{ ,for } x>x'. \tag{8.55b}$$

The boundary conditions are:

$$\text{(i)} \quad I_1(0,s) = -\frac{G_1}{Z_1} = -\frac{1}{Ls}\frac{\partial G_1}{\partial x}, \quad \text{and} \tag{8.56a}$$

$$\text{(ii)} \quad I_2(\ell_p,s) = \frac{G_2}{Z_2} = -\frac{1}{Ls}\frac{\partial G_2}{\partial x}, \tag{8.56b}$$

where L=inductance of the overhead line, H/m. The four constants in (8.55) can be evaluated from the properties of the Green's function and the boundary conditions of the overhead line. These constants are given as follows:

$$A_1 = -\frac{c}{2s}\left[\beta_2 e^{s(x'-2\ell_p)/c} + e^{-sx/c}\right]F_1, \tag{8.57a}$$

$$B_1 = -\frac{c}{2s}\left[\beta_1\beta_2 e^{s(x'-2\ell_p)/c} + \beta_1 e^{-sx'/c}\right]F_1, \tag{8.57b}$$

$$A_2 = -\frac{c}{2s}\left[\beta_2 e^{s(x'-2\ell_p)/c} + \beta_1\beta_2 e^{-s(x'+2\ell_p)/c}\right]F_1, \quad \text{and} \tag{8.57c}$$

$$B_2 = -\frac{c}{2s}\left[e^{sx'/c} + \beta_1 e^{-sx'/c}\right]F_1, \tag{8.57d}$$

$$\text{where,} \quad F_1 = \sum_{n=0}^{\infty}(\beta_1\beta_2)^n e^{-2sn\ell_p/c}\ ; \ \beta_1 = \frac{Z_1-Z_o}{Z_1+Z_o}\ ; \ \beta^2 = \frac{Z_2-Z_o}{Z_2+Z_o}\ ; \ Z_o = \sqrt{\frac{L}{C}}, \quad \text{and}$$

C=capacitance to ground of overhead line, F/m. The inducing voltage is given by (8.14). It should be noted that F_1 represents multiple reflections at both ends of the line.

8.7 Estimation of Flashover Rates by Nearby Lightning Strokes [53]

Most outages on overhead power distribution lines are caused by lightning strokes to nearby ground. In assessing the severity of the induced voltage, the significant parameters are:

(i) line height, (ii) distance of strike point from line, (iii) amplitude and front time of return-stroke current, (iv) return-stroke velocity, and (v) cloud height.

The designer of an electric distribution system needs to know the outage rate of an overhead power distribution line for a selected critical flashover voltage, V_{50}, to meet the reliability criteria set for the system. A method is outlined which will facilitate estimating the outage rate of an overhead power distribution line caused by lightning strokes to nearby ground. The data are given in the form of curves with the line flashover rate ***vs.*** V_{50} of line for 8-, 10- and 12-meter high lines situated in a region of ground flash density, n_g of 10 lightning flashes per km^2 per year. The data can easily be converted to any other ground flash density, because the flashover rates are directly proportional to n_g.

If the perpendicular distance of the leader stroke from the line is greater than a critical distance, the leader will strike the ground; otherwise, it will strike the line causing a direct stroke. This critical perpendicular distance y_{o1} of the stroke from the line can be calculated, once the striking distance r_s is known, as follows (Fig. 8.11):

$$y_{o1} = \sqrt{r_s^2 - (r_s - h_p)^2} \quad \text{for } r_s > h_p, \text{ and} \tag{8.58a}$$

$$y_{o1} = r_s, \quad \text{for } r_s \leq h_p \; . \tag{8.58b}$$

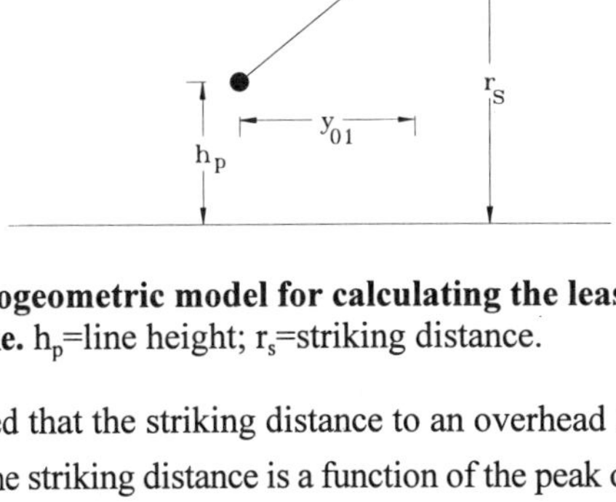

Fig. 8.11 Electrogeometric model for calculating the least distance of ground strike. h_p=line height; r_s=striking distance.

Here it is assumed that the striking distance to an overhead line is the same as that to a ground plane. The striking distance is a function of the peak of the return-stroke current. Several empirical equations have been proposed. The most widely used equation is:

$$r_s = 8\, I_p^{0.65} \;, \quad (m) \tag{8.59}$$

where I_p = peak of return-stroke current in kA. y_{o1} in (8.58) is the shortest distance of a stroke of a given return-stroke peak current, I_p which will result in a flash to ground. y_{o1} is then a function of I_p.

The return-stroke current, I_p is varied from 1 kA to 200 kA in steps of 0.5 kA. The current front time, t_f is varied from 0.5 μs to 10.5 μs in steps of 0.5 μs. At each current level, the shortest possible distance of the stroke y_{o1} is calculated from (8.58) and (8.59). Starting at t_f=0.5 μs, the induced voltage is computed as a function of time and compared with the given V_{50} of the line. If the V_{50} is not exceeded, then the next higher level of current is chosen. If the V_{50} is exceeded, then the perpendicular distance y_o of the stroke from the line is increased, the induced voltage is recalculated and compared with the V_{50} of the line. y_o is progressively increased until the induced voltage does not exceed the line V_{50}. This distance is called y_{o2}. For the selected I_p and t_f, the induced voltage will then exceed the V_{50} and cause line flashover, if the lightning stroke hit the ground along a 100 km length of the line within an area A given by (Fig.8.12):

$$A = 0.2\,(y_{o2} - y_{o1}) \quad km^2 \;. \tag{8.60}$$

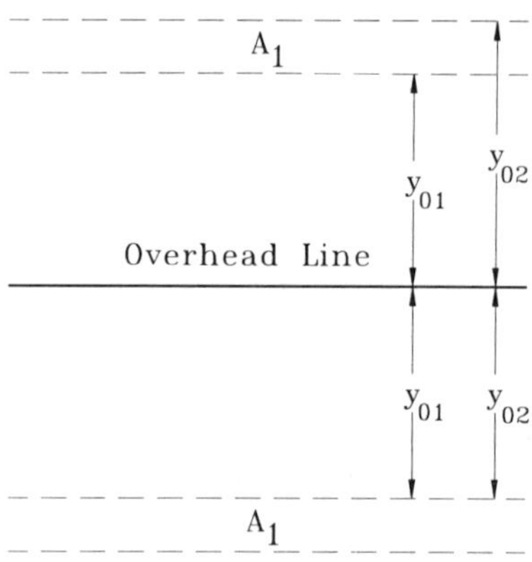

Fig. 8.12 Schematic of areas of lightning ground flash to cause line flashover.
$A = 2A_1 = 0.2(y_{o2}\text{-}y_{o1})$ km^2.

The joint probability density function $p(I_p,t_f)$ is then calculated for the selected I_p-t_f combination. It was assumed that the probability density functions, $p(I_p)$ and $p(t_f)$, follow log-normal distribution, and the statistical data were taken from references 54 and 55. If n_g is the ground flash density for that region, then the expected number of line flashovers per 100 km per year for that particular I_p-t_f combination will be:

$$nf_o = p(I_p,t_f)\,\Delta I_p\,\Delta t_f\,n_g A \ , \tag{8.61}$$

where ΔI_p = current step, and Δt_f = front-time step. The front time, t_f, is then increased by Δt_f=0.5 μs to the next step and nf_o for the same current but with the new t_f computed and added to the previous nf_o. Once t_f=10.5 μs is reached, the return-stroke current is increased by ΔI_p=0.5 kA, and the whole procedure repeated until the limits I_p=200 kA and t_f=10.5 μs are reached. The cumulative nf_o will then give the total number of expected line flashovers per 100 km per year for the selected line $V_{50.}$

The results of such a computation are shown in Fig. 8.13 for three line heights (h_p=8,10,12 m), situated in a region of ground flash density n_g=10 ground flashes per km^2 per year. This figure can be used to estimate line flashover rate for any other n_g, because the line flashover rate is directly proportional to n_g. For any other line height, the flashover rate can be approximated by assuming that it is directly proportional also to the line height.

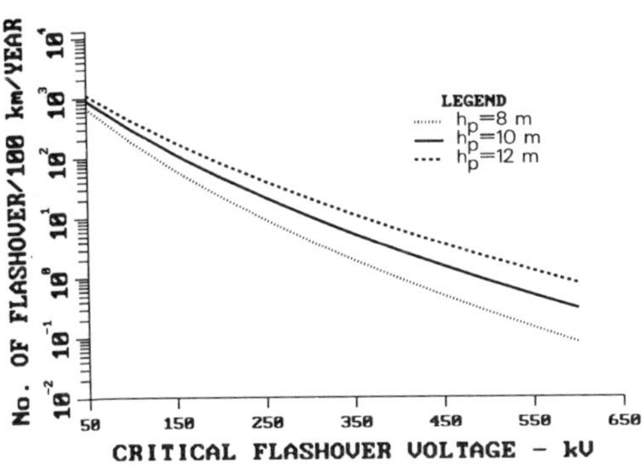

Fig. 8.13 Flashover rate of single-conductor overhead line.
Line heights, h_p=8,10,12 m; Cloud height, h_c= 3 km;
Ground flash density, n_g=10/km^2/year; Striking distance, r_s=8$(I_p)^{0.65}$.

The computations were carried out with the assumptions that the per-unit return-stroke velocity is 0.3 (β=0.3) and the cloud height, h_c=3 km. The striking distance is another important parameter which needs to be known more accurately. The correlation coefficient, ρ (between I_p and t_f) was assumed to be 0.47 [54,55]. It has been shown that the correlation coefficient is another sensitive parameter which needs to be known accurately [53].

Figure 8.14 shows the effectiveness of the shield wire in reducing the outage rate caused by lightning-induced voltages. The shield wire was placed one meter below the phase conductor. For shielding against direct lightning strokes, the shield wire is placed above the phase conductor to intercept the lightning stroke. For low-voltage power distribution systems, the majority of the direct strokes, even if intercepted by the shield wire result in backflashover, caused by voltage drop across the footing impedance. Therefore, shield wires are not usually used. However, the neutral conductor is installed below the phase conductor. The neutral conductor, if properly grounded, would reduce the outage rate caused by lightning-induced voltages.

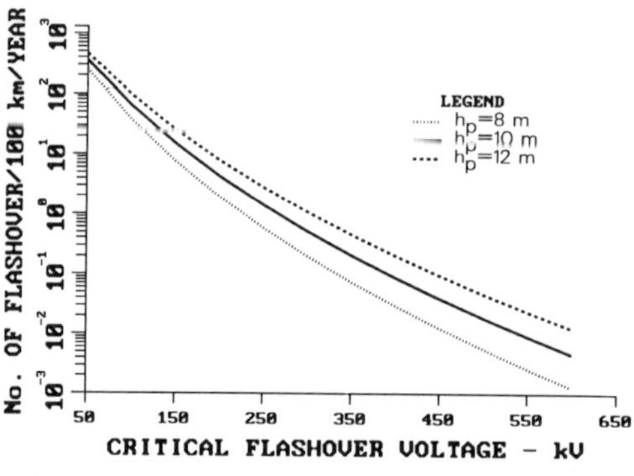

Fig. 8.14 Flashover rate of shielded single-conductor overhead line caused by lightning-induced voltages.

h_p=8,10,12 m; h_s=7,9,11 m; r_p=1.0 cm; r_s=0.5 cm

h_c=3 km; β=0.3; n_g=10.

8.8 References to Chapter 8

1. C.L. Fortescue, "Direct Strokes - Not Induced Surges - Chief Cause of High-Voltage Line Flashover," ***The Electric Journal***, Vol. 27, pp. 459-462, 1930.

2. B. Grunner-Nielson, et al., "Lightning Arresters in Low Voltage Networks," The Danish Association of Electricity Supply Undertakings and the Danish Association of Mutual Fire Insurance Companies, Copenhagen, 1957.

3. S. Rusck, "Lightning Over-Voltages and Over-Voltage Protection in Low-Voltage Networks," ***ASEA Journal***, Vol. 31, pp. 75-81,1958.

4. A.G. Greve, "Overvoltage Protection in Low-Voltage Networks," ***ASEA Journal***, Vol. 34, pp. 34-38, 1961.

5. Technical Group on Lightning-Proofing and Distribution Lines, Report of Central Research Institute of Electric Power Industry, Japan, Vol. 13, No. 4, 1963.

6. R.H. Golde, "Lightning Performance of British High-Voltage Distribution Systems," ***Proc.IEE*** (G.B.), Vol. 113, pp. 601-610, 1966.

7. Working Group Report: Lightning Protection of Distribution Networks. The Electricity Council, London, 1973.

8. ***Proc. Conference on Lightning and The Distribution System***. IEE Conference Publication No. 108, London, 1974.

9. K.W. Wagner, "Electromagnetic Oscillations on Overhead Lines and Cables," (in German), Leipzig, 1908.

10. L.V. Bewley, "Traveling Waves Due to Lightning," ***AIEE Trans.***, Vol. 48, pp. 1050-1064, 1929.

11. V. Aigner, "Lightning-Induced Overvoltage and its Relation to Backflashover," ***Elektrotechnische Zeitschrift***, Vol. 56, pp. 497-500, 1935.

12. H. Norinder, "Some Recent Tests on the Investigation of Induced Surges," ***CIGRE***, Paris, France, Report No. 303, 1939.

13. C.F. Wagner and G.D. McCann, "Induced Voltage on Transmission Lines," ***AIEE Trans.***, Vol. 61, pp. 916-930, 1942.

14. S. Szpor, "A New Theory of the Induced Voltages," ***CIGRE***, Paris, France, Report No. 308, 1948.

15. R.H. Golde, "Lightning Surges on Overhead Distribution Lines Caused by Indirect and Direct Lightning Strokes," ***AIEE Trans.***, Vol. 73, Part III, pp. 437-446, 1954.

16. R. Lundholm, "Induced Overvoltage-Surges on Transmission Lines and their Bearing on the Lightning Performance at Medium Voltage Networks," ***Trans. Chalmers University of Technology***, Gothenburg, Sweden, No. 188, 1957.

17. S. Rusck, "Induced Lightning Over-Voltages on Power-Transmission Lines with Special Reference to the Over-Voltage Protection of Low-Voltage Networks," ***Trans. Royal Inst. of Tech.***, Stockholm, Sweden, No. 120, 1958.

18. J. Pepet-Lepine, "Contribution to the Calculation of the Overvoltages Induced by Lightning on the Electric Power Transmission Lines," (in French), ***Rev. Gen. Elec.***, Vol. 69, pp. 591-600, 1960.

19. G. Owa, "Study of Induced Lightning Surges and their Frequency of Occurrence," ***Jour. IEE of Japan***, Vol. 84, pp. 44-55, 1964.

20. S.B. Griscom, "The Prestrike Theory and Other Effects in the Lightning Stroke," ***AIEE Trans.***, Vol. 77, Part III, pp. 919-933, 1958.

21. F.W. Chapman, "Atmospheric Disturbances due to Thundercloud Discharges, Part I," ***Proc. Phys. Soc. of London***, Vol. 51, pp. 876-894, 1939.

22. B.F.J. Schonland, D.B. Hodges and H. Collens, "Progressive Lightning V - A Comparison of Photographic and Electrical Studies of the Discharge Process," ***Proc. Royal Soc.***,(G.B.), Vol. 166, Series A, pp. 56-75, 1938.

23. F.R. Perry, "The Measurement of Lightning Voltages and Currents in South Africa and Nigeria, 1935-1937," ***Jour. IEE*** (G.B.), Vol. 88, Part II, pp. 69-87, 1941.

24. F.R. Perry, G.H. Webster and P.W. Baguley, "The Measurement of Lightning Voltages and Currents in Nigeria (Part 2, 1938-1939)," ***Jour. IEE*** (G.B.), Vol. 89, Part II, pp. 185-203, 1942.

25. P.P. Barker, R.T. Mancao, D.J. Kvaltine and D.E. Parrish, " Characteristics of Lightning Surges Measured at Metal Oxide Distribution Arresters", ***IEEE Trans. on Power Delivery***, Vol. 8, pp. 301-310, 1993.

26. K. Uehara and G. Ohwa, "Investigation of Lightning Damages on Distribution Lines," ***IEEE Trans. on Power Apparatus and Systems***, Vol. PAS-87, pp. 1018-1025, 1968.

27. P. Chowdhuri and E.T.B. Gross, "Voltage Surges Induced on Overhead Lines by Lightning Strokes," ***Proc. IEE*** (G.B.), Vol. 114, pp. 1899-1907,1967.

28. P. Chowdhuri and E.T.B. Gross, "Voltages Induced on Overhead Multiconductor Lines by Lightning Strokes," ***Proc. IEE*** (G.B.), Vol. 116, pp. 561-565, 1969.

29. G. Cornfield: Correspondence on reference 26, ***Proc. IEE*** (G.B.), Vol. 119, pp. 893-894, 1972.

30. E. Cinieri and A. Fumi, "On the Overvoltage Induced by Lightning Strike on Electrical Power Lines," (in Italian), ***L'Energia Elettrica***, Vol. 55, pp. 1-7, 1978.

31. E. Cinieri and A. Fumi, "The Effect of the Presence of Multiconductors and Ground Wires on the Atmospheric High Voltages Induced on Electrical Lines," (in Italian), ***L'Energia Elettrica***, Vol. 56, pp. 3-9, 1979.

32. A.C. Liew and S.C. Mar, "Extension of the Chowdhuri-Gross Model for Lightning Induced Voltage on Overhead Lines," ***IEEE Trans. on Power Delivery***, Vol. PWRD-1, pp. 240-247, 1986.

33. M.K. Haldar and A.C. Liew, "Alternative Solution for the Chowdhuri-Gross Model of Lightning-Induced Voltages on Power Lines," ***Proc. IEE*** (G.B.), Vol. 135, Part C, pp. 324-329,1988.

34. C. Singarajah, "Surges Induced on Overhead Transmission Lines by Indirect Lightning Strokes, with Particular Reference to Nigeria," Ph.D. Thesis, University of London, September, 1971.

35. J.R. Wait: ***Electromagnetic Waves in Stratified Media***. MacMillan, New York, N.Y., 1962.

36. H. Koga, T. Motomitsu and M. Taguchi, "Lightning Surge Waves Induced on Overhead Lines," ***Trans. IECE of Japan***, Vol. E62, pp. 216-223, 1979.

37. M.J. Master, "Lightning Induced Voltages on Power Lines: Theory and Experiment," Ph.D. Dissertation, University of Florida, 1982.

38. F. de la Rosa, "Effects of Close Lightning on Electric Power Distribution Lines," Doctoral Thesis, Uppsala University, 1985.

39. A. Agrawal, H. Price and S.H. Gurbaxani, "Transient Response of Multiconductor Transmission Lines Excited by a Nonuniform Electromagnetic Field," ***IEEE Trans. on Electromagnetic Compatibility***, Vol. EMC-22, pp. 119-129, 1980.

40. H. Price and A.K. Agrawal, "The Response of a Transmission Line Illuminated by Lightning-Induced Electromagnetic Fields", ***IEEE Trans. on Electromagnetic Compatibility***, Vol. EMC-22, pp. 150-156, 1980.

41. G. Diendorfer, "Induced Voltage on an Overhead Line due to Nearby Lightning", ***IEEE Trans. on Electromagnetic Compatibility***, Vol. 32, pp. 292-299, 1990.

42. C.A. Nucci, F. Rachidi, M.V. Ianoz and C. Mazzetti, "Lightning-Induced Voltages on Overhead Lines", ***IEEE Trans. on Electromagnetic Compatibility***, Vol. 35, pp. 75-86, 1993.

43. V. Cooray, "Calculating Lightning-Induced Overvoltages in Power Lines: A Comparison of Two Coupling Models", ***IEEE Trans. on Electromagnetic Compatibility***, Vol. 36, pp. 179-182, 1994.

44. C.A. Nucci, F. Rachidi, M. Ianoz and C. Mazzetti, "Comparison of Two Coupling Models for Lightning-Induced Overvoltage Calculations", ***IEEE Trans. on Power Delivery***, Vol. 10, pp. 330-339, 1995.

45. P.M. Morse and H. Feshbach: ***Methods of Theoretical Physics***, Vol. 1. McGraw-Hill, New York, N.Y., 1950.

46. P.A.M. Dirac: ***Principles of Quantum Mechanics***. Oxford University Press, New York, N.Y., 1947.

47. P. Chowdhuri, "Analysis of Lightning-Induced Voltages on Overhead Lines," ***IEEE Trans. on Power Delivery***, Vol. 4, pp. 479-492, 1989.

48. P. Chowdhuri, "Parametric Effects on the Induced Voltages on Overhead Lines by Lightning Strokes to Nearby Ground," ***IEEE Trans. on Power Delivery***, Vol. 4, pp. 1185-1194, 1989.

49. P. Chowdhuri, "Lightning-Induced Voltages on Multiconductor Overhead Lines", ***IEEE Trans. on Power Delivery***, Vol. 5, pp. 658-667, 1990.

50. S. Ramo and J.R. Whinnery: ***Fields and Waves in Modern Radio***, Second Edition. New York: John Wiley & Sons, 1964.

51. E. Cinieri and F. Muzi, "Lightning Induced Overvoltages. Improvement in Quality of Service in MV Distribution Lines by Addition of Shield Wires", ***IEEE Trans. on Power Delivery***, Vol. 11, pp. 361-372, 1996.

52. P. Chowdhuri, "Response of Overhead Lines of Finite Length to Nearby Lightning Strokes", ***IEEE Trans. on Power Delivery***, Vol. 6, pp. 343-351, 1991.

53. P. Chowdhuri, "Estimation of Flashover Rates of Overhead Power Distribution Lines by Lightning Strokes to Nearby Ground," ***IEEE Trans. on Power Delivery***, Vol. 4, pp. 1982-1989, 1989.

54. R.B. Anderson and A.J. Eriksson, "Lightning Parameters for Engineering Applications," ***Electra***, No. 69, pp. 65-102, 1980.

55. A.J. Eriksson,"Notes on Lightning Parameters for System Performance Estimation," CIGRE Note 33-86 (WG33-01) IWD, July 15, 1986.

CHAPTER 9
FUNDAMENTAL CONCEPTS OF SWITCHING TRANSIENTS

9.1 Introduction

A switching operation in an electrical system involves either applying voltage (and current) or removing voltage (and current) from the system. All electrical circuits and systems contain inductance, capacitance and resistance. These could be design parameters, or they could be inherent (stray). When current flows through an inductance, it produces magnetic flux. Any effort to change the magnetic flux (i.e., the current) will be opposed by the inductance, which is manifested by the generation of a counter e.m.f. in the inductance in such a direction as to keep the magnetic flux (and the current) in the inductance constant. The higher the rate of change of current, the higher the counter e.m.f. Theoretically, the counter e.m.f. can be infinite if the current through the inductance is snapped off instantaneously. However, as every electrical system can withstand a maximum level of voltage, the system dielectric eventually breaks down, causing either a temporary or a permanent damage. In practice, all electrical systems have capacitances. When the current is interrupted in an electrical system, the current through the inductance flows through the capacitance, thus transferring the inductive energy from the inductance to the capacitance as capacitive energy. The energy originally stored in the inductance is transferred back and forth between the inductance and the capacitance until the stored energy is finally dissipated by the system resistance.

In a system with large extent, such as a power transmission line, where the inductance and the capacitance are distributed, traveling voltage and current waves are generated during a switching operation. Dangerously high voltage can be generated by multiple reflections at the ends of the line.

9.2 Basic Simulation of Switching

It is mostly assumed in transient analysis that the system is linear. This means that the effects (responses) of various causes (forcing functions) can be algebraically added when the various causes exist simultaneously.

A voltage exists across the contacts of an open switch. This voltage becomes zero when the switch is closed. This closing operation can be simulated by injecting a voltage across the switch contacts in such a way that the algebraic sum of the voltage which would have existed across the switch prior to switching and the injected voltage on switch closing remains zero while the switch remains closed [1]. The change in current through any branch of the network in which the switch was closed is calculated by assuming the injected voltage across the switch contacts as the source voltage, and short circuiting all other voltage sources and open circuiting all current sources. This is the transient component of the current. To this transient component of current is (algebraically) added the current through that branch which would have existed if the switch were not closed. This sum is the total current through that branch after the switch is closed. If the current through that branch was zero prior to the closing of the switch, then the transient component is the total current through that branch. The voltage across that branch is the voltage drop across the impedance of that branch caused by the flow of the total current. The simulation of the closing of a switch is schematically shown in Fig. 9.1.

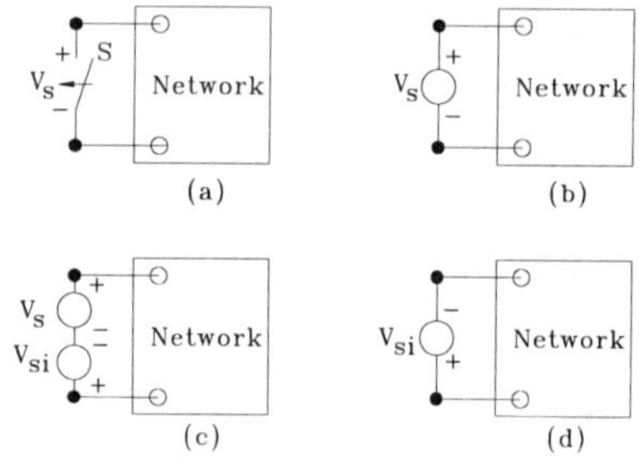

Fig. 9.1 Simulation of the closing of a switch.
(a) switch open; (b) switch replaced by equivalent voltage source;
(c) cancelled voltage at the switch on closing; (d) injected voltage at the switch.

Similarly, the opening of a switch can be simulated by injecting a current through the switch contacts such that the algebraic sum of this injected current and the current which would have existed if the switch were not opened remains zero while the switch is in the open position. The current and the voltage in any branch of the switched network is calculated by placing a current source equal to the injected current across the switch contacts. This is shown schematically in Fig. 9.2.

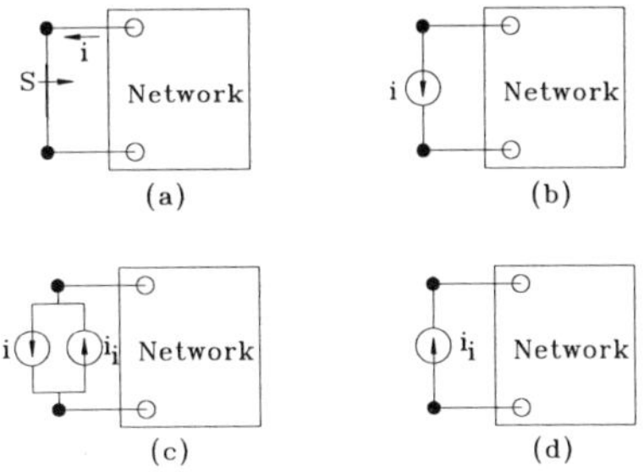

Fig. 9.2 Simulation of the opening of a switch.
(a) switch closed; (b) switch replaced by equivalent current source;
(c) cancelled current at the switch on opening; (d) injected current at the switch.

9.3 Examples of Simple Switching

9.3.1 Closing of a switch

A simple example of switch closing is shown in Fig. 9.3. Prior to switch closing, the current through and the voltage across R_2 are:

$$i_{2o} = \frac{V_o}{R_1 + R_2}, \text{ and } V_{2o} = V_o \frac{R_2}{R_1 + R_2}. \tag{9.1}$$

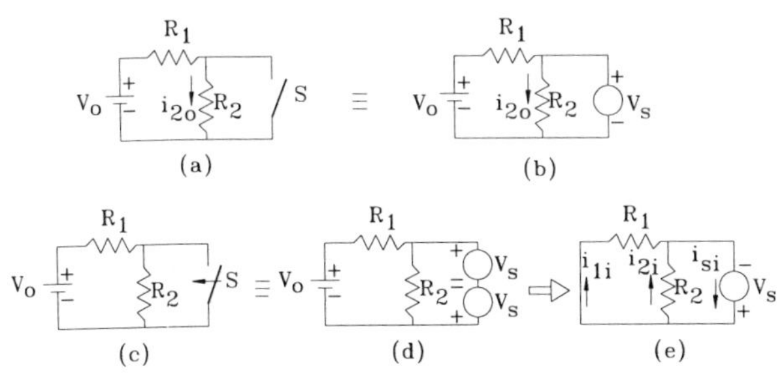

Fig. 9.3 Example of switch closing.

V_{2o} is also the voltage, V_s, across the open switch, S, with the polarity shown in Fig. 9.3b. When S is closed (Fig. 9.3c), V_s is nullified as shown in Fig. 9.3d. By short-circuiting V_o, the switch current, i_{si}, and the current, i_{2i}, through R_2 are given by (Fig. 9.3e):

$$i_{si} = V_s \frac{R_1 + R_2}{R_1 R_2} = \frac{V_o}{R_1}. \tag{9.2}$$

$$i_{2i} = \frac{V_s}{R_2} = \frac{V_o}{R_1 + R_2}. \tag{9.3}$$

The initial current, i_{so}, through S prior to switching was obviously zero. Therefore, the total current, i_{sT}, through S and the total current, i_{2T}, through R_2 after the closure of S are:

$$i_{sT} = i_{so} + i_{si} = 0 + \frac{V_o}{R_1} = \frac{V_o}{R_1}, \text{ and} \tag{9.4}$$

$$i_{2T} = i_{2o} - i_{2i} = \frac{V_o}{R_1 + R_2} - \frac{V_o}{R_1 + R_2} = 0. \tag{9.5}$$

The final results could have been predicted without the injected voltage method. However, this simple example shows the strength of the method.

9.3.2 Opening of a switch

Using the same circuit as in Fig. 9.3, if the switch, S, is initially in the closed position (Fig. 9.4), the initial currents through R_1, R_2 and S are given by:

$$i_{1o} = i_{So} = \frac{V_o}{R_1}, \quad i_{2o} = 0. \tag{9.6}$$

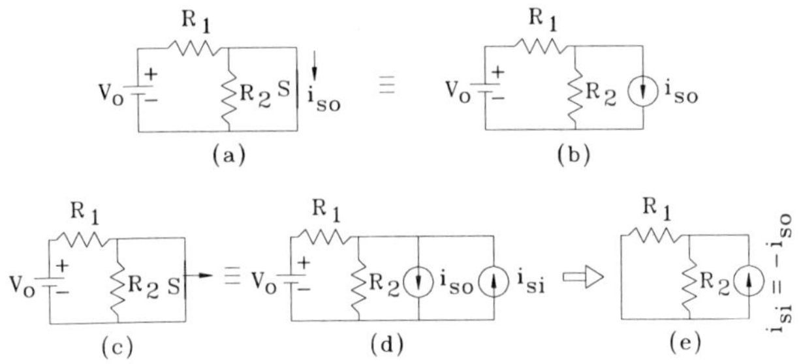

Fig. 9.4 Example of switch opening.

When S is opened, a current, i_{si}, is injected through the switch contacts such that $i_{si} = -i_{so}$, as shown in Fig. 9.4e. The currents, i_{1i} and i_{2i}, through R_1 and R_2 due to i_{si} are:

$$i_{1i} = i_{si}\frac{R_2}{R_1 + R_2} = V_o\frac{R_2}{R_1(R_1 + R_2)}, \text{ and} \tag{9.7}$$

$$i_{2i} = i_{si}\frac{R_1}{R_1 + R_2} = \frac{V_o}{R_1 + R_2}. \tag{9.8}$$

The total currents through R_1, R_2 and S after the switch is opened are given by:

$$i_{1T} = i_{1o} + i_{1i} = \frac{V_o}{R_1} - V_o\frac{R_2}{R_1(R_1+R_2)} = \frac{V_o}{R_1+R_2}, \tag{9.9}$$

$$i_{2T} = i_{2o} + i_{2i} = 0 + \frac{V_o}{R_1+R_2} = \frac{V_o}{R_1+R_2}, \text{ and} \tag{9.10}$$

$$i_{ST} = i_{So} + i_{Si} = i_{So} - i_{So} = 0. \tag{9.11}$$

These results can also be anticipated from simple circuit analysis.

Complex circuits which contain inductances and capacitances can be analyzed by the injected voltage (for circuit closing) and the injected current (for circuit opening) methods applying Laplace transform. Some of these problems are left as exercises at the end of this chapter.

It was assumed in Fig. 9.4 that the direct current through the switch was interrupted instantaneously (Fig. 9.5a). In actual circuits, inductance is always present. During the process of opening the contacts of the switch, an arc is formed across the opening contacts. This slows down the rate of decrease of the current. In Fig. 9.5b, if the arc voltage, V_a, is constant, then:

$$V_o = Ri + L\frac{di}{dt} + V_a . \tag{9.12}$$

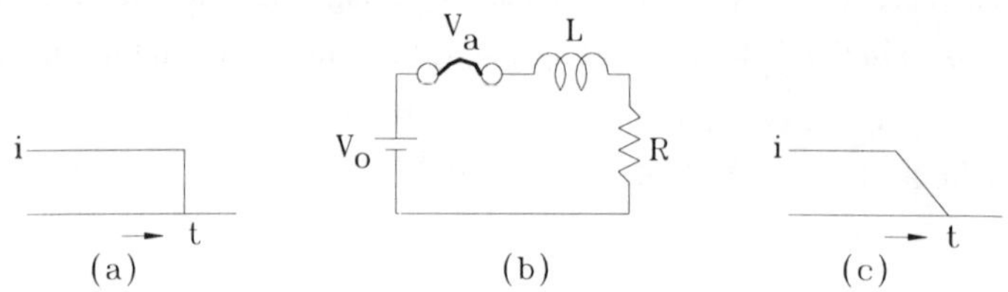

Fig. 9.5 Current interruption in an inductive circuit.

Neglecting the resistive drop:

$$\frac{di}{dt} = \frac{V_o - V_a}{L} . \tag{9.13}$$

The above equation shows that the current can only be interrupted if the arc voltage, V_a, is greater than the source voltage, V_o. In that case, the current will decrease to zero as a ramp function, as shown in Fig. 9.5c. The injected current through the switch should also be a ramp function of opposite polarity. This ramp-function injected current can be handled the same way as discussed in Chapter 1.

In an ac system, an arc is formed across the switch contacts during opening. The arc is quenched when the arc current goes through its cyclic current zero. The injected current at current zero should also be a sinusoidal current but 180° phase shifted from the current which is interrupted (Fig. 9.6).

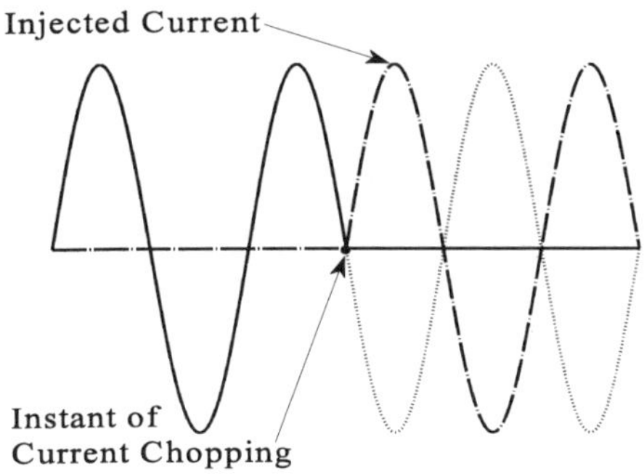

Fig. 9.6 Current interruption in an ac system.

Under certain conditions, the arc quenching medium in the switch may be too strong for the current to be interrupted. In that case, the current may be interrupted before its natural current zero (Fig. 9.7). This is called current chopping. Current chopping can generate very high switching transient voltage.

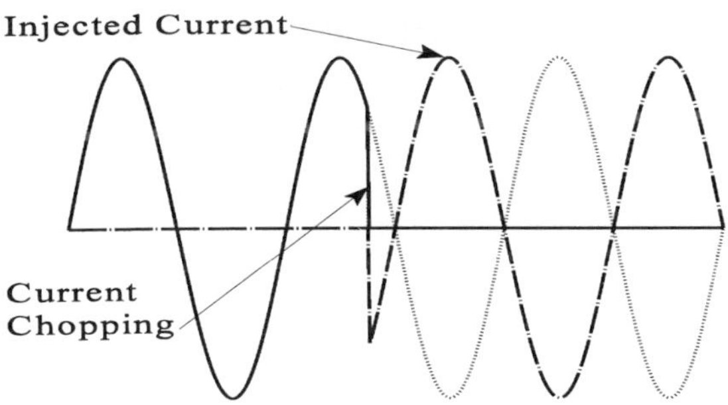

Fig. 9.7 Current chopping in an ac system.

9.4 Circuit Breaker Recovery Voltage [2]

When the contacts of a circuit breaker (or a switch) part to interrupt the current, an arc is formed bridging the gap between the parting contacts. At current zero, the arc is quenched, and the dielectric medium between the contacts starts to recover its dielectric strength. Concurrently, a transient voltage is developed across the open contacts. This voltage, known as the transient recovery voltage (V_{tr}) of the circuit breaker, is given by:

$$V_{tr} = i_i Z, \tag{9.14}$$

where Z is the impedance of the network as seen from the open contacts of the circuit breaker (Fig. 9.8), and i_i is the injected current between the contacts.

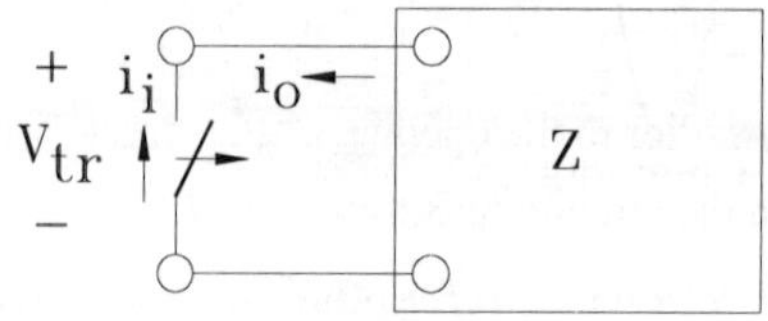

Fig. 9.8 Transient recovery voltage across opening circuit-breaker contacts.

As an example, consider the electrical system shown in Fig. 9.9a. The equivalent circuit is shown in Fig. 9.9b. L_1 is the total inductance to the left of the circuit breaker; this includes the inductances of the step-up transformer, T_1, and of the generator, G. C_1 is the capacitance of the bushing of the breaker pole 1 and also that of the bushing of the transformer, T_1. L_2 is the total inductance to the right of the circuit breaker, including that of the line and the transformer, T_2. C_2 includes the bushing capacitance of the pole 2 and that of the line.

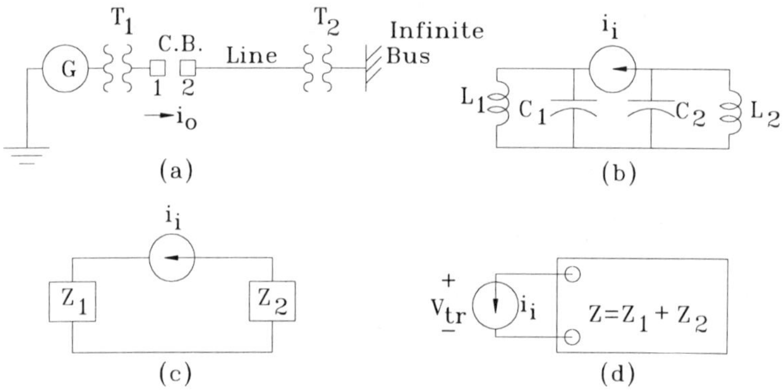

Fig. 9.9 Recovery voltage across the circuit breaker in a transmission system.
G=generator; CB=circuit breaker;
T_1=step-up transformer; T_2=step-down transformer.

The two LC-branches can be expressed in Laplace transform as:

$$Z_1 = \frac{s}{C_1(s^2 + \omega_1^2)} ; \; Z_2 = \frac{s}{C_2(s^2 + \omega_2^2)}, \tag{9.15}$$

$$\text{where,} \; \omega_1^2 = \frac{1}{L_1 C_1} ; \; \omega_2^2 = \frac{1}{L_2 C_2}. \tag{9.16}$$

Assuming the circuit current prior to the opening of the circuit breaker to be $I=I_o \sin \omega t$, the injected current through the circuit breaker is $I_i=-I=-I_o \sin \omega t$. The transient recovery voltage across the switch after the circuit current is interrupted is given in Laplace transform by:

$$V_{tr}(s) = I_i(Z_1 + Z_2) \tag{9.17}$$

$$= -I_o \omega \left[\frac{s}{C_1(s^2+\omega^2)(s^2+\omega_1^2)} + \frac{s}{C_2(s^2+\omega^2)(s^2+\omega_2^2)} \right]. \tag{9.18}$$

Taking inverse Laplace transform,

$$V_{tr}(t) = -I_o \omega \left[\frac{\cos\omega t - \cos\omega_1 t}{C_1(\omega_1^2-\omega^2)} + \frac{\cos\omega t - \cos\omega_2 t}{C_2(\omega_2^2-\omega^2)} \right]. \tag{9.19}$$

A plot of the transient recovery voltage across the circuit breaker is shown in Fig. 9.10, for L_1=10 mH, L_2=20 mH, C_1=1 nF, C_2=100 nF, and I_o=10 kA.

The rate of rise of the transient recovery voltage (r.r.t.r.v) is an extremely important parameter in the successful operation of an electrical system. During current interruption when the arc persists between the opening contacts, the space between them is ionized. After current interruption, the space between the opening contacts of the circuit breaker gradually recovers its dielectric strength. However, the transient recovery voltage also appears across the opening contacts immediately after the arc is quenched. The current interruption will be successful if the rate of dielectric recovery of the space between the contacts is faster than the rate of rise of the transient voltage across the

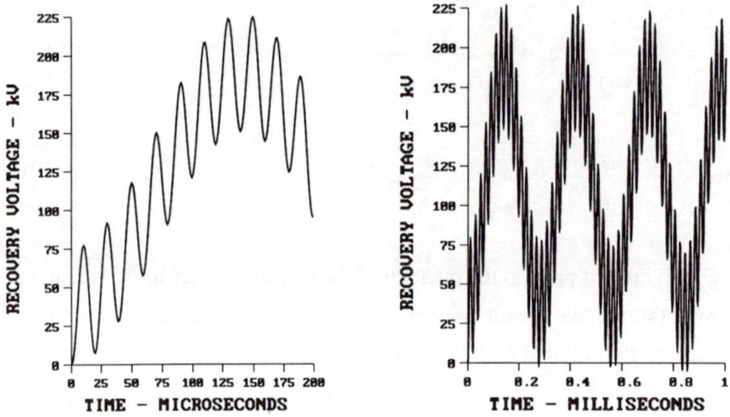

(a) (b)

Fig. 9.10 Transient recovery voltage across the circuit breaker in Fig. 9.9.
L_1=10 mH; L_2=20 mH; C_1=1 nF; C_2=100 nF; I_o=10 kA.
Note: The same transient is shown in (a) and (b) but on different time scales.

circuit breaker. If the transient recovery voltage across the open circuit breaker is higher than the voltage withstand strength of the dielectric space between the contacts, the arc will be reignited. This may produce even higher transient recovery voltage when the second arc is quenched. Circuit breakers have been blown apart by repeated reignitions, and system components damaged by excessive currents or voltage or both. A successful current interruption depends upon the circuit breaker design as well as the system parameters. In selecting a circuit breaker, this interaction between the characteristics of the circuit breaker and those of the system must be taken into consideration.

Going back to (9.19), generally, $\omega_1 \gg \omega$, $\omega_2 \gg \omega$, and because of the losses (which have been neglected in this simple analysis) in the system, the terms - $\cos \omega_1 t$ and $\cos \omega_2 t$ - are damped out after some time. Therefore, the steady-state expression of (9.19) can be written as:

$$V_{tr} = -I_o\omega(L_1 + L_2)\cos\omega t = -V_o\cos\omega t. \tag{9.20}$$

This is the steady-state system voltage which appears across the open circuit breaker.

9.5 References to Chapter 9

[1] M.F. Gardner and J.L. Barnes: ***Transients in Linear Systems***, Vol.1. John Wiley & Sons, New York, 1942.

[2] H.A. Peterson: ***Transients in Power Systems***. John Wiley & Sons, New York, 1951.

Problems for Chapter 9

1.

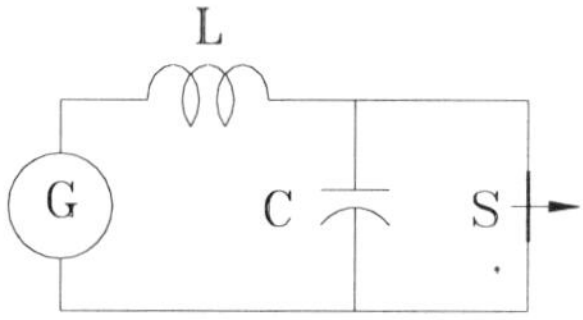

For the circuit shown above, the capacitance is initially short-circuited by the switch, S. The switch is opened to interrupt the circuit current at its normal current zero. The generator voltage is given as $V_g = V_o \cos\omega t$. Plot the recovery voltage across the switch, S, for a time period of 25 μs if L=10 μH, C=0.1 μF and V_o=100 kV.

2.

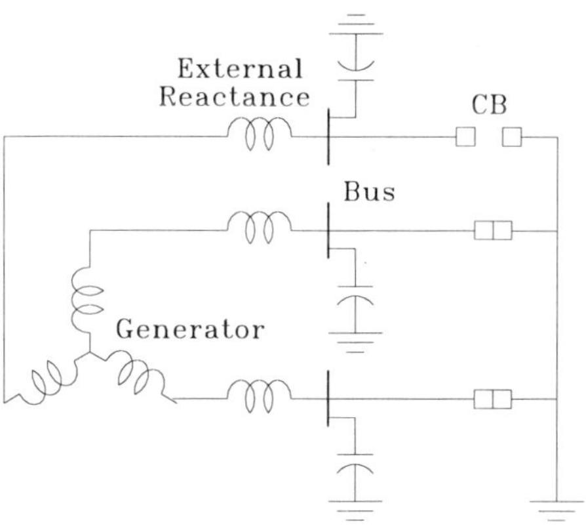

A three-phase fault occurs on an electrical system with isolated neutral as shown above. The total inductance per phase (generator and external reactance) is L. The bus has a capacitance to ground per phase of C farads. Find the recovery voltage across the circuit breaker, CB, for the first phase to clear, for current interruption at normal current zero.

3.

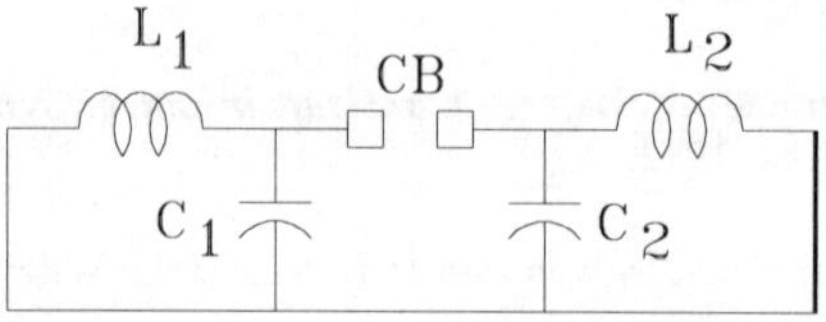

A fault occurs on a transmission line. The circuit breaker, CB, interrupts the fault current at its normal current zero. Find an expression for the recovery voltage across the circuit breaker, assuming the equivalent for the network under faulted condition as shown above.

4.

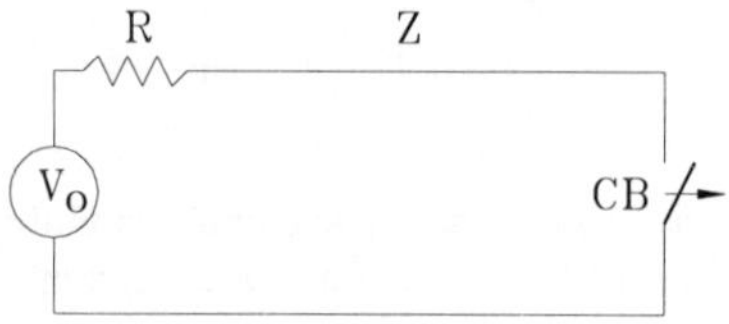

A dc voltage source, V_o, feeds the short circuit on a transmission line through a resistor, R, as shown above. The short circuit is removed by opening the circuit breaker, CB, at $t = 0$. If the surge impedance of the line is Z, the wave travel time on the line is τ and $Z/R = 4$, draw the voltage profile (in terms of V_o) at the open end of the line for a period $t = 5\tau$.

5.

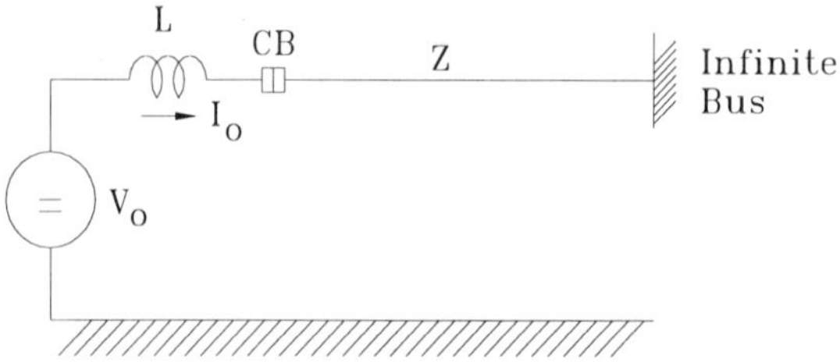

The above figure shows the schematic of a dc converter supplying dc power to a remote infinite bus through a dc reactor in series with a long transmission line of surge impedance, Z. During interruption, the circuit breaker, CB, interrupts the current instantaneously. Find an expression for the total voltage on the line after current interruption.

CHAPTER 10
SWITCHING SURGE PHENOMENA

10.1 Current Suppression or Chopping

Current suppression or chopping occurs when the current in a circuit is forced to zero. In an ac circuit, current zero occurs cyclically. When the contacts of a switch part to open a circuit, an arc is formed across the parting contacts. The arc is quenched when the current goes through its natural zero. The function of the arc-quenching medium is to restore the dielectric strength of the space between the contacts so that the arc is not reignited when the voltage builds up across the open contacts. However, if the arc-quenching medium is too strong, the current may be brought to zero prematurely, and a current chopping will occur. In a dc circuit, as there is no natural current zero, current chopping always occurs during interruption of direct current.

10.1.1 Current chopping in dc systems

Let us consider a simple parallel LC-circuit, where a steady direct current, i_o, is flowing initially (Fig. 10.1a). If this current is abruptly switched off, then the recovery voltage across the switch, S, can be calculated by injecting a step-function current, $-i_o$, through the switch (Fig. 10.1b). The circuit impedance, looking from the open switch, is:

$$Z(s) = \frac{Ls}{LCs^2 + 1}. \qquad (10.1)$$

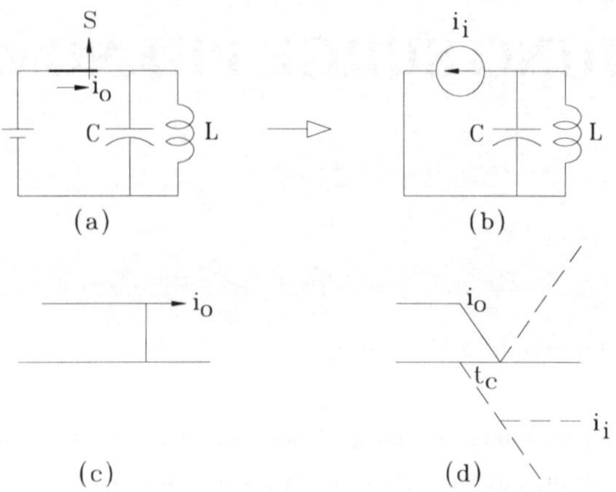

Fig. 10.1 Current chopping in a dc circuit.

The recovery voltage across the switch is given by:

$$V_r(s) = i(s)Z(s) = \frac{i_o}{C}\frac{1}{s^2 + \frac{1}{LC}}. \tag{10.2}$$

Taking inverse Laplace transform,

$$V_r(t) = i_o\sqrt{\frac{L}{C}}\sin\frac{t}{\sqrt{LC}}. \tag{10.3}$$

Instead of interrupting the current abruptly (Fig. 10.1c), if it is brought to zero linearly as a ramp (Fig. 10.1d), then the injected current at the switch can be simulated as:

$$i_i = -[\alpha_1 t u(t) - \alpha_1 (t - t_c) u(t - t_c)], \tag{10.4}$$

where $\alpha_1 = i_o/t_c$. The Laplace transform of (10.4) is:

$$i_i(s) = -\left[\frac{\alpha_1}{s^2} - \frac{\alpha_1}{s^2}e^{-t_c s}\right]. \tag{10.5}$$

The recovery voltage across the switch, in Laplace transform, is then given by:

$$V_r(s) = -\frac{\alpha_1}{C}\left[\frac{1}{s(s^2+\frac{1}{LC})} - \frac{e^{-t_c s}}{s(s^2+\frac{1}{LC})}\right]. \quad (10.6)$$

Taking the inverse Laplace transform,

$$V_r(t) = -\frac{i_o L}{t_c}\left[(1-\cos\frac{t}{\sqrt{LC}}) - (1-\cos\frac{t-t_c}{\sqrt{LC}})u(t-t_c)\right]. \quad (10.7)$$

Figure 10.2a shows the recovery voltage across a switch for an instantaneous current chop, and Fig. 10.2b shows the recovery voltage for a ramp chop at two different ramp rates. It should be noticed that the slower the chopping rate, the lower the recovery voltage will be.

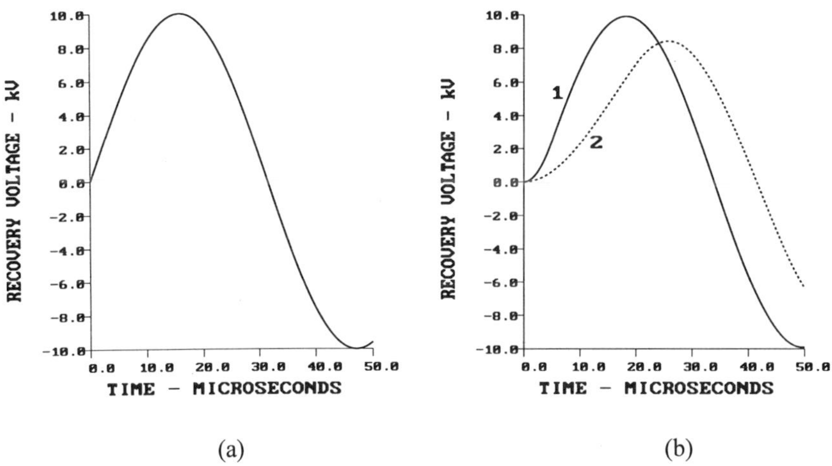

Fig. 10.2 Recovery voltage across a switch in a dc circuit.
i_o = 1 kA; L = 100 μH; C = 1 μF.
(a) instantaneous current chop; (b) ramp chop of current;
1: t_c=5 μs, 2: t_c=20 μs.

10.1.2 Current chopping in ac systems

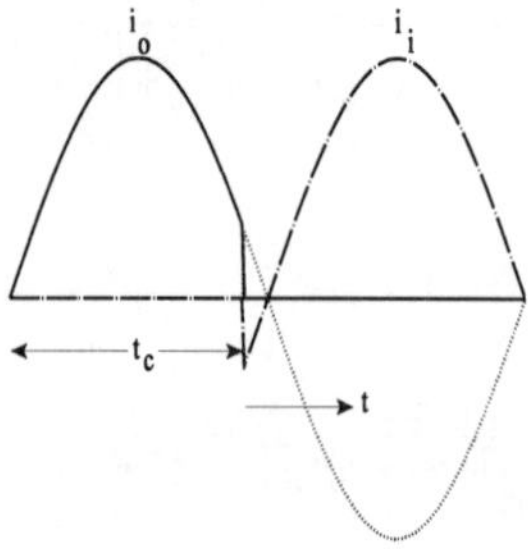

Fig. 10.3 Current chopping in an ac system.

In an ac system, the circuit breaker generally interrupts a current at its natural current zero. However, if the strength of the arc quenching medium is too strong, the current may be brought to zero prematurely before its natural current zero. This was discussed in Section 9.3.2. As an example, current chopping can occur during interruption of the magnetizing current of a transformer at no load. The magnetizing current is much smaller (about 5 percent) than the full-load current. The circuit breaker is designed to interrupt fault currents which can be as high as 10 times the rated full-load current of the transformer. The strength of the arc quenching medium of the circuit breaker would force the magnetizing current prematurely to zero, thus generating high switching transients in the system. Redrawing Fig. 9.7 in Fig. 10.3, the injected current can be simulated as:

$$i_i = -i_o \sin\omega(t + t_c), \qquad (10.8)$$

where t_c = time interval (in seconds) between the forced current zero and the previous natural current zero.

The Laplace transform of the injected current is:

$$i_i(s) = -i_o\left[\frac{\omega}{s^2+\omega^2}\cos\omega t_c + \frac{s}{s^2+\omega^2}\sin\omega t_c\right]. \qquad (10.9)$$

The recovery voltage across the switch is:

$$V_r(s) = i_i(s)Z(s), \qquad (10.10)$$

where Z(s) is the impedance of the circuit looking from the switch. Applying this to the circuit of Fig. 10.1,

$$V_r(s) = \frac{i_o}{C}\left[\frac{s\,\omega\cos\omega t_c}{(s^2+\omega^2)(s^2+\omega_o^2)} + \frac{s^2\sin\omega t_c}{(s^2+\omega^2)(s^2+\omega_o^2)}\right]. \qquad (10.11)$$

Taking inverse Laplace transform,

$$V_r(t) = \frac{i_o}{C(\omega_o^2-\omega^2)}\left[\omega\cos\omega(t+t_c) - A_o\sin(\omega_o t-\phi)\right], \qquad (10.12)$$

where $A_o = \sqrt{\omega^2\cos^2\omega t_c + \omega_o^2\sin^2\omega t_c}$, and $\tan\phi = \frac{\omega}{\omega_o}\cot\omega t_c$. The recovery voltage of the circuit in Fig. 10.1 at two different t_c are shown in Fig. 10.4.

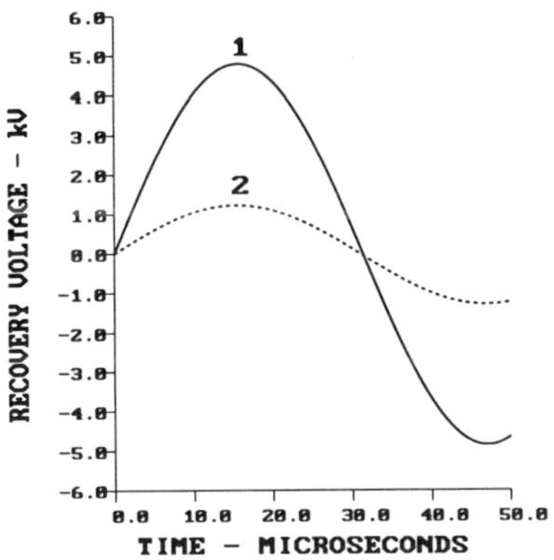

Fig. 10.4 Recovery voltage caused by current chopping in an ac circuit.
Instant chop; $i_o = 1$ kA; L = 100 μH; C = 1 μF.
1: $t_c = 7$ ms; 2: $t_c = 8$ ms.

10.2 Compound Transients

A compound switching transient occurs when a switch is operated before the transient caused by the previous operation of a switch has subsided. A simple example of a compound transient in a dc circuit is shown in Fig. 10.5a.

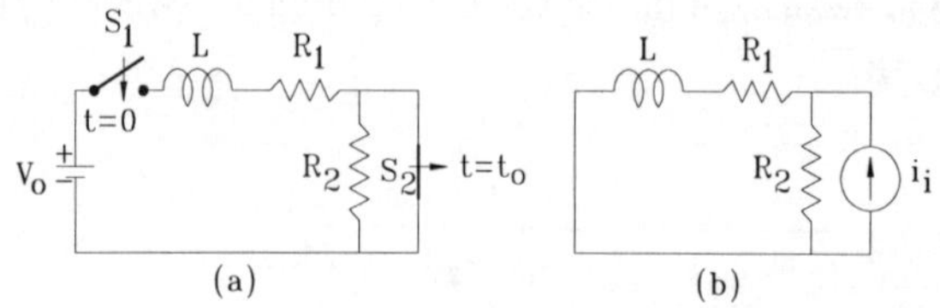

Fig. 10.5 An example of a compound transient in a dc circuit.

The switch, S_1, is closed at time, t=0. The switch, S_2, is opened at time, $t=t_o$, before the current transient caused by the closing of S_1 has been dissipated. The recovery voltage across S_2 is to be determined.

The current through S_1, L, R_1 and S_2 is given, in Laplace transform, as

$$i(s) = \frac{V_o}{Ls\left(s + \frac{R_1}{L}\right)}, \text{ for } 0 \le t \le t_o. \tag{10.13}$$

$$\text{In time domain: } i(t) = \frac{V_o}{R_1}\left(1 - e^{-\frac{R_1}{L}t}\right). \tag{10.14}$$

The injected current at S_2 will be $i_i = -i$ for $t \ge t_o$ (Fig. 10.5b). Shifting the origin of time to t_o:

$$i_i(\tau) = -\frac{V_o}{R_1}\left[1 - e^{-\frac{R_1}{L}(\tau + t_o)}\right]. \tag{10.15}$$

$$\text{In Laplace transform: } i_i(s) = -\frac{V_o}{R_1}\left[\frac{1}{s} - \frac{e^{-\frac{R_1}{L}t_o}}{s + \frac{R_1}{L}}\right]. \tag{10.16}$$

The impedance seen from open S_2 will be:

$$Z_{eq} = \frac{R_2(Ls + R_1)}{R_1 + R_2 + Ls}. \tag{10.17}$$

The recovery voltage across S_2 will be:

$$V_r(s) = i_i(s)Z_{eq} = -\frac{V_o R_2}{R_1}\left[\frac{s+\frac{R_1}{L}}{s\left(s+\frac{R_1+R_2}{L}\right)} - \frac{e^{-\frac{R_1}{L}t_o}}{s+\frac{R_1+R_2}{L}}\right]. \tag{10.18}$$

In time domain, with $\tau = t - t_o$:

$$V_r(\tau) = -V_o\frac{R_2}{R_1}\left[\frac{R_1}{R_1+R_2} + \frac{R_2 e^{-\frac{R_1+R_2}{L}\tau}}{R_1+R_2} - e^{-(\frac{R_1}{L}t_o + \frac{R_1+R_2}{L}\tau)}\right]. \tag{10.19}$$

An example of a compound transient in an ac circuit is shown in Fig. 10.6. The inductance, L, represents a short overhead line; C is the stray capacitance at the end of the open line. A short circuit occurs across C at a time t = 0. The fault subsequently clears at the first current zero of the fault current at t = t_o. What is the recovery voltage across the air gap which cleared the fault ?

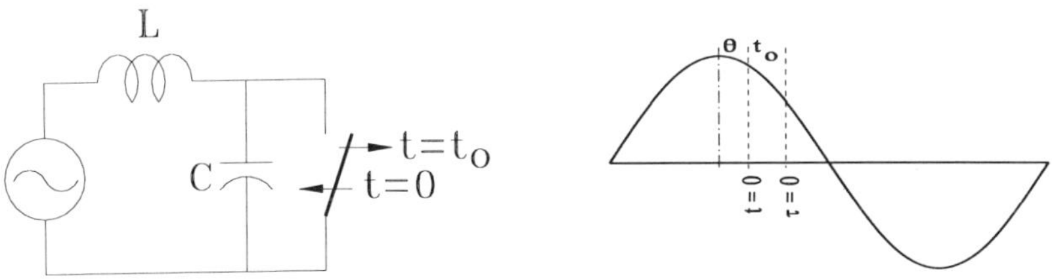

(a) (b)

Fig. 10.6 An example of a compound switching transient in an ac circuit.

If the fault occurs at a time when the instantaneous voltage is θ radians after the peak has occurred (Fig. 10.6b), then the generated voltage can be expressed as:

$$V_g = V_o\cos(\omega t + \theta), \tag{10.20}$$

where the reference of time (t=0) starts when the fault occurs at θ. Looking from the position of the fault, the circuit impedance will consist of L and C connected in parallel. Therefore, in Laplace transform, the equivalent impedance of the circuit looking from the faulted position is

$$Z_{eq}(s) = \frac{s}{C(s^2 + \omega_o^2)}, \tag{10.21}$$

where $\omega_o = \frac{1}{\sqrt{LC}}$.

The voltage across C, prior to the short circuit, is

$$V_c(s) = V_g(s)\frac{\frac{1}{Cs}}{Ls+\frac{1}{Cs}} = \frac{V_o}{LC(s^2+\omega_o^2)}\left[\frac{s\cos\theta}{s^2+\omega^2} - \frac{\omega\sin\theta}{s^2+\omega^2}\right]. \tag{10.22}$$

The fault current, in Laplace transform,

$$i(s) = \frac{V_c(s)}{Z_{eq}(s)}. \tag{10.23}$$

Inverting,

$$i(t) = \frac{V_o}{\omega L}[\sin(\omega t + \theta) - \sin\theta], \tag{10.24}$$

where time starts (t=0) at the instant of short circuit when the generated voltage is θ radians past the voltage peak (Fig.10.6b). It should be noticed in (10.24) that the fault current has a dc component, $-V_o\sin\theta/\omega L$. This dc component vanishes if the fault occurs at the instant the voltage is at its peak (θ=0), and it is maximum at $\theta=\pm0.5\pi$. It should also be noticed that the dc component is positive if θ is negative, and vice versa. The current lags the voltage by 90° in an inductive circuit, such as shown in Fig. 10.6a when C is shorted out. Therefore, the current will be at its peak when the voltage is going through zero. However, the current has to start from zero and cannot abruptly change to its expected value instantaneously (constant flux linkage theory). Therefore, if a short circuit occurs at the instant the voltage is going through its cyclic zero, and the short-circuit current cannot instantaneously jump to its expected value, a dc component of current is induced which is equal, but of opposite polarity, to the expected short-circuit current. This is clearly shown in Fig. 10.7 [1].

If the short circuit is interrupted at first current zero of the fault current at a time $t=t_o$, then $i(t_o) = 0$. Equating (10.24) to 0 at $t=t_o$,

$$t_o = \frac{\pi - 2\theta}{\omega} . \tag{10.25}$$

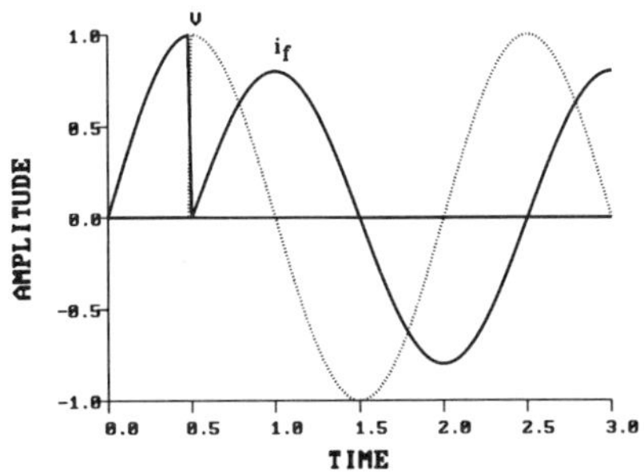

(a)

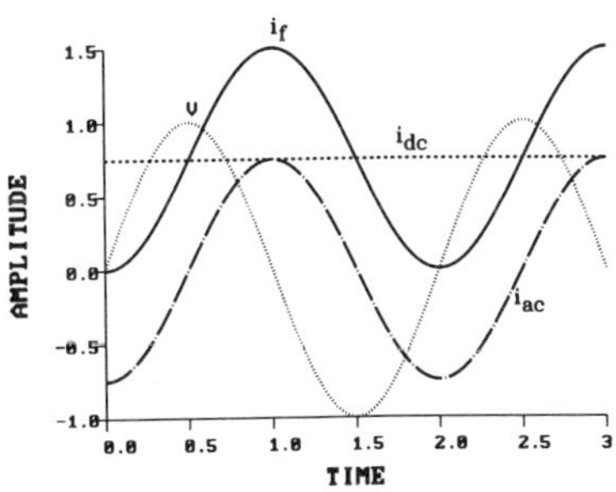

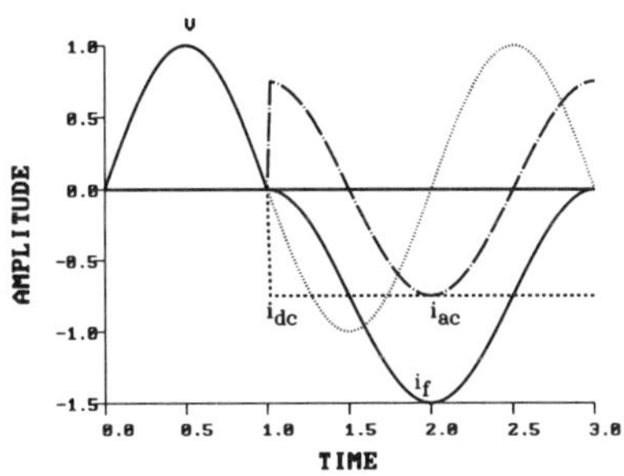

(b) (c)

Fig. 10.7 Effect of instant of fault occurrence on dc component of fault current [1].

$$i(\tau) = \frac{V_o}{\omega L}[\sin(\omega\tau + \omega t_o + \theta) - \sin\theta]. \quad (10.26)$$

Knowing that the injected current for current interruption at the short circuit, i_i, is equal but of opposite polarity to the fault current, i, the recovery voltage across the fault-clearing gap is given by:

$$V_r(s) = i_i(s) Z_{eq} = \frac{V_o}{\omega LC} \frac{s}{(s^2+\omega_o^2)} \left[\frac{\omega\cos\theta}{s^2+\omega_o^2} - \frac{s\sin\theta}{s^2+\omega_o^2} + \frac{\sin\theta}{s} \right]. \quad (10.27)$$

Inverting and simplifying,

$$V_r(t) = \frac{V_o \omega_o^2}{\omega_o^2 - \omega^2} \left[\cos(\omega t - \theta) - \frac{\cos\theta \cos(\omega_o t - \alpha)}{\cos\alpha} \right], \quad (10.28)$$

$$\text{where} \quad \tan\alpha = \frac{\omega}{\omega_o}\tan\theta. \quad (10.29)$$

As usually, $\omega_o \gg \omega$, then (10.28) can be simplified to the following:

$$V_r(t) \approx V_o[\cos(\omega t - \theta) - \cos\theta \cos\omega_o t]. \quad (10.30)$$

It should be noticed in (10.30) that the amplitude of the natural-frequency oscillations is proportional to $\cos\theta$. It was shown in (10.24) that the dc component of the fault current is proportional to $\sin\theta$. Therefore, it can be concluded that the higher the dc component of the fault current, the lower is the amplitude of the natural-frequency component of the recovery voltage. In the limit when $\theta = \pm 0.5\pi$, the high-frequency oscillations in the recovery voltage disappear. The effect of the fault current transient is shown in Fig. 10.8.

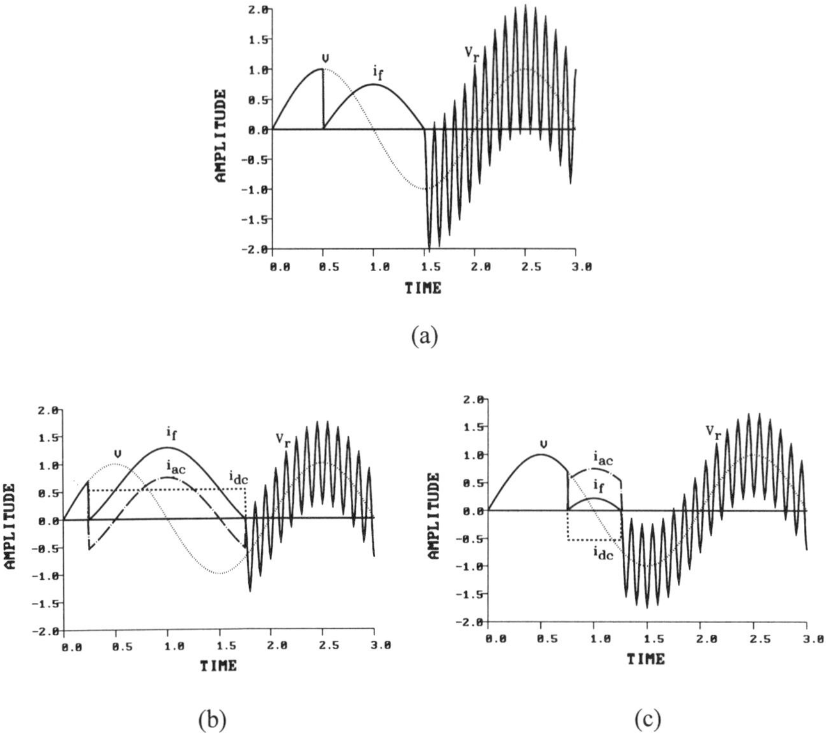

Fig. 10.8 Effect of fault current transient on recovery voltage [1].

10.3 Switching Surges in Capacitive Circuits

Figure 10.9a shows an LC-circuit where the switch is initially in the closed position. Similar to Fig. 10.6, L could represent a short overhead line. However, unlike the C in Fig. 10.6, the C in Fig. 10.9 is relatively large, such as a power-factor correction capacitor bank.

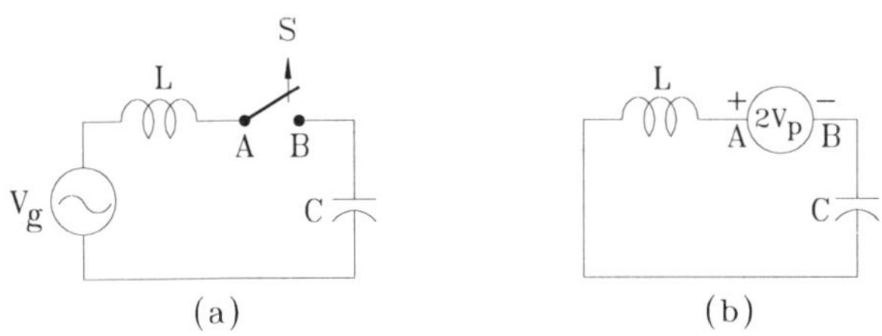

Fig. 10.9 Current interruption in a lumped capacitive circuit.

Figure 10.9 is essentially a capacitive circuit. With S closed, and a small L, the capacitor voltage, V_c, will be about the same as the generator voltage, V_g, and the capacitor current will lead the voltage by 90 degrees, being zero when the voltage is at its peak. If the switch opens the circuit at current zero, the capacitor will be isolated from the rest of the circuit with the peak voltage, V_p, across its terminals. Although pole B of the switch is held at V_p, pole A of the switch will follow the instantaneous voltage of the generator. The switch voltage will be $2V_p$ half-a-cycle later. A transient current will flow if the switch reignites (restrikes) at this instant. This transient current can be calculated by injecting a voltage, $-2V_p$, across the open switch. This high-frequency current is calculated by dividing the injected voltage by the impedance seen from the open switch:

$$i_s = -2V_p\sqrt{\frac{C}{L}}\sin\omega_o t, \tag{10.31}$$

where $\omega_o = \dfrac{1}{\sqrt{LC}}$.

The capacitor voltage due to this current can be calculated to be:

$$V_c(s) = \frac{i_s(s)}{Cs} = -\frac{2V_p}{LCs(s^2 + \omega_o^2)}. \tag{10.32}$$

Inverting the Laplace transform,

$$V_c(t) = -2V_p(1 - \cos\omega_o t). \tag{10.33}$$

This voltage will reach its peak value of $-4V_p$ when i_s reaches zero. The total capacitor voltage will be the algebraic sum of this transient component and the initial capacitor voltage ($+V_p$). Therefore, the total voltage across C will be $-3V_p$ when the high-frequency current, i_s, goes through its cyclic zero and the switch isolates the capacitor again from the rest of the circuit.

After half-a-cycle of 60 Hz, when the voltage of pole A is V_p, the voltage across the open switch will be $-4V_p$. If the switch restrikes at this instant, an even higher high-frequency current will flow because of the higher injected voltage ($+4V_p$) across the switch. This current flow will charge up the capacitor to still higher voltage ($+5V_p$). Such repetitive restrikes would build up very high voltages in the system.

10.4 Switching Surges in Distributed-Constant Systems

The transmission line cannot be represented by an inductance, as in Fig. 10.9, if it is long, particularly when the rate of rise of the transient recovery voltage is high. In that case, the transmission line is represented by its surge impedance. Figure 10.10a represents a long open-circuited transmission line which is connected to a generator through a switch installed at the generator end of the line. The inductance, L, in this circuit represents the generator leakage inductance, and Z represents the surge impedance of the line.

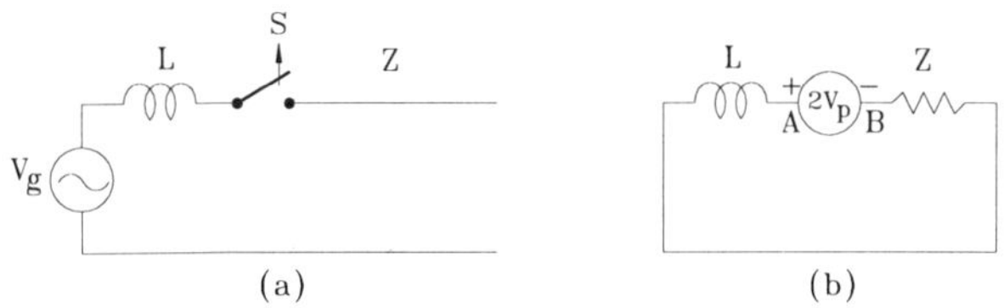

Fig. 10.10 Current interruption in a distributed-constant transmission line.

Under steady-state conditions, the open-circuited line acts like a capacitive circuit. If the switch, S, interrupts the capacitive current at its normal current zero, the line will be left isolated, charged with the peak value, V_p, of the generator voltage, just as in Fig. 10.9. The voltage across S will rise from the initial value of zero, at the instant of current interruption, to $2V_p$, half-a-cycle of 60 Hz later. If the switch restrikes at this instant, this condition can be mathematically simulated by injecting a voltage, $-2V_p$, across the open switch. The similarity with Fig. 10.9 ends here. The current which will be injected to the line at B will be given by:

$$i_f(s) = -\frac{2V_p}{sZ_{eq}} = -\frac{2V_p}{s(Ls + Z)}. \tag{10.34}$$

In the time domain,

$$i_f(t) = -\frac{2V_p}{Z}(1 - e^{-\frac{Z}{L}t}). \tag{10.35}$$

This will also be the current flowing through the switch. The forward current, i_f, will produce a voltage, V_f (=i_f Z), on the line, and both i_f and V_f will travel along the line with the velocity of light, and be reflected at the open end of the line. The total voltage at point B will be:

$$V_B = V_p + i_f(t)Z = -V_p + 2V_p e^{-\frac{Z}{L}t} \quad \text{for } 0 \le t \le 2\tau, \tag{10.36}$$

where τ = travel time of the waves from B to the open end of the line.

The reflected current wave will be equal to $-i_f$, and the reflected voltage wave will be equal to $+V_f$. The reflected waves will travel back towards S, where part of the reflected waves will be reflected back again towards the open end of the line, and part will be transmitted through S. The transmission coefficients of the current and the voltage waves at B are:

$$a_{ct} = \frac{2Z}{Ls+Z} \; ; \; a_{vt} = \frac{2Ls}{Ls+Z}. \tag{10.37}$$

The total current at S will be:

$$i_s(s) = -\frac{2V_p}{Ls(s+\frac{Z}{L})} + \frac{2Z}{L(s+\frac{Z}{L})}\frac{2V_p}{Ls(s+\frac{Z}{L})}e^{-2\tau s}. \tag{10.38}$$

In time domain: $i_s(t)=0$ for $t<2\tau$, else

$$i_s(t) = -\frac{2V_p}{Z}\left[1-e^{-\frac{Z}{L}t}\right] + \frac{4V_p}{Z}\left[1-\left(1+\frac{Z}{L}(t-2\tau)\right)e^{-\frac{Z}{L}(t-2\tau)}\right]. \tag{10.39}$$

The current component which was reflected from the open end being of opposite polarity to the initial switch current, the total switch current will eventually be brought to zero.

As long as S is conducting, the traveling-voltage waves at the point B will be:

$$V_B(s) = -\frac{2V_pZ}{Ls(s+\frac{Z}{L})} - \frac{2Ls}{L(s+\frac{Z}{L})}\frac{2V_pZ}{Ls(s+\frac{Z}{L})}e^{-2\tau s}. \tag{10.40}$$

The first term in (10.40) is caused by i_f, and the second term is the component which is transmitted across the switch, after being reflected at the open end of the line. Inverting to time domain, and remembering that the initial voltage at B on the interruption of the 60-Hz current was V_p, the *total* voltage at B is:

$$V_B(t) = V_p - 2V_p(1 - e^{-\frac{Z}{L}t}) - \frac{4V_p Z}{L}(t - 2\tau)e^{-\frac{Z}{L}(t-2\tau)} \tag{10.41}$$

This is the case as long as S continues to conduct. By the time the reflected waves from the open end of the line arrive at the switch, the first component of the switch current of (10.39) would attain its steady-state value of $(-2V_p/Z)$. Changing the reference time to the instant of arrival of the reflected waves at S, the terms $(t - 2\tau)$ should be replaced by t. Thus,

$$i_s(t) = \frac{2V_p}{Z}\left[1 - 2\left(1 + \frac{Z}{L}t\right)e^{-\frac{Z}{L}t}\right], \tag{10.42}$$

where time is counted from the instant of arrival of the reflected waves from the open end of the line. Equation (10.42) becomes zero when $Zt/L \approx 1.68$. Injecting a current $i_i = -i_s$ at the switch, the voltage component at the point B due to i_i will be given by:

$$\Delta V_B = -2V_p\left[1 - 2\left(1 + \frac{Z}{L}(t + t_o)\right)e^{-\frac{Z}{L}(t+t_o)}\right], \tag{10.43}$$

where $t_o = 1.68L/Z$, and t is the new reference time starting from the instant of current zero. The total voltage at B will be the algebraic sum of (10.41) and (10.43):

$$V_B(t) \approx -3V_p + 4V_p e^{-\frac{Z}{L}(t+t_o)}. \tag{10.44}$$

A second restrike at S may occur when the 60-Hz voltage reverses sign. It can be shown that the voltage at B will then build up to about $+5V_p$. The successive restrikes will keep increasing the voltage to higher values.

10.5 References to Chapter 10

[1] H.A. Peterson: ***Transients in Power Systems***. John Wiley & Sons, New York, 1951.

Problems for Chapter 10

1. For Problem 2 of Chapter 9, find the recovery voltage across the circuit breaker, CB, for the first phase to clear, for current interruption before the natural zero of the fault current at a time, t_o, after the initiation of the fault.

2. For the same fault condition as in Problem 3 of Chapter 9, calculate the recovery voltage if the circuit breaker clears the fault before natural current zero of the fault current at a time t_o after the initiation of the fault.

3.

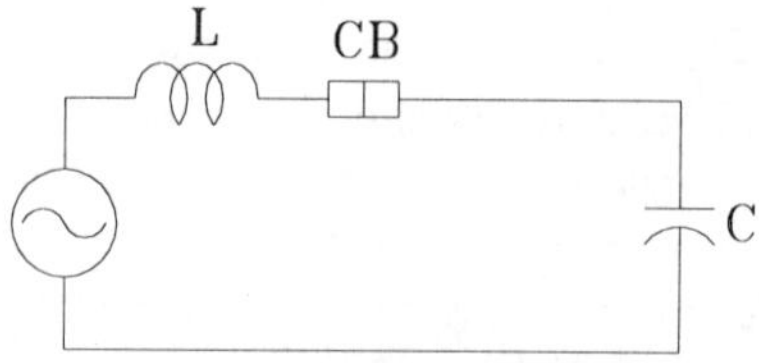

A lumped LC-circuit is energized from an ac voltage source as shown in the above figure. The circuit breaker, CB, interrupts the current at normal current zero. CB restrikes when the voltage across it reaches peak half a cycle later. Calculate the total voltage across the capacitor, C, after the first restrike.

4.

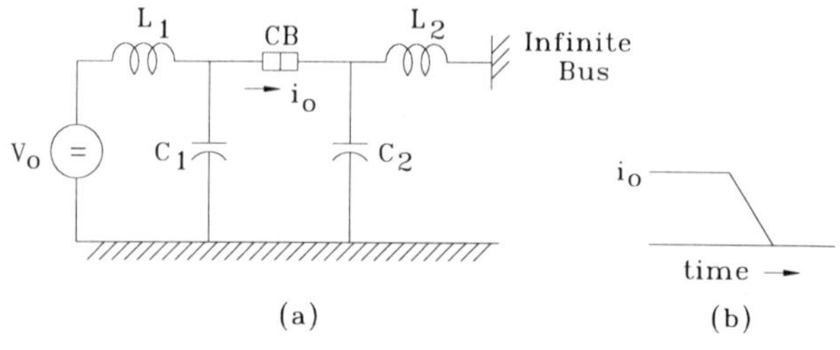

Figure a shows the schematic of a dc transmission system connected to an infinite bus. During interruption of current by the circuit breaker, CB, the current goes to zero linearly as shown in Fig. b. Find an expression for the recovery voltage across the circuit breaker.

5.

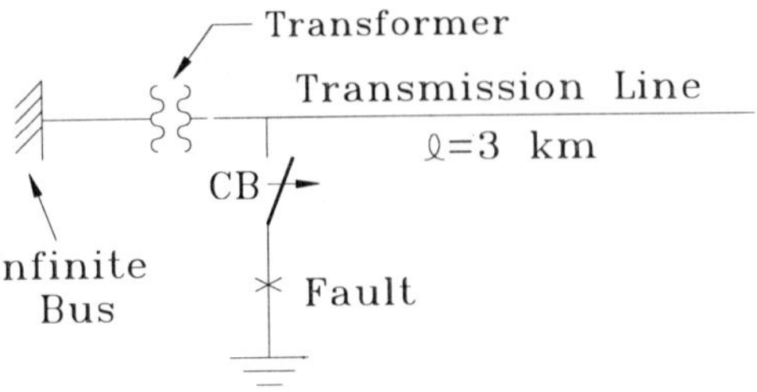

The above figure shows a transmission line ($\ell = 3$ km) supplied through a grounded-neutral transformer from an infinite bus. A feeder line is connected to the line-side of the transformer. The circuit breaker, CB, clears a fault on this feeder line at normal current zero. If the leakage inductance of the transformer is L and the surge impedance of the transmission line is Z, calculate the switching surge voltage initially impressed on the transformer when the circuit breaker opens. Calculate also the switching surge voltage on the transformer after the first reflection from the far end (open circuit) of the transmission line has reached the transformer terminal. Neglect the transformer capacitance to ground. Assume that near current zero (when the circuit is interrupted), the short-circuit current can be represented as $I_o \sin\omega t \approx I_o \omega t$.

6.

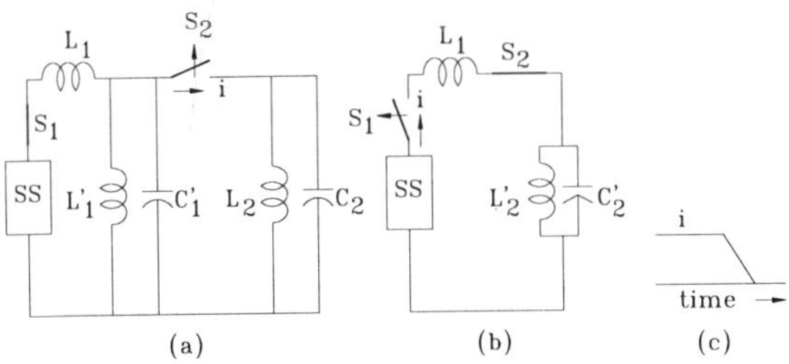

Figure a above is the simplified schematic of a propulsion system when the on-board dc circuit breaker opens. Figure b is the simplified schematic of the same system but now the dc circuit breaker of the substation opens. Figure c shows that the current goes to zero linearly. Find an expression for the recovery voltage across the circuit breaker for each case in terms of the Laplace transform. Assume the impedance of the substation to be negligible.

CHAPTER 11
SYSTEM PERFORMANCE UNDER SWITCHING SURGES

11.1 Introduction

The insulation of an electric power system consists of both internal and external insulation. The internal insulation is inside a power apparatus, such as the circuit breaker, transformer, capacitor and rotating machine. The external insulation consists essentially of air gaps, such as the air space across bushings, insulator strings, phase conductors, and across phase conductors and towers. The external insulation is recoverable (self-restoring); once the arc, formed by the breakdown of an air gap, is quenched the insulation can withstand voltage without any deterioration of its strength. An internal insulation is damaged once it breaks down; it is non-self-restoring. Because of its recoverable nature, the external insulation is used as the sacrificial lamb in the coordination of insulation of electric power systems by choosing its withstand level to be lower than that of the internal insulation. However, to maintain system reliability, the number of flashovers of the external insulation and the subsequent outages caused by short circuits of the power-frequency voltage must not exceed a maximum limit set by the system designers.

The power-system insulation is stressed by two types of transient voltages: (i) lightning, and (ii) switching. The duration of the lightning voltage surges is in the microseconds regime, whereas the duration of the switching voltage surges is in the milliseconds regime, although submicrosecond switching surges are also present.

11.2 Critical Flashover Voltage of External Insulation under Switching Surges

The air gaps, generally used for external insulation in electric power systems, can be broadly classified into: (i) rod-to-rod, (ii) rod-to-plane, and (iii) conductor-to-conductor gaps, as shown in Fig. 11.1. The insulation strength of an external insulation is measured by its critical flashover (CFO) level, V_{50}. The CFO is defined as the voltage level at which the insulation has a 50 percent probability of breakdown. The withstand level is generally estimated by subtracting 3σ (σ=standard deviation) from the CFO. The positive-polarity CFO is generally lower than the negative-polarity CFO, and the rod-to-plane gap has the lowest CFO. The CFO of an external insulation under lightning voltages is linearly proportional to the air-gap length. The switching-surge CFO, on the other hand, shows a saturation effect at longer gaps [1-3]. The CFOs for a rod-to-plane gap under various types of voltages as a function of the gap length are shown in Fig. 11.2. It has been found that the switching-surge CFO reaches a minimum at a critical front time of the switching surge voltage (Fig. 11.3) [4-6]. The CFO under switching surges could be lower than the power-frequency sparkover voltage of the air gap. Therefore, protection against switching surges becomes critical for very high voltage systems (above 345 kV). The international standard for the switching surge is the 250/2500-μs voltage wave [7,8].

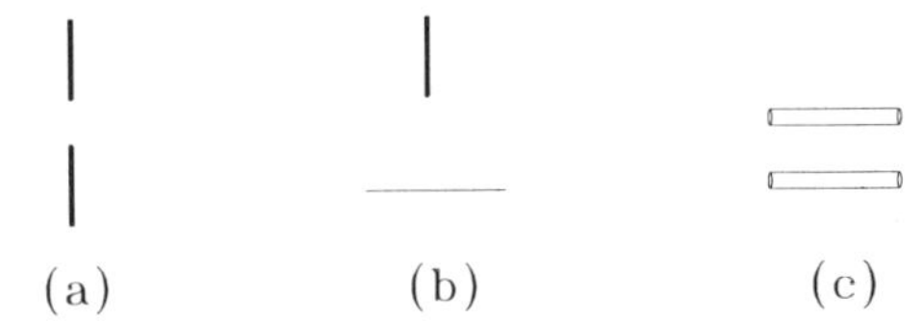

Fig. 11.1 Three commonly used air gaps in electrical power systems.
(a) rod-rod gap; (b) rod-plane gap; (c) conductor-conductor gap

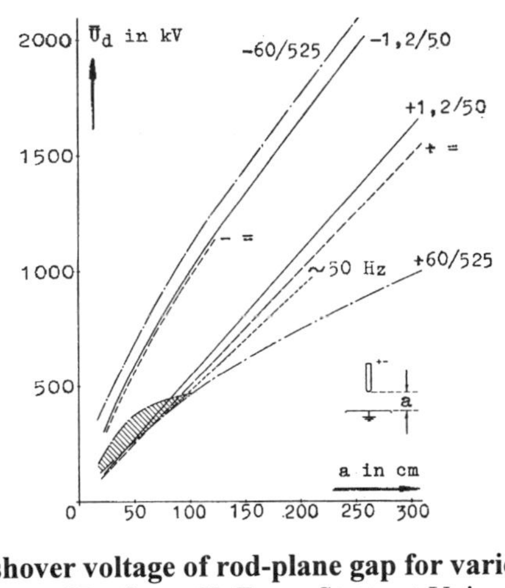

Fig. 11.2 Critical flashover voltage of rod-plane gap for various waveforms [3].
(Courtesy of Professor K. Feser, Stuttgart University)

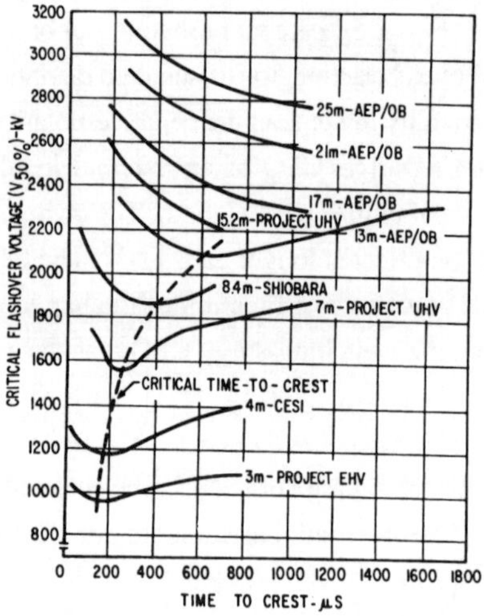

Fig. 11.3 Critical switching-surge flashover voltages for rod-plane air gaps [6]. (Copyright © 1987. Electric Power Research Institute. EPRI EL-2500. *Transmission Line Reference Book: 345 kV and Above, Second Edition.* Reprinted with Permission.)

Based on tests, Paris and Cortina [2] have suggested the following empirical relation between the positive-polarity CFO (V_{50}) and the gap length, for gap lengths less than 10 m:

$$V_{50} = k500d^{0.6}, \tag{11.1}$$

where V_{50} is in kV, the gap length (d) is in meters, and $k = \dfrac{V_{50}}{V_{50} \text{ of rod-plane gap}}$.

Feser [3], however, has proposed the following empirical equation:

$$V_{50} = k[100\sqrt{50d + 1} - 250] . \tag{11.2}$$

A third expression has been proposed which extends the V_{50} *vs.* d relation to longer gap lengths (2 m < d < 15 m) [5]:

$$V_{50} = k \frac{3400}{1 + \dfrac{8}{d}} . \tag{11.3}$$

For d>13 m, the V_{50} *vs.* d characteristics exhibit a fairly linear trend which can be approximated as [6]:

$$V_{50} = k(1400 + 55d), \text{ for } 13 \text{ m} < d < 30 \text{ m}. \tag{11.4}$$

For negative-polarity switching surges, the following empirical equation seems to fit experimental data well for 2 m ≤ d ≤ 15 m [6]:

$$V_{50} = 1180\, d^{0.45}. \tag{11.5}$$

Table 11.1 shows the gap factors (k) for various electrode arrangements [2].

Table 11.1: Gap Factors for Various Types of Air Gaps [2]

Gap Type	Gap Configuration	k
Rod–Plane		1.00
Rod–Structure (under)		1.05
Conductor–Plane		1.15
Conductor–Window		1.20
Conductor–Structure (under)		1.30
Rod–Rod (h=3 m/6 m)	h	1.30/1.40
Conductor–Rod (h=3 m/6 m)	h	1.65/1.90

11.3 Physical Mechanism of Air Breakdown

The breakdown of air gaps under impulse voltages has been extensively investigated, although all aspects of the breakdown phenomenon are not yet completely understood. Excellent reviews of the present knowledge on the breakdown phenomenon are given in [7-10]. Figure 11.4 shows a simplified sketch of the breakdown process of a rod-rod air gap under a switching surge, the upper rod being of positive polarity.

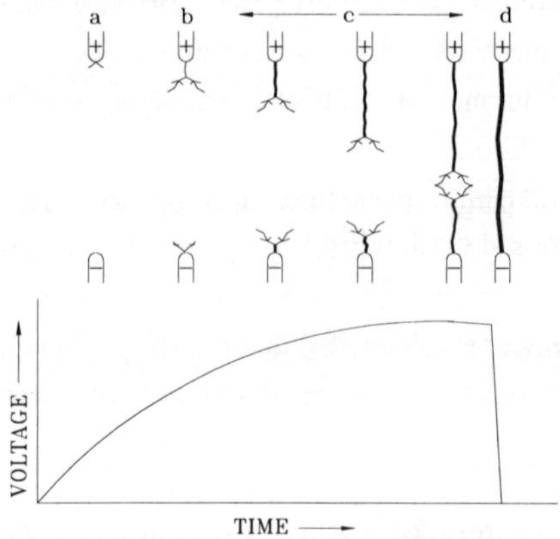

Fig. 11.4 Simplified sketch of the breakdown phenomenon of a rod-rod air gap. (a) avalanche start; (b) streamer formation; (c) leader growth; (d) breakdown.

The atmosphere contains electrons and ions, which are created by natural ionization sources, such as the cosmic rays. The electrons, because of their light weight, are accelerated by the electric field in the gap, thus gaining kinetic energy.The drifting electrons lose their energy to neutral molecules by collision. The level of energy transfer increases as the voltage across the electrodes increases. If the energy transfer is high enough, electrons from the excited molecules may be detached. The newly freed electrons, in turn, take part in the collision process, exciting and ionizing other molecules. Each electron creates new electrons at the rate of α electrons per unit distance travelled. However, during the same distance travelled, electrons are attached to oxygen molecules at the rate of η electrons per unit length. (α-η) represents the net electron multiplication rate.The ionization process is sustained if (α-η) is positive. This condition is reached when the electric field exceeds a critical level (ca. 26 kV/cm), and an electron avalanche is created. A number of electrons, n_o, which are accelerated in the gap at a distance, x, from the anode, would create n electrons in its train on its arrival at the positive electrode:

$$n = n_0 e^{\int_0^x (\alpha - \eta) dx} \quad . \tag{11.6}$$

The electron avalanche leads to complete breakdown of a gap which is stressed by uniformly distributed electric field. However, the electric field is nonuniformly distributed in most practical gaps. In that case, the electron avalanche is initially confined around one or both electrodes. The space charge of the ions creates its own electric field, enhancing and diminishing the original applied electric field throughout the gap. The space-charge field plays a significant role in the breakdown of gaps stressed by nonuniform electric field. If the number of ions, n, in (11.6) exceeds 10^8, the magnitude of the space-charge field is about the same as that of the applied field, and initiates a streamer. The streamer is a highly ionized channel with a voltage gradient of about 5 kV/cm, and propagates at a velocity of about 10 cm/μs. The current flowing through the core of the streamer produces heat, decreasing the gas density and increasing ionization. Eventually, the streamer 'ignites' into a leader which has a very low voltage gradient across it. This means that the positive electrode is practically thrust into the gap, thus increasing the electric field farther in the gap, repeating the cycle of avalanche, streamer and leader. Ultimately, the streamers emanating from the leader tip reach the other electrode, bridging the gap and causing sparkover of the gap. For a rod-plane gap, the streamers and the leaders proceed from the rod toward the plane. But in a rod-rod gap (Fig. 11.4), ionizations start from both electrodes. However, the electrons near the cathode advance into regions of reducing field. Therefore, the ionization near the cathode is less extensive and requires a higher field for streamer propagation. Sparkover in a rod-rod gap occurs when the streamers from both electrodes meet. The voltage at breakdown, V_b, and the gap spacing, d, are given by [10]:

$$V_b = E_\ell^+ \ell_\ell^+ + E_s^+ \ell_s^+ + E_\ell^- \ell_\ell^- + E_s^- \ell_s^-, \text{ and} \tag{11.7}$$

$$d = \ell_\ell^+ + \ell_s^+ + \ell_\ell^- + \ell_s^-, \tag{11.8}$$

where $E_\ell^+, E_\ell^-, \ell_\ell^+, \ell_\ell^-$ are the electric fields and lengths of positive and negative leaders at the instant of breakdown, and the corresponding symbols with subscript s refer to the streamers. For a rod-plane gap, as $\ell_\ell^- = \ell_s^- = 0$ for positive rod, (11.7) and (11.8) can be rewritten as:

$$V_b = E_s^+ d - \ell_\ell^+ (E_s^+ - E_\ell^+). \tag{11.9}$$

Comparing (11.7) and (11.9), it can be seen that the breakdown voltage of a rod-plane gap with positive rod, where no negative corona exists, is lower than that of a rod-rod gap where a significant negative corona exists. In fact, the breakdown voltage of a rod-plane gap with positive rod is the lowest of any gap stressed by nonuniform electric field. The breakdown voltage of the rod-plane gap with positive rod is taken as the reference i.e., gap factor, k=1.

The critical breakdown voltage level of a nonuniform-field gap, discussed in Section 11.2, can be explained from the foregoing description of the breakdown process. More space charge is formed in a gap for a longer front time of the applied impulse voltage. More space charge means higher electric field in the remainder of the gap, and hence lower breakdown voltage. However, more ions are drifted away from the gap for longer front times, which tends to lower the space-charge field.Up to the critical front time, the buildup of the space charge predominates; therefore, the breakdown voltage level diminishes with increasing front times. Above the critical front time, the loss of ions due to drift predominates, increasing the breakdown voltage level with longer front times. For a longer gap, smaller percent of the total ions are lost; therefore, the critical front time for longer gaps increases.

Breakdown of air gaps stressed by negative-polarity impulse voltages has been less studied because it has higher breakdown voltage and also because of its complexity. However, the same main predischarge phases have been identified for both polarities of applied voltages [9].

11.4 Effects of Atmospheric Conditions on Switching-Surge CFO

Atmospheric temperature, pressure and humidity affect the breakdown voltage level of external insulation. Test results under any atmospheric conditions can be converted to that for standard atmospheric conditions [11]. The standard atmospheric condition is defined as [11]:

Temperature, $t_o = 20°C$
Pressure, $p_o = 760$ mmHg
Humidity, $h_o = 11$ g of water vapor per cubic meter

The air density correction factor, k_d is given by

$$k_d = \frac{p}{p_o}\frac{273 + t_o}{273 + t}, \qquad (11.10)$$

where p and t are the atmospheric pressure and temperature under test conditions. The actual test results can be corrected to the standard atmospheric conditions as:

$$V_{50}(\text{standard}) = \frac{V_{50}(\text{test})}{k_d} . \tag{11.11}$$

The effects of humidity on the breakdown voltage of air gaps under impulse voltages have been extensively investigated and reviewed [9,10,12-25]. Essentially, there are three physical processes which are influenced by water vapor [23,24]. (i) Water molecules, being electronegative in nature, attach to electrons, thus removing the electrons from the ionization process. Therefore, an increase in humidity tends to increase the breakdown voltage level. (ii) The high absorption of uv radiation by water molecules reduces the length of the electron avalanche in front of the streamer head. This results in higher breakdown voltage level for higher absolute humidity for positive streamer discharges in air gaps. (iii) The relaxation time of vibration of the nitrogen molecules is reduced by the water vapor, thus supporting the leader inception voltage.

Standards [11,26] provide humidity correction factors which were determined from tests performed by various groups of researchers. For engineering applications, the following were suggested [24]. Only positive streamer predischarges are significantly influenced by humidity. Therefore, no humidity correction should be applied for gaps with uniform electric fields, and for negative rod-plane gaps. For a rod-plane gap under positive-polarity stress, a correction of 1 percent per g/m^3 should be applied.

11.5 Estimation of the Performance of External Insulation under Switching Surges

The high cost of equipment failure and/or power outages requires cost effective as well as reliable design of insulation coordination. The insulation strength must be correlated with the stress applied by the expected switching surges. The magnitude and the waveshape of the switching surges vary at random. However, unlike lightning, the system is the source of the switching surges, i.e., the switching surges are system specific. Field tests, to collect statistical data on the switching surges, require the system to be built first. This is impractical. Therefore, the statistical data on the switching surges for a specific system are collected by extensive computation either by a transient network analyzer or by a digital computer. The flashover strength of an external insulation fluctuates from one surge to the next even under identical test conditions. The flashover strength also varies with the waveshape of the applied surge as well as with the weather conditions.

The simplest method of designing an external insulation is to choose its withstand strength at a specified margin above the maximum amplitude of the expected switching surge in the system. The withstand strength of an external insulation is estimated from laboratory test data. The withstand strength is usually assumed as the critical flashover voltage level (V_{50}) minus three times the standard deviation ($V_{50} - 3\sigma_i$). The expected switching surge level is estimated from either field tests or from analytical studies.

As both the insulation strength and the system switching surge magnitude are statistical in nature, an optimized design can be achieved by a statistical design procedure. Assuming normal (Gaussian) probability distribution, the probability that a surge of magnitude between V and V+ΔV will appear is given by $f_s(V)dV$, where f_s is the probability density function of the expected surge, given by [27-29]:

$$f_s(V) = \frac{1}{\sqrt{2\pi}} e^{-\frac{z_s^2}{2}} , \tag{11.12}$$

$$\text{where} \quad z_s = \frac{V - V_{ms}}{\sigma_s} , \text{ and} \tag{11.13}$$

V_{ms}, σ_s = mean and standard deviation of the system switching surges.The probability, $P_i(V)$, that the insulation will flash over at the voltage, V, is given by:

$$P_i(V) = \frac{1}{\sqrt{2\pi}} \int_{-\infty}^{z_i} e^{\frac{z_i^2}{2}} dz_i , \tag{11.14}$$

$$\text{where} \quad z_i = \frac{V - V_{50}}{\sigma_i} . \tag{11.15}$$

The probability that both events will occur simultaneously, i.e., that the switching surge, V, will be impressed across the insulation and that the insulation will flash over is the product of (11.12) and (11.14). The probability that the insulation will flash over across the entire range of the expected switching surges is then given by the integral of the product [29]:

$$P = \frac{1}{\sqrt{2\pi}} \int_{-\infty}^{z} e^{-\frac{z^2}{2}} dz = 0.5\left[1 + \operatorname{erf}\left(\frac{z}{\sqrt{2}}\right)\right], \tag{11.16}$$

$$\text{where} \quad z = \frac{V_{ms} - V_{50}}{\sqrt{\sigma_s^2 + \sigma_i^2}}. \tag{11.17}$$

The probability of occurrence of either polarity of the switching surge is the same. As the withstand voltage level of an insulator under a positive-polarity switching surge is lower than that under the negative-polarity surge, (11.16) should be divided by 2. Therefore, the probability of a flashover under the positive-polarity surge is:

$$P_1(V) = 0.25\left[1.0 + \operatorname{erf}\left(\frac{z}{\sqrt{2}}\right)\right]. \tag{11.18}$$

The probability of insulation flashover,$P_1(V)$, in (11.18) can be computed by evaluating the error function from mathematical tables. Some representative values of $P_1(V)$ as a function of z are shown in Table 11.2.

$P_1(V)$ in (11.18) is the probability that one insulator string will flash over under a positive-polarity switching surge. If there are n insulators which are exposed to the same switching surge, then the probability of their flashover will be:

$$P_n(V) = 1.0 - \left[1.0 - P_1(V)\right]^n. \tag{11.19}$$

As an example, consider the case of a line where 100 insulator strings are subjected to the same switching surge. If the mean of the expected surge in the system is V_s = 1 MV with a standard deviation of σ_s = 0.06 p.u. (=60.0 kV), and for the insulators, V_{50} = 1.2 MV, and σ_i = 0.05 p.u. (=60 kV), then z = -2.357. From Table 11.2, $P_1(V)$ = 4.6055E-03. The probability of failure of one of the 100 insulator strings will be $P_n(V)$ = 0.3697. Table 11.3 shows how the probability of failure of one of the 100 insulator strings would vary by increasing the V_{50} of the insulator strings.

A simplified statistical method may be used during the preliminary design of the external insulation system [28,29]. In this method, two statistical quantities are defined - the statistical overvoltage, V_s, and the statistical withstand voltage, V_w. The probability of encountering a switching surge in the system to exceed V_s is 2%, and the probability that the apparatus will withstand V_w is 90%. The ratio V_w/V_s is the statistical safety factor.

Table 11.2: Probability, $P_1(V)$, *vs.* z

z	$P_1(V)$	z	$P_1(V)$	z	$P_1(V)$	z	$P_1(V)$
0.0	0.2500	-1.0	7.93E-02	-2.0	1.14E-02	-3.0	6.75E-04
-0.1	0.2301	-1.1	6.78E-02.	-2.1	8.93E-03	-3.1	4.84E-04
-0.2	0.2104	-1.2	5.75E-02	-2.2	6.95E-03	-3.2	3.44E-04
-0.3	0.1910	-1.3	4.84E-02	-2.3	5.36E-03	-3.3	2.42E-04
-0.4	0.1723	-1.4	4.04E-02	-2.4	4.10E-03	-3.4	1.68E-04
-0.5	0.1543	-1.5	3.34E-02	-2.5	3.10E-03	-3.5	1.16E-04
-0.6	0.1371	-1.6	2.74E-02	-2.6	2.33E-03	-3.6	7.96E-05
-0.7	0.1210	-1.7	2.23E-02	-2.7	1.73E-03	-3.7	5.39E-05
-0.8	0.1059	-1.8	1.80E-02	-2.8	1.28E-03	-3.8	3.62E-05
-0.9	9.20E-02	-1.9	1.44E-02	-2.9	9.33E-04	-3.9	2.41E-05

Table 11.3: $P_n(V)$ *vs.* V_{50}

$V_s = 1$ MV, $\sigma_s = 0.06$ p.u., $\sigma_i = 0.05$ p.u.
Number of insulator strings, $n_i = 100$
Number of positive-polarity switching surges per year, $n_s = 100$

V_{50}, MV	Single Insulator (n=1)		100 Insulators (n=100)	
	P_1	$N_f(1)$	P_n	$N_f(100)$
1.2	4.6055E-02	4.6E+01	0.3697	3.70E+02
1.3	1.7384E-04	1.74E-01	1.7232E-02	1.72E+01
1.4	3.5912E-06	3.59E-03	3.5757E-04	3.58E-01
1.5	4.4703E-08	4.47E-05	5.9605E-06	5.96E-03

N_f = Number of insulator flashovers per 10 years

11.6 Performance of Internal Insulation under Switching Surges

Because of its non-self-restoring character, a deterministic approach is taken in the design of the internal insulation of a power apparatus, contrary to the statistical approach taken for the design of the external insulation. A basic switching impulse level (BSL) is assigned to the power apparatus, similar to the BIL for lightning voltages. The power apparatus is required to withstand the assigned BSL during specified switching-surge tests. Lightning arresters are connected across the apparatus to ensure that the applied switching surge is held below the assigned BSL of the apparatus with a margin of safety. The selection of lightning arresters for reliable operation will be discussed in Chapter 15.

11.7 Control of Switching Surges

Unlike lightning, switching surges are generated within the electrical system by the interaction between the switching device and the electrical characteristics of the system. Therefore, it is possible to control the switching-surge severity by the modification of the switching device as well as the system electrical parameters.

Insertion of a resistor either in series or in parallel with the circuit breaker (Fig. 11.5) is the most effective method in reducing the severity of a switching surge. This method is almost universally used. In a closing operation, S_2 is closed first, then S_1 is closed. S_1 is opened first, and then S_2 for an opening operation.

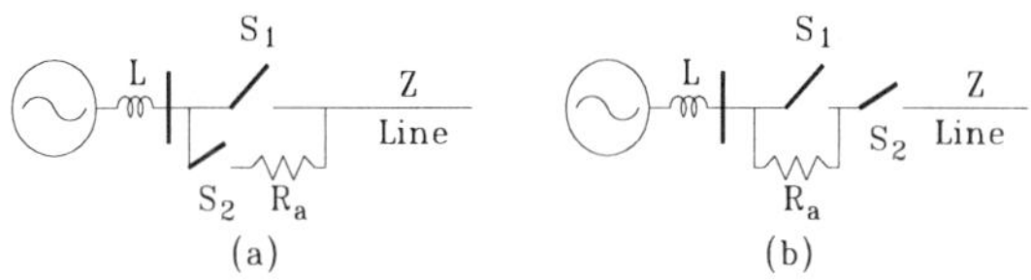

Fig. 11.5 Insertion of resistor during breaker operation.
(a) parallel and (b) series insertion of resistor.

When an open-circuited line is closed by closing S_2 first, the voltage across S_2 is cancelled, and is distributed among L, R_a and the surge impedance of the line, Z. The traveling voltage on the line will thus be lower than that for the case when no resistor is inserted during the switching operation. After the traveling voltage wave returns to

R_a after being reflected at the open end of the line, with S_1 still open, the transmitted voltage at R_a will be attenuated at a faster rate, $(R_a+Z)/L$, than the case, Z/L, with no insertion of the resistor. Therefore, the severity of the closing switching surge will be significantly lower with the insertion of the resistor than that without it. Similarly, for an opening operation, S_1 opens first, thus inserting the resistor, R_a, in the circuit. The injected current across S_1 will be equal but of opposite polarity to the cancelled current. Some of this injected current will be bypassed through the resistor, R_a. Therefore, lower current, hence lower voltage will be transmitted along the line than the case without the resistor. Similar to the closing operation, the voltage reflected from the open end, on arrival at R_a will be attenuated at a faster rate than the case without the insertion resistor.

The insertion of resistor during opening operation also slows down the rate of rise of the recovery voltage (rrrv) across the opening contacts, thus reducing the possibility of a restrike. The transient recovery voltage is capacitively distributed among the individual breaks of a multi-break circuit breaker. Because of stray capacitances, the voltage distribution is not uniform. The voltage distribution can be equalized by connecting discrete capacitors across each break and/or dividing the insertion resistor equally across each individual break.

11.7 References to Chapter 11

1. T. Udo, "Switching Surge and Impulse Sparkover Characteristics of Large Gap Spacings and Long Insulator Strings", ***IEEE Trans. on Power Apparatus and Systems***, Vol. 84, pp. 304-309,1965.

2. L. Paris and R. Cortina, "Switching and Lightning Impulse Discharge Characteristics of Large Air Gaps and Long Insulator Strings", ***IEEE Trans. on Power Apparatus and Systems***, Vol. PAS-87, pp. 947-957, 1968.

3. K. Feser, "The Influence of the Voltage Waveform on the Breakdown of Air Gaps with Inhomogeneous Field", (in German), ***Energie und Technik***, Vol. 22, pp. 319-324, 1970.

4. L. Paris, "Influence of Air Gap Characteristics on Line-to-Ground Switching Surge Strength", ***IEEE Trans. on Power Apparatus and Systems***, Vol. PAS-86, pp. 936-947, 1967.

5. G. Gallet, G. Leroy, R. Lacey and I. Kromer, "General Expression for Positive Switching Impulse Strength Valid up to Extra Long Air Gaps", ***IEEE Trans. on Power Apparatus and Systems***, Vol. PAS-94, pp. 1989-1993, 1975.

6. R. Cortina, E. Gargabnati, A. Pigini, G. Sartorio and L. Thione, "Switching Impulse Strength of Phase-to-Earth UHV External Insulation - Research at the 1000 kV Project", ***IEEE Trans. on Power Apparatus and Systems***, Vol. PAS-104, pp. 3161-3168, 1985.

7. B. Jones and J.N. Ross, "The Physics of the Long Spark", ***CEGB Research*** (U.K.), No. 8, pp. 27-36, 1978.

8. E. Kuffel and W.S. Zaengl, "Electrical Breakdown in Gases, Liquids and Solids", Chapter 5 of ***High Voltage Engineering Fundamentals***. Permagon Press, 1984.

9. CIGRE Committee 33.07 Report, "Guidelines for the Evaluation of the Dielectric Strength of External Insulation", 1992.

10. N.L. Allen, "Fundamental Aspects of Air Breakdown", Chapter 14 in ***High Voltage Engineering and Testing***, edited by H.M. Ryan. Peter Peregrinus Ltd/IEE, Stevenage, Herts, U.K., 1994.

11. ***IEEE Standard Techniques for High Voltage Testing***. IEEE Standard 4-1995, The Institute of Electrical and Electronics Engineers, Piscataway, NJ., 1995.

12. F.D. Fielder, "Humidity Effects on Flashover of High-Voltage Apparatus", ***The Electric Journal***, Vol. 32, pp. 543-547, 1935.

13. T. Harada, Y. Aihara and Y. Aoshima, "Influence of Humidity on Lightning and Switching Impulse Flashover Voltages", ***IEEE Trans. on Power Apparatus and Systems***, Vol. Pas-90, pp. 1433-1442, 1971.

14. J. Kucera, T.W. Liao and A.F. Rohlfs, "Atmospheric Correction Factors for High Voltage Testing", ***Electra***, No. 21, pp. 74-84, 1972.

15. W. Büsch, "Air Humidity, An Important Factor for UHV Design", ***IEEE Trans. on Power Apparatus and Systems***, Vol. PAS-91, pp. 2086-2093, 1972.

16. T.E. Allibone and D. Dring, "Influence of Humidity on the Breakdown of Sphere and Rod Gaps under Impulse Voltages of Short and Long Wavefronts", ***IEE Proc.*** (G.B.), Vol. 122, pp. 235-238, 1975.

17. C.T. Phelps and R.F. Griffiths, "Dependence of Positive Streamer Propagation on Air Pressure and Water Vapour Content", ***Jour. Applied Physics***, Vol. 47, pp. 2929-2934, 1976.

18. N.L. Allen and D. Dring, "Effect of Humidity on the Properties of Corona in a Rod-Plane Gap under Positive Impulse Voltages", ***Proc. Royal Society*** (G.B.), Vol. A396, pp. 285-295, 1984.

19. N.L. Allen, "The Effect of Humidity on Avalanche Growth and Streamer Initiation", ***Jour. Physics***, D Applied Physics, Vol. 18, pp. 47-52, 1985.

20. K. Feser and A. Pigini, "Influence of Atmospheric Conditions on the Dielectric Strength of External Insulation", ***Electra***, No. 112, pp. 83-95, 1987.

21. N.L. Allen and M. Boutlendj, "Study of Electric Fields Required for Streamer Propagation in Humid Air", ***IEE Proc.-A*** (G.B.), Vol. 138, pp. 37-43, 1991.

22. CIGRE Task Force 33.07.03 Report, "Humidity Influence on Non-Uniform Field Breakdown in Air", ***Electra***, No. 134, pp. 62-90, 1991.

23. J. Schmid, "Influence of Absolute Humidity on the Electrical Breakdown in Air", ***European Trans. on Electrical Power Engineering***, Berlin, Germany, Vol. 2, pp. 327-331, 1992.

24. IEEE Task Force 15.09 Report, "Review of Research on Nonstandard Lightning Voltage Waves", ***IEEE Trans. on Power Delivery***, Vol. 9, pp. 1972-1981, 1994.

25. IEEE Task Force 15.09 Report, "Bibliography of Research on Nonstandard lightning Voltage Waves", ***IEEE Trans. on Power Delivery***, Vol. 9, pp. 1982-1990, 1994.

26. ***High-Voltage Test Techniques***. IEC Standard 60-1, Geneva, Switzerland, 1989.

27. IEEE Committee Report, "Guide for Application of Insulators to Withstand Switching Surges", ***IEEE Trans. on Power Apparatus and Systems***, Vol. PAS-94, pp. 58-67, 1975.

28. ***Insulation Co-ordination Part 2: Application Guide.*** IEC Standard 71-2, Geneva, Switzerland, 1976.

29. K.J. Lloyd and L.E. Zaffanella, "Insulation for Switching Surges", Chapter 11, ***Transmission Line Reference Book 345 kV and Above***, EPRI EL-2500, Second Edition Revised. Electric Power Research Institute, Palo Alto, CA., 1987.

CHAPTER 12
PERFORMANCE OF POWER APPARATUS UNDER TRANSIENT VOLTAGES

12.1 Introduction

Transformers, reactors and rotating machines behave in a complex manner when impressed by a fast-rising transient voltage. Sometimes, they fail even when impressed by a transient voltage of level below their BIL rating [1-11]. Therefore, it would be beneficial to understand their behavior under transient voltages.

An extensive critique and bibliography on the surge performance of transformers and rotating machines have been published by Abetti [12,13]. It became clear from the beginning of the 20th century that the response of windings of transformers, coils and rotating machines to transient voltages, such as lightning and switching surges, was quite different from that under power frequencies. It was realized that the voltage distribution along a winding under a steep-front transient was highly nonuniform, and that excessive voltage buildup occurs at the end turns of a winding. From early on, engineers were involved in the design of windings to alleviate stress concentrations along part of the winding. Research was also progressing in the understanding of the natural or resonant frequencies of windings.

It was recognized that the initial voltage distribution along the winding caused by a steep-front transient voltage was due to the (stray) capacitive network of the winding. There was no agreement on the theory of the subsequent voltage oscillations. The research on the subsequent oscillations followed two distinct directions:

1. ***Traveling-wave*** theory was used in the earlier attempts to explain the nonuniform voltage distribution in a winding. According to this theory, the winding behaves like a transmission line. A part of an oncoming traveling voltage wave will penetrate the winding, and travel from the line end to the neutral along the winding turns. According to this traveling-wave theory, most of the transmitted voltage will occur across the first two entrance turns if the front time of the voltage wave is comparable with the travel time of the wave along the turn. Rudenberg [14], refining this traveling-wave theory, proposed that the winding behaves like a waveguide when a transient voltage is applied across it, blocking the passage of all frequencies above a critical frequency. The components of an oncoming voltage wave which have frequencies higher than the critical frequency of the winding will pile up at the entrance of the winding, producing a stress concentration there. The traveling-wave theory ignores the capacitive coupling between winding parts. It also ignores the magnetic coupling between winding parts. These electrostatic and magnetic couplings between winding parts, which are physically close but electrically remote, make the concept of a wave traveling along the winding implausible.

2. According to the ***standing-wave*** theory, the capacitance network of the winding dominates the initial voltage distribution if the winding is impressed with a steep-front voltage wave, and that the inductance network of the winding eventually dominates [15]. The transition from the capacitive voltage distribution to the inductive voltage distribution takes place through damped oscillations. The initial voltage distribution is nonuniform because of the structure of the capacitive network - the series capacitances between turns separated by the capacitances of the winding parts to the ground (i.e., core and tank wall).

Although both theories have ardent advocates, the performance of coils and transformers under transient voltages can be adequately explained by the standing-wave theory. On the other hand, traveling-wave phenomena can be detected on large rotating machines with single-turn slots for the stator.

12.2 Transient Behavior of Transformers and Reactors

The dielectric components of a coil can be electrically stressed in two ways if a transient voltage is impressed across its terminals. First, the voltage distribution along the coil will be highly nonuniform if the transient voltage has a steep front. This will concentrate the electrical stresses at the end turns, causing a dielectric failure between adjacent turns at these locations. Second, the coil may resonate at the frequency of one of its space harmonics if the applied transient voltage contains such a frequency component. Under this condition, an internal point of the coil may attain a voltage peak much higher than the applied voltage. It may cause dielectric failure between parts of the coil, even when the peak amplitude of the applied voltage is below the BIL of the coil, and a surge protector is installed across the coil terminals.

The inductances of the coil parts are dominant in the distribution of voltage across the coil under power-frequency operation. However, the capacitances between various parts of the winding, e.g., turn-to-turn, turn-to-ground capacitances, play a significant role under transient conditions. Such a complex LC-network of a winding is shown in Fig. 12.1, where L is the leakage inductance per unit length of the winding, C_s is the total series capacitance (between turns), C_g is the total capacitance to ground (core), and ℓ is the winding length.

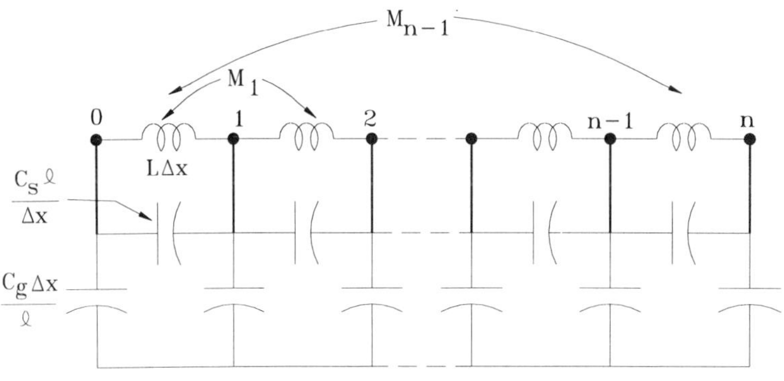

Fig. 12.1 Schematic representation of a winding under transient condition.

12.2.1 Initial voltage distribution

The coil behaves like a capacitive network (Fig. 12.2) initially when impacted by a steep-front voltage. For a unit step-function input voltage, the voltage to ground, V(x), of a point on the coil at an axial distance, x, from the origin can be found by analyzing Fig. 12.2 as follows. The origin (x=0) can be taken either at the line end or the neutral end of the winding.

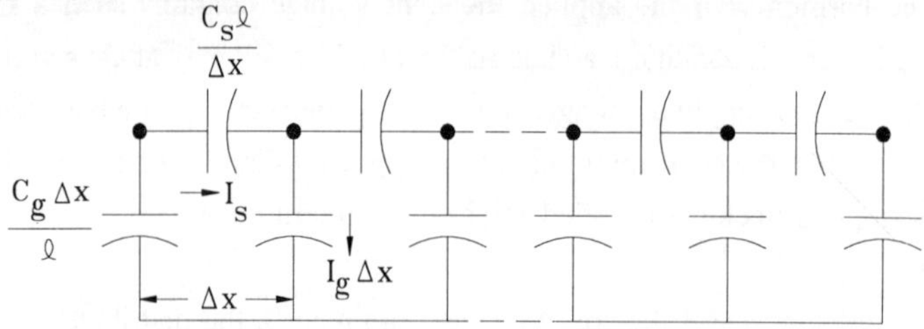

Fig. 12.2 Equivalent circuit of a winding for initial voltage distribution.

$$I_g \Delta x = \Delta I_s = \frac{\partial I_s}{\partial x} \Delta x. \tag{12.1}$$

However, $$I_g \Delta x = \frac{C_g \Delta x}{\ell} \frac{\partial V}{\partial t}, \text{ and} \tag{12.2}$$

$$I_s = \frac{C_s \ell}{\Delta x} \frac{\partial}{\partial t}(\Delta V). \tag{12.3}$$

For $\Delta x \to 0$: $$I_g = \frac{\partial I_s}{\partial x} = \frac{C_g}{\ell} \frac{\partial V}{\partial t}, \text{ and} \tag{12.4}$$

$$I_s = C_s \ell \frac{\partial^2 V}{\partial x \, \partial t}. \tag{12.5}$$

From (12.5): $$\frac{\partial I_s}{\partial x} = C_s \ell \frac{\partial}{\partial t} \left(\frac{\partial^2 V}{\partial x^2} \right). \tag{12.6}$$

Equating (12.4) and (12.6), and integrating with respect to t, one gets:

$$C_s \ell \frac{\partial^2 V}{\partial x^2} - \frac{C_g}{\ell} V = 0 \tag{12.7}$$

The solution of (12.7) is

$$V(x) = A_1 \exp(\gamma x) + A_2 \exp(-\gamma x), \tag{12.8}$$

$$\text{where } \gamma = \frac{1}{\ell}\sqrt{\frac{C_g}{C_s}} = \frac{\alpha}{\ell}. \tag{12.9}$$

The constants A_1 and A_2 can be evaluated from the boundary conditions.

12.2.1.1 Winding neutral grounded

With the origin (x=0) at the line end, V(0) = 1 p.u. and V(ℓ) = 0

$$A_1 = -\frac{\exp(-\alpha)}{2\sinh\alpha}\ ;\ A_2 = \frac{\exp(\alpha)}{2\sinh\alpha}, \text{and} \tag{12.10}$$

$$V(x) = \frac{\sinh\alpha(1 - \frac{x}{\ell})}{\sinh\alpha}. \tag{12.11}$$

The final voltage distribution is a straight line drawn from V(0)=1 on the voltage ordinate at x/ℓ=0 (line end) to V(ℓ)=0 at x/ℓ=1 (neutral end) of the coil. The passage from the initial to the final voltage distribution is accompanied by intermediate oscillations. These oscillations may be assumed to oscillate around the final voltage distribution as the axis and with an amplitude equal to the difference between the final and the initial distributions, just like a simple pendulum. Therefore, the envelope of the maximum voltages along the coil can be drawn by adding the difference between the final and the initial voltage distributions at each point along the coil to the final voltage distribution. This is shown in Fig. 12.3 for two values of α.

If the neutral-end of the coil is chosen as x=0, then V(0)=0 and V(ℓ)=1, and

$$V(x) = \frac{\sinh(\alpha\frac{x}{\ell})}{\sinh\alpha}. \tag{12.12}$$

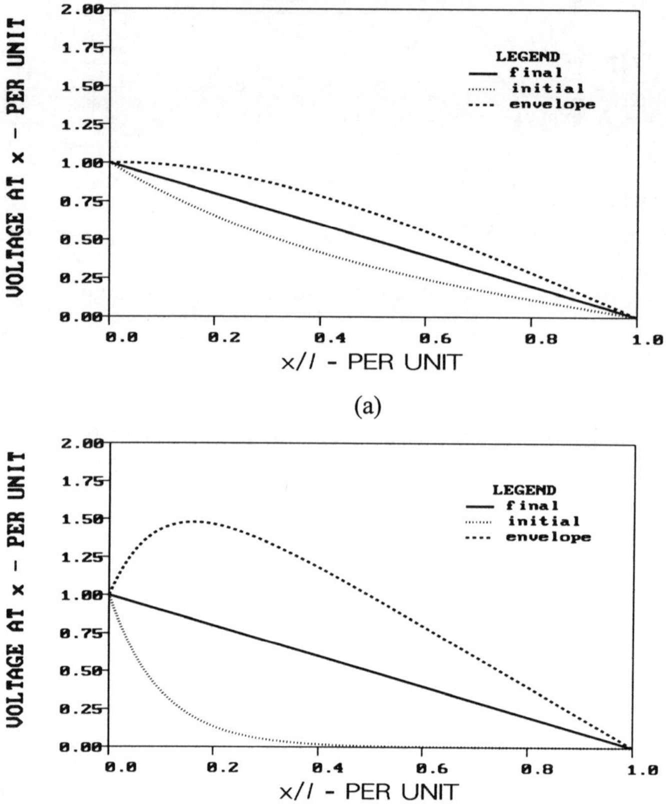

(a)

(b)

Fig. 12.3 Transient voltage distribution in a coil with grounded neutral.
solid line: final; dotted line: initial; dashed line: envelope.
(a) α=2.0; (b) α=10.0.

12.2.1.2 Winding neutral isolated

If the line end is chosen as the origin (x=0), then $I_s(\ell) = 0$ i.e., $\partial V/\partial x = 0$ at $x = \ell$ (neutral end), and $V(0) = 1$. Then:

$$V(x) = \frac{\cosh\alpha(1 - \frac{x}{\ell})}{\cosh\alpha}. \tag{12.13}$$

If the neutral end is chosen as x = 0, then

$$V(x) = \frac{\cosh(\alpha\frac{x}{\ell})}{\cosh\alpha}. \tag{12.14}$$

The transient voltage distribution along a winding with an isolated neutral is shown in Fig. 12.4 for two values of α and the line end at x=0.

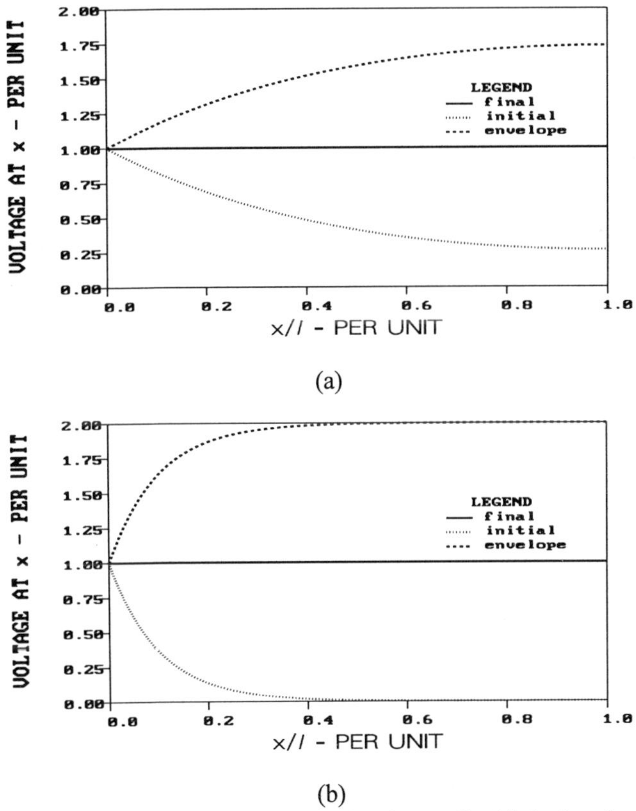

Fig. 12.4 Transient voltage distribution in a coil with isolated neutral.
solid line: final; dotted line: initial; dashed line: envelope.
(a) α=2.0; (b) α=10.0.

12.2.1.3 Transient voltage gradient along winding

From (12.11), ***for the grounded neutral***, the voltage gradient, g, is given by:

$$g = -\frac{\partial V}{\partial x} = \frac{\alpha}{\ell}\frac{\cosh\alpha(1-\frac{x}{\ell})}{\sinh\alpha}. \tag{12.15}$$

The voltage gradient is maximum at x = 0, which is given by:

$$g_{max} = \frac{\alpha}{\ell}\coth\alpha \simeq \frac{\alpha}{\ell} = \alpha\bar{g}, \tag{12.16}$$

where $\bar{g} = \frac{1}{\ell}$ = per-unit uniform voltage gradient.

From (12.13), ***for the isolated neutral***, the maximum voltage gradient (i.e., at the line end) is given by:

$$g_{max} = (\alpha\bar{g})\tanh\alpha \simeq \alpha\bar{g}. \tag{12.17}$$

The oscillations along the winding length form a standing-wave pattern, a pattern of maximum and minimum voltages along the winding. These standing-wave patterns can be broken down into space harmonics. Three of the space harmonics are shown in Fig. 12.5.

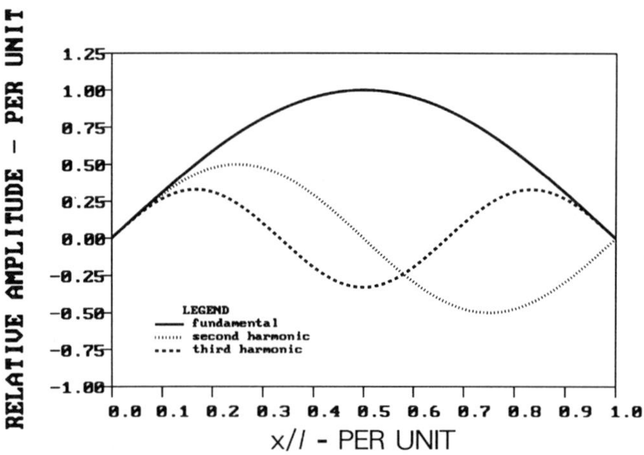

Fig. 12.5 Space harmonics along a winding with grounded neutral.

These space harmonics vary in amplitude in space, i.e., along the winding length; the fundamental is maximum at the midpoint of the winding, the second harmonic is maximum at the 1/4 and 3/4 points of the winding, and so on. Each space harmonic also oscillates in time at a specific frequency. These specific frequencies of the space harmonics are called the natural frequencies of the coil. If the coil is excited by an external voltage source at any of its natural frequencies, then it will resonate internally at that frequency. For instance, if the coil is excited by an external voltage source at the natural frequency of its fundamental space harmonic, then the midpoint of the winding will have the highest voltage. Actual tests on coils have shown that the midpoint voltage of the coil may exceed 20 times the peak of the externally applied voltage if the frequency of the external voltage source coincides with the fundamental natural frequency of the coil [16].

12.2.2 Natural frequencies and transient voltages in windings

The calculation of the natural frequencies of a coil is difficult and controversial. The difficulty arises from the assumptions regarding the inductance and the series capacitance elements. For instance, Wagner [17] calculated the natural frequencies of coils by neglecting the mutual inductances between the coil parts. According to his analysis, a coil would have a critical natural frequency above which no oscillations can be sustained by the coil. According to the analysis by Blume and Boyajian [18], which takes the mutual inductances into account, no such critical frequency exists. As the mutual inductance is an important parameter in the transient analysis of coils, we have used the Blume-Boyajian equation for calculating the natural frequencies of uniform, single-layer coils. This is:

$$\omega_n = \frac{\alpha_n^2}{\sqrt{L c_g \left(1 + \dfrac{\alpha_n^2 c_s}{c_g}\right)}}, \qquad (12.18)$$

where ω_n = n-th natural angular frequency of coil, radians/s
$\alpha_n = n\pi/\ell$, (n=1,2,3....) ; $L = \mu_o (N_c/\ell)^2 (2\pi R_m/h)$;
$\mu_o = 4\pi x 10^{-7}$; N_c = total number of turns in coil
ℓ = axial length of coil ; R_m = mean radius of a turn
$h = b + \ell/3n$, (n=1,2,3....) ; b = radial thickness of coil
c_g = capacitance to ground per unit length of coil, and
c_s = series capacitance per unit length of coil.

The natural frequencies of complex coils, where c_g and c_s are not constant along the coil length, are determined by considering the winding as a lumped ladder network as shown in Fig. 12.1. Transient voltages along the winding can also be found by extending the same analysis. Once the winding has been divided into an n-section ladder network, a convenient way of solving the network is by the state variable method [19,20]. In the general case of circuit analysis, a loop or node equation may be a second-order integro-differential equation. Higher order equations would pose even greater difficulties. The objective of the state space or state variable analysis is to describe an nth-order system by a set of n simultaneous first-order equations. This set of n simultaneous equations is then replaced by a single first-order matrix equation.

The state equations of an n-section ladder network can be written as [21,22]:

$$\frac{dX(t)}{dt} = A_1 X(t) + A_2 V(t), \text{ and} \tag{12.19}$$

$$U(t) = A_3 X(t) + A_4 V(t), \tag{12.20}$$

where $X(t)$ = state vector (2nx1),
$U(t)$ = output vector of node voltages (nx1),
$V(t)$ = applied impulse voltage (a scalar), and
A_1, A_2, A_3 and A_4 = matrices of constant coefficients.

A_1 is a 2nx2n matrix whose eigenvalues are the natural frequencies of the winding. A_2 is a 2nx1 column matrix, A_3 is a nx2n matrix, and A_4 is a nx1 column matrix. The matrices A_1, A_2, A_3 and A_4 are:

$$A_1(2n,2n) = \begin{bmatrix} -C_{n1}^{-1}G_{n1} & I \\ -C_{n1}^{-1}L_{n1}^{-1} & 0 \end{bmatrix}, \tag{12.21}$$

$$A_2(2n,1) = \begin{bmatrix} C_{n1}^{-1}(G_{n1}C_{n1}^{-1}C_m - G_m) \\ \text{-----------------} \\ C_{n1}^{-1}(L_{n1}^{-1}C_{n1}^{-1}C_m - L_m^{-1}) \end{bmatrix}, \tag{12.22}$$

$$A_3(n,2n) = [I \mid 0], \quad \text{and} \tag{12.23}$$

$$A_4(n,1) = -C_{n1}^{-1} C_m . \tag{12.24}$$

For an n-mesh circuit, C_{n1},G_{n1} and L_{n1}^{-1} are the nxn-matrices which remain when the m-th row and the m-th column are removed from C_n, G_n and L_n^{-1}. Node m is where the input voltage is applied. C_m, G_m and L_m^{-1} are the m-th column of C_n, G_n and L_n^{-1} with the m-th row removed, and

$C_n(n+1,n+1)$ = nodal capacitance matrix including the input node,
$G_n(n+1,n+1)$ = nodal conductance matrix including the input node,
$L_n^{-1}(n+1,n+1)$ = inverse nodal matrix of inductances including the input node, and
I and 0 = identity and null matrices, respectively.

The elements of the G_n-matrix are the conductances across the series and ground capacitances of the lumped ladder network. These simulate the dielectric losses in the insulation. They can be highly nonlinear and frequency dependent. Their primary effect is a damping of the oscillations. When the conductance terms are neglected, the expressions for A_1 and A_2 are simplified:

$$A_1(2n,2n) = \left[\begin{array}{c|c} 0 & I \\ \hline -C_{n1}^{-1} L_{n1}^{-1} & 0 \end{array}\right], \quad \text{and} \tag{12.25}$$

$$A_2(2n,1) = \left[\begin{array}{c} 0 \\ \hline C_{n1}^{-1}(L_{n1}^{-1} C_{n1}^{-1} C_m - L_m^{-1}) \end{array}\right]. \tag{12.26}$$

The voltage at each node, U(t), for any type of applied voltage, V(t), can be determined from (12.20), once the state vector X(t) is found from (12.19). Recognizing that (12.19) is a matrix differential equation, it can be solved either by any of the classical methods or by Laplace transform. Assuming a unit step-function applied voltage, the solution of (12.19) is

$$X(t) = \left[\exp(A_1 t) - I\right] A_1^{-1} A_2 \ . \tag{12.27}$$

The exponential function in (12.27) is also a matrix function. It is evaluated in closed form by invoking Sylvester's theorem [23]:

$$\exp(A_1 t) = \sum_{r=1}^{m} \exp(\alpha_r t) \prod_{s=1,s\neq r}^{m} \frac{\alpha_s I - A_1}{\alpha_s - \alpha_r}, \tag{12.28}$$

where $\alpha_r(r=1,m)$ are m distinct eigenvalues of A_1, which were already evaluated to determine the natural frequencies of the winding.

The node voltages for any applied transient voltage, other than the unit step-function voltage, can be found by applying Duhamel's theorem (convolution integral) to (12.27).

12.2.2.1 An example

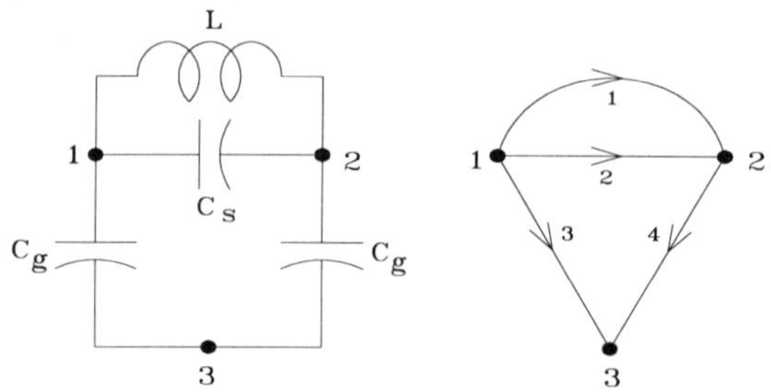

Fig. 12.6 Analysis of a one-mesh circuit.

The incidence matrix is given by:

$$[A]_{2,4} = \begin{bmatrix} 1| & 1 & 1 & 0 \\ -1| & -1 & 0 & 1 \end{bmatrix} = \left[A_L | A_C\right]$$

$$\left[A_L\right] = \begin{bmatrix} 1 \\ -1 \end{bmatrix} ; \left[A_L^t\right] = [1 \ \ -1]$$

$$[A_C] = \begin{bmatrix} 1| & 1 & 0 \\ -1| & 0 & 1 \end{bmatrix} = [A_L \mid I]; \; [A_C^t] = \begin{bmatrix} 1 & -1 \\ 1 & 0 \\ 0 & 1 \end{bmatrix} = \begin{bmatrix} A_L^t \\ ---- \\ I \end{bmatrix}$$

$$[\Gamma_n] = [A_L][\Gamma_b][A_L^t] = \frac{1}{L}\begin{bmatrix} 1 & -1 \\ -1 & 1 \end{bmatrix}$$

$$[\Gamma_{n1}] = \frac{1}{L}$$

$$[C_n] = [A_C]\begin{bmatrix} C_s & 0 & 0 \\ 0 & C_g & 0 \\ 0 & 0 & C_g \end{bmatrix}[A_C^t] = \begin{bmatrix} (C_s+C_g) & -C_s \\ -C_s & (C_s+C_g) \end{bmatrix}$$

$$[C_{n1}] = (C_s+C_g) \text{ , and } [C_{n1}^{-1}] = \frac{1}{C_s+C_g}$$

$$[C_{n1}^{-1}][\Gamma_{n1}] = \frac{1}{(C_s+C_g)L}$$

$$[A_1]_{2,2} = \begin{bmatrix} 0 & 1 \\ \dfrac{-1}{(C_s+C_g)L} & 0 \end{bmatrix}$$

The eigenvalues of A_1 are found from:

$$\begin{vmatrix} -\lambda & 1 \\ \dfrac{-1}{(C_s+C_g)L} & -\lambda \end{vmatrix} = 0$$

$$\lambda = \pm \frac{j}{\sqrt{(C_s+C_g)L}}$$

To compute A_2, A_3 and A_4: $L_m = -L$; $\Gamma_m = -1/L$; $C_m = -C_s$

From (12.22),

$$[A_2]_{2,1} = \begin{bmatrix} 0 \\ \dfrac{1}{(C_s+C_g)L}\left(1 - \dfrac{C_s}{(C_s+C_g)}\right) \end{bmatrix}$$

$$[A_3]_{1,2} = [1 \quad 0]$$

$$[A_4]_{1,1} = \frac{C_s}{C_s+C_g}$$

12.2.3 Modeling of transformer as a system component

As was stated before, a winding behaves initially like a capacitance. Its long-term behavior is that of an inductance. It is a complex network between these two stages.

We shall estimate the value of this equivalent capacitance for the initial stage. If V_o is the voltage applied at the line-end of the transformer, and I_o is the line-end current, then the equivalent capacitance, C_e, of the transformer is given by:

$$I_o = C_e \frac{\partial V}{\partial t}. \qquad (12.29)$$

From (12.5), the initial line-end current is also given by:

$$I_o = C_s \ell \frac{\partial}{\partial t}\left(\frac{\partial V}{\partial x}\right). \qquad (12.30)$$

Equating (12.29) and (12.30), and integrating with respect to time,

$$C_e V = C_s \ell \frac{\partial V}{\partial x}. \qquad (12.31)$$

For ***grounded neutral***, from (12.11), at x=0:

$$\frac{\partial V}{\partial x} = -\frac{\alpha}{\ell} V \coth\alpha \approx -\frac{\alpha}{\ell} V. \qquad (12.32)$$

From (12.31) and (12.32):

$$C_e = C_s \alpha = \sqrt{C_g C_s}. \qquad (12.33)$$

It can be similarly shown that the equivalent capacitance of the winding, C_e, for the case of ***isolated neutral*** is also given by (12.33). The terminal capacitance to ground must be added to (12.33) to obtain the total equivalent capacitance of the winding. Most of this terminal capacitance comes from the capacitance of the terminal bushing to ground. In fact, this terminal capacitance may be significantly larger than that obtained from (12.33).

The charging time of C_e can be estimated by assuming that the charging time depends upon the magnitude of C_e, as well as the surge impedance of the line to which it is connected. If V_f is the oncoming step-function voltage wave, V_t is the voltage transmitted to the transformer, and Z is the surge impedance of the connected line, then

$$V_t = 2V_f\left[1 - \exp\left(-\frac{t}{ZC_e}\right)\right]. \qquad (12.34)$$

For a given winding [22], if C_g=900 pF, and C_s=9 pF, then C_e=90 pF. If the surge impedance of the connected line is Z=500 Ω, then the time constant of (12.34) is 45 ns. It will take 0.1036 μs to charge this capacitor to 90 percent of its final voltage. This is indeed a very small time. It should, however, be borne in mind that the additional terminal capacitance would make the actual charging time much larger.

The impedance of a winding to an external transient voltage during the intermediate time range has not been rigorously analyzed because of its complexity. Surge impedance as a function of the harmonic voltages has been proposed [24,25]. However, its applications are not practical nor are they accurate. The measured values show that the surge impedance of a winding is on the order of a few thousand ohms. Therefore, for all practical purposes, the winding termination can be considered to be an open circuit.

12.2.4 Computation of winding parameters

It should be obvious that the larger the number of meshes in the equivalent circuit (Fig. 12.1), the greater will be the accuracy of results. It should also be pointed out that increasing the number of meshes increases the accuracy of estimating the higher order space harmonics. The first three space harmonics dominate the transient voltage distribution in a coil. Therefore, increasing the number of meshes arbitrarily to large numbers is not advisable. The practical approach is to divide a winding into a convenient number of meshes, and compute the natural frequencies. Then, increase the number of meshes by convenient steps, recalculate the natural frequencies and compare the results until the differences are within acceptable limits. For instance, a 100-turn winding can be initially divided into five meshes, each with 20 turns. The next iteration should be with 10 meshes with 10 turns each, etc..

12.2.4.1 Winding inductances

For the calculation of the inductance of an iron-cored winding, it is generally assumed that the magnetic flux is confined in the air space, because the time required for the magnetic flux to diffuse into the ferromagnetic material is large compared to the duration of the transient. However, it must be cautioned that several studies have found that the ferromagnetic core does affect the inductance even under transient conditions, and that this influence of the ferromagnetic core is frequency dependent, i.e., the inductance is dependent on the waveshape of the applied voltage. Many papers have been published on the methods of calculating the inductance of a coil [26-31]. The method described in [27] has proven to be adequate.

12.2.4.2 Winding capacitances

The basic capacitances, in general, can be computed by assuming either an equivalent parallel-plate, a cylindrical-symmetry configuration, or more accurately by methods proposed in [32]. For parallel-plate equivalence:

$$C = \frac{kA}{36\pi d} 10^{-9} \quad \text{(Farad)}, \tag{12.35}$$

where k = dielectric constant of the medium between the two plates,
A = area of the plates, m^2
d = distance between the plates, m.

For electrodes with cylindrical symmetry, the capacitance is given by:

$$C = \frac{k}{18 \ln \frac{r_2}{r_1}} 10^{-9} \quad \text{(Farad/meter)}, \tag{12.36}$$

where r_2 and r_1 are the radii of the outer and the inner electrodes, respectively.

In many practical situations, the space between the two electrodes is filled with mixed dielectrics. These mixed dielectrics could be solid dielectrics of different dielectric constants, or solid and liquid, such as in an oil-filled transformer, or solid and gas, such as in a gas-filled transformer. In that case, the capacitance between two parallel electrodes with two dielectrics is:

$$C = \frac{k_1 A}{36\pi (d_1 + \frac{k_1}{k_2} d_2)} 10^{-9} = \frac{k_1 A}{36\pi d_1'} 10^{-9}, \tag{12.37}$$

where d_1 and d_2 are the thicknesses of the two dielectrics of dielectric constants k_1 and k_2. d_1' is the equivalent spacing filled with the dielectric k_1 which would give the same capacitance. If the space is filled with n different dielectric materials, then the equivalent space filled with the dielectric k_1 is:

$$d_1' = d_1 + \frac{k_1}{k_2} d_2 + \frac{k_1}{k_3} d_3 + \cdots + \frac{k_1}{k_n} d_n . \tag{12.38}$$

Similar relations hold for the capacitance of cylindrical symmetry with mixed dielectrics.

For calculating the capacitances to ground, the total ground capacitance is calculated, and then it is divided by the number of meshes to get the ground capacitance per mesh. However, the computation of the series capacitance is not that simple because of its complex network structure. This will be discussed next.

12.2.4.3 Computation of the series capacitance

Many researchers have proposed methods of computing the equivalent series capacitance of a winding [33-41]. As each method is quite different from the others, the computation of the natural frequencies of windings becomes uncertain.

A method proposed by Chowdhuri is discussed here [41]. The capacitive network for computing the equivalent series capacitance of a winding is shown in Fig. 12.7. The first section of Fig. 12.7a is shown in Fig. 12.8 to facilitate mathematical analysis.

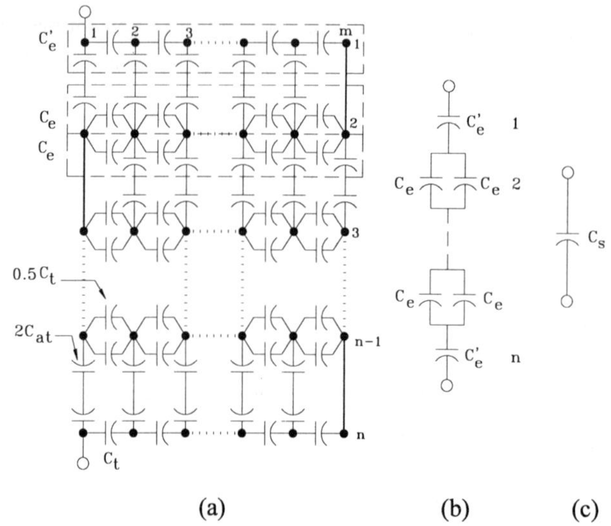

Fig. 12.7 Series-capacitance network of a winding.
(a) Complete network; (b) Reduced string; (c) Equivalent series capacitance.
C_{at}=capacitance between turns of adjacent coils; C_t=inter-turn capacitance.
within a coil; m=number of turns in a coil; n=number of coils in a winding.

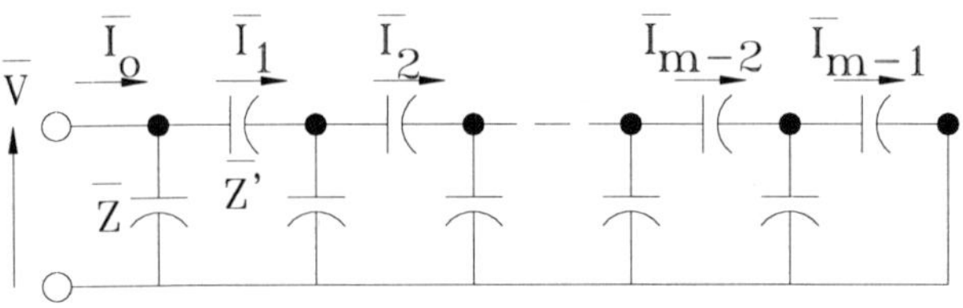

Fig. 12.8 Ladder network to calculate C_e' or C_e of Fig. 12.7.
$Z=1/2sC_{at}$; s=Laplace operator
For C_e': $Z'=1/sC_t$. For C_e: $Z'=2/sC_t$.

Applying a step-function voltage, V, at the input of the network in Fig. 12.8, and using Laplace transform,

$$\bar{V} = \bar{Z}\bar{I}_o - \bar{Z}\bar{I}_1, \tag{12.39}$$

$$0 = -\bar{Z}\bar{I}_o + (2\bar{Z} + \bar{Z}')\bar{I}_1 - \bar{Z}\bar{I}_2,$$

. . . .

. . . .

$$0 = -\bar{Z}\bar{I}_{m-3} + (2\bar{Z} + \bar{Z}')\bar{I}_{m-2} - \bar{Z}\bar{I}_{m-1}, \quad \text{and} \tag{12.40}$$

$$0 = -\bar{Z}\bar{I}_{m-2} + (\bar{Z} + \bar{Z}')\bar{I}_{m-1}. \tag{12.41}$$

Except for the two end sections, the Laplace transform of currents in the neighboring sections are connected by the difference equation,

$$-\bar{Z}\bar{I}_r + (2\bar{Z} + \bar{Z}')\bar{I}_{r+1} - \bar{Z}\bar{I}_{r+2} = 0. \tag{12.42}$$

The solution of (12.42) is obtained by assuming [40]

$$\bar{I} \propto \mu^r. \tag{12.43}$$

Substituting for I_r in (12.42) from (12.43), and using the boundary conditions prescribed in (12.39) and (12.41), a general equation for I_r can be derived. I_o is then found by putting r = 0 in the equation for I_r. Then,

$$\text{Input impedance, } \bar{Z}_e = \frac{1}{C_e s} = \frac{\bar{V}}{\bar{I}_o}. \tag{12.44}$$

After algebraic manipulation, it can be shown that, in Fig. 12.7b,

$$C_e' = C_{at}[1 + \coth(m - 1)\theta' \coth\frac{\theta'}{2}], \tag{12.45}$$

$$C_e = C_{at}[1 + \coth(m - 1)\theta \coth\frac{\theta}{2}], \tag{12.46}$$

$$\theta' = \text{arccosh}(1 + \frac{C_{at}}{C_t}), \text{ and} \tag{12.47}$$

$$\theta = \text{arccosh}(1 + \frac{2C_{at}}{C_t}), \tag{12.48}$$

where C_{at} = capacitance between turns of adjacent coils,
C_t = inter-turn capacitance within a coil,
m = number of turns in a coil, and
n = number of coils.

For shielded windings, C_e' does not exist [38], and the series-capacitance network of Fig. 12.7 will consist of n series-connected branches, each branch comprising two parallel-connected capacitances each of value C_e.

The total equivalent series capacitance of an unshielded n-coil winding will then be given by:

$$C_s = \frac{2C_e C_e'}{4C_e + (n - 2)C_e'}. \tag{12.49}$$

For a shielded winding,

$$C_s = \frac{2C_e}{n}. \tag{12.50}$$

12.2.5 Surge transfer through a transformer [24,25,43-52]

When a step-function voltage is applied to the high-voltage winding of a transformer, the voltage which appears across the low-voltage winding is composed of four voltage components, as shown in Fig. 12.9 [43]:

1. An electrostatic voltage component is transferred to the low-voltage winding. Its amplitude depends on the ratio of the capacitance between the two windings and that to ground.

2. The space harmonics in the high-voltage winding induce corresponding oscillations in the low-voltage winding. This induction process is both electrostatic and magnetic, and is dependent on the distributed constants and the turns ratio of the two windings.

3. A free oscillation is also generated in the low-voltage winding, the magnitude of which depends upon the distributed constants of the low-voltage winding.

4. A voltage which exponentially rises to peak and subsequently decays exponentially is generated in the low-voltage winding by magnetic induction. This voltage component is directly proportional to the turns ratio, and is a simple function of the short-circuit inductance of the transformer and the surge impedances of the external circuits.

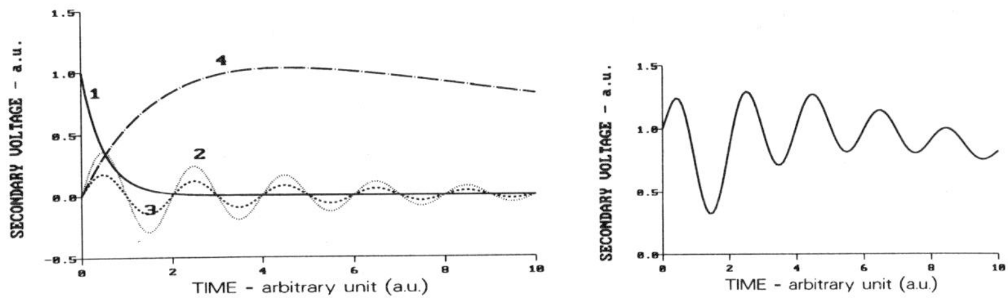

(a) (b)

Fig. 12.9 Secondary terminal voltage and its four components. [43]
(a) component voltages; (b) total voltage.

Rotating machines connected to the low-voltage winding of a transformer have occasionally been damaged by lightning surges traveling along the transmission line which is connected to the high-voltage side of the transformer. For this reason, most of the work has been devoted to unit-connected generators and transformers. Because of the complexity of the coupled-circuit network, data were collected either by actual tests or by analog computers. The difficulty of estimating the magnitude of the transferred surge is enhanced by the uncertainty in computing the inductance and capacitance elements of the coupled network.

The electrostatic component is of very short duration (fraction of a microsecond). Moreover, the electrostatic component and the intermediate oscillations are highly damped out if the secondary winding is connected to a line or a load. Neglecting the electrostatic component and the intermediate oscillations, the surge transfer to the secondary terminals can be calculated easily with reference to Fig. 12.10.

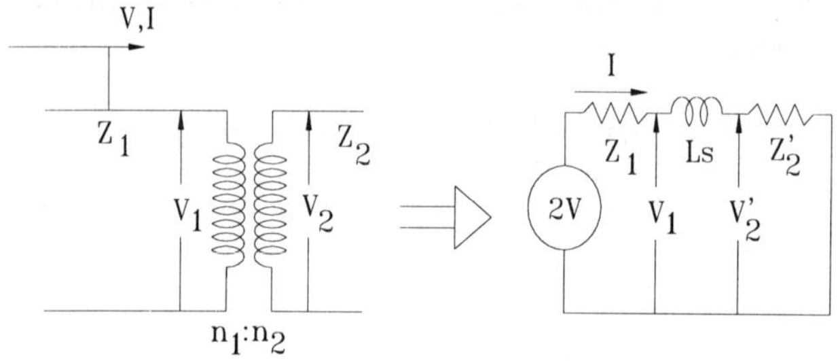

Fig. 12.10 Surge transfer to the secondary side for a single-phase transformer.

If a traveling voltage wave, V, travels along the line on the primary side of the transformer, then the voltage equations will be:

$$2\bar{V} = Z_1\bar{I} + \bar{I}Ls + Z_2'\bar{I}, \tag{12.51}$$

$$\bar{I} = \frac{2\bar{V}}{Z_1 + Ls + Z_2'}, \tag{12.52}$$

$$\bar{V}_2' = \bar{I} Z_2' = \frac{2\bar{V}}{Z_1 + Ls + Z_2'} Z_2'. \quad (12.53)$$

$$\bar{V}_2 = \frac{2\bar{V}}{n} \frac{Z_2'}{Z_1 + Ls + Z_2'}. \quad (12.54)$$

In time domain:

$$V_2(t) = \frac{2V}{n} \frac{Z_2'}{Z_1 + Z_2'} \left[1 - \exp\left(- \frac{Z_1 + Z_2'}{L} t \right) \right]. \quad (12.55)$$

Here, n = transformer turns ratio (n_1:n_2), and $Z'_2 = n^2 Z_2$.

12.2.6 Surge transfer through a three-phase transformer

12.2.6.1 Grounded Y to grounded Y

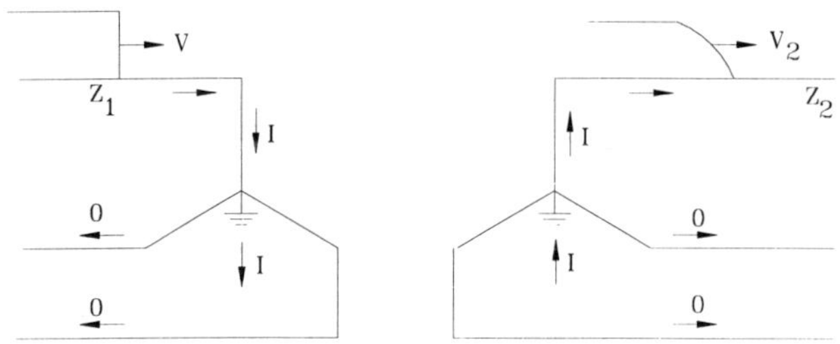

Fig. 12.11 Surge transfer through grounded Y to grounded Y transformer.

$$2\bar{V} = \bar{I}(Z_1 + Ls + Z_2'), \quad (12.56)$$

$$\bar{V}_2' = Z_2' \bar{I}, \quad \text{and} \quad (12.57)$$

$$V_2(t) = \frac{2V}{n} \frac{Z_2'}{Z_1 + Z_2'} \left[1 - \exp\left(- \frac{Z_1 + Z_2'}{L} t \right) \right]. \quad (12.58)$$

12.2.6.2 Grounded Y to isolated Y

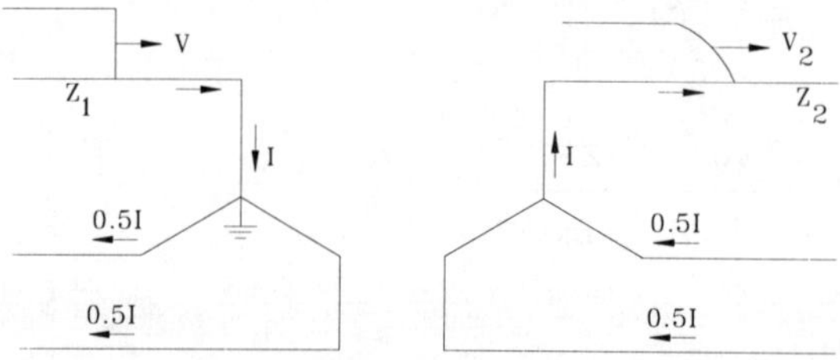

Fig. 12.12 Surge transfer through grounded Y to isolated Y transformer.

$$2\bar{V} = \bar{I}(1.5Z_1 + 1.5Ls + 1.5Z_2'), \text{ and} \tag{12.59}$$

$$V_2(t) = \frac{2V}{n}\frac{2}{3}\frac{Z_2'}{Z_1 + Z_2'}\left[1 - \exp\left(-\frac{Z_1 + Z_2'}{L}t\right)\right]. \tag{12.60}$$

12.2.6.3 Isolated Y to delta

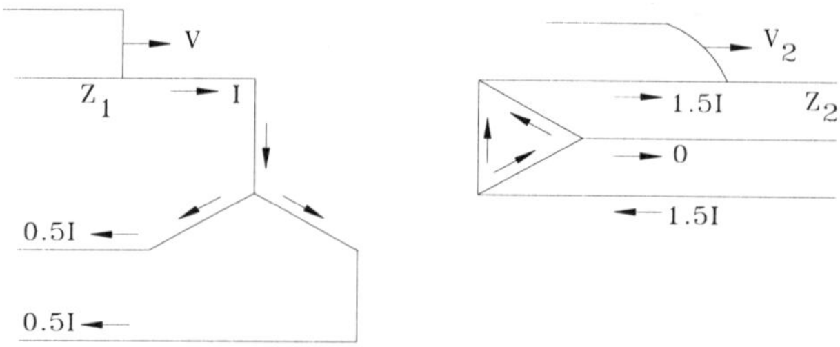

Fig. 12.13 Surge transfer through isolated Y to delta.

$$2\bar{V} = \bar{I}(1.5Z_1 + 1.5Ls + 2\times1.5\times1.5^2 Z_2'), \tag{12.61}$$

$$V_2(t) = \frac{2V}{n}\frac{Z_2'}{Z_1 + 4.5Z_2'}\left[1 - \exp\left(-\frac{Z_1 + 4.5Z_2'}{L}t\right)\right]. \tag{12.62}$$

12.2.6.4 Grounded Y to delta

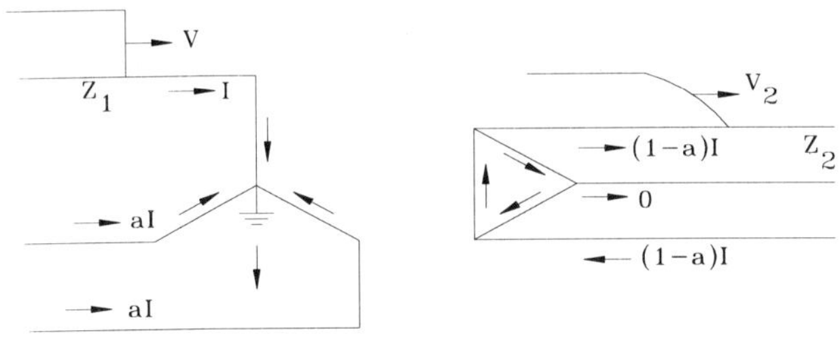

Fig. 12.14 Surge transfer through grounded Y to delta.

$$2\bar{V} = \bar{I}(Z_1 + Ls - aZ_1 - asL + (1 - a)2Z_2'), \tag{12.63}$$

$$V_2(t) = \frac{2V}{n}\frac{Z_2'}{Z_1 + 2Z_2'}\left[1 - \exp\left(-\frac{Z_1 + 2Z_2'}{L}t\right)\right]. \tag{12.64}$$

12.2.7 Protection of transformers against transient voltages

Static plates at the line end of a transformer have been used to effectively increase the series capacitances of the end turns; this tends to equalize the voltage stress concentration at the end turns. Static or graded shields have also been used to buck the current flowing through the ground capacitances of the winding. Both methods reduce α, and tend to equalize the transient voltage distribution in coils. These design innovations also reduce the amplitudes of oscillations of the space harmonics. However, if the applied voltage is oscillatory and contains a frequency which coincides with one of the natural frequencies, especially the fundamental, then dangerous oscillations can still be generated within the winding. This has caused field failures of transformers [1-3,5-10]. Switching surges are generally oscillatory, and the fundamental natural frequencies of high-voltage transformers seem to coincide with the oscillations of some of the switching surges. In recent years, most high-voltage transformer failures have been due to internal oscillations caused by switching surges. The fundamental natural frequency of a transformer can be roughly estimated by the following equation [1]:

$$f = 325\frac{MVA^{0.275}}{kV^{0.95}} \quad (kHz), \tag{12.65}$$

where MVA and kV are the power and voltage ratings of the transformer. Failure of high-voltage transformers from internal resonances caused by oscillatory lightning voltages have also been reported [10]. The frequency of oscillations of lightning voltage is much higher than that of a switching surge, except those generated by vacuum or gas insulated switchgear. According to (12.65), the distribution transformers would have higher natural frequencies than those of high-voltage transformers. Therefore, distribution transformers are more susceptible to internal resonances when impacted by a lightning voltage.

Internal oscillations in transformers caused by switching surges can be minimized by reconfiguring the switching scheme so that the switching-surge frequency does not coincide with the natural frequency of the transformer. However, lightning-caused internal oscillations cannot be prevented by this remedy. Installation of lightning arresters at the transformer terminal will not prevent the internal resonance, because the voltages at the internal points can be many times higher than the terminal voltage limited by the arrester. One possible solution is to install nonlinear resistor blocks inside the transformer tank connecting the 1/4, 1/2 and 3/4 points on the winding to the tank. In the past, nonlinear resistors have been installed inside the transformer tank at transition points, such as tap changers.

12.3 Transient Behavior of Rotating Machines

Small and large rotating machines behave differently under transient voltages. Small machines, which have many turns or coils in the same stator slot, will have high series capacitances between turns and coils as well as high mutual inductances. Therefore, these machines would behave like transformers under transient conditions. Larger machines, which have fewer and longer turns per slot, will have smaller series capacitances; the mutual inductances will be smaller as well. In that case, these machines behave like transmission lines.

The traveling-wave theory on transmission lines is well known. Rudenberg's traveling-wave theory of coils [14] was, therefore, the natural extension of its predecessor. According to this theory, the voltage to ground of any point inside the winding is:

$$V(x,t) = A\, e^{j\omega(t - \frac{x}{v})} . \tag{12.66}$$

This is a traveling wave oscillating in time with angular frequency ω, and traveling along the winding with a velocity v given by:

$$v = \sqrt{\frac{\ell}{LC_g} - \frac{C_s \omega^2}{C_g}}, \tag{12.67}$$

where C_s = total series capacitance, C_g = total capacitance to ground, and L = total winding inductance. This velocity becomes zero when

$$\frac{\ell}{LC_g} = \frac{C_s}{C_g}\omega_c^2, \text{ or } \omega_c = \sqrt{\frac{\ell}{LC_s}} . \tag{12.68}$$

This frequency, ω_c, at which the velocity of propagation is zero, is called the critical frequency; at higher frequencies, the velocity of propagation becomes imaginary. Therefore, the voltage cannot penetrate into the winding at such frequencies. An oncoming transient voltage can be represented by a series of sinusoidal waves of different frequencies by Fourier's theorem. The concept of the critical frequency can therefore be interpreted by saying that the oncoming voltage wave splits into two parts upon arrival at the terminal of the coil: one part comprised of all the component sinusoidal waves with angular frequencies from ω=0 to $\omega=\omega_c$ penetrates through the winding; the second part comprised of the components with angular frequencies higher than ω_c cannot penetrate through the winding but establishes an exponentially decaying standing wave along the winding. This exponentially decaying standing wave is the same as the initial distribution according to the standing-wave theory. The traveling-wave part of the voltage wave travels back and forth inside the winding according to the transmission-line traveling-wave theory. But the analysis of traveling waves becomes more complicated than that for the transmission line because of the different velocities of propagation of the different harmonic waves. The solution of the traveling wave analysis should then be superimposed on the solution of the decaying exponential waves to get the complete solution.

The usual method of finding the surge impedance of a machine is to connect a resistor from the neutral to ground, and pulse the winding at the line end. The value of the neutral resistor which does not produce any reflection is the surge impedance. However, it has been found that the initial response of a generator winding to a fast-front transient produces an impedance which varies with time [44]. Initially, the magnetic flux is confined in the air space inside the slot. Therefore, the inductance is small. Consequently, the surge impedance is small. With time, as the magnetic flux penetrates the ferromagnetic material, the inductance increases and, consequently, the surge impedance increases. Measured surge impedances of machines, ranging in MVA rating from 35 to 125 MVA, varied from about 165 to 85 ohms, respectively [42]. It was reported that high-frequency (1 kHz) measurements of machine inductance and capacitance would lead to fairly accurate values of the surge impedance ($Z=\sqrt{(L/C)}$). The velocity of propagation of the surge in the slot portion of the winding was between 13.7 and 15.2 m per microsecond for all machines [44]. If a fast-front transient voltage is applied across the terminals of a rotating machine, most of the wavefront will reside across the first few turns of the machine, thus overstressing these turns. The wavefront slopes off and the amplitude is attenuated as the wave penetrates along the winding, because of the eddy currents generated in the slots surrounding the coils.

12.4 Test Techniques

12.4.1 Impulse testing of transformers

Power transformers are usually tested with the standard 1.2/50-μs lightning test voltage at their full BIL (basic impulse insulation level), specified by the industry standards [53,54]. For single-phase transformers, the impulse voltage is applied across the two terminals of one winding while both the terminals of the other winding are grounded either directly or through resistors. This practice protects the non-impulsed winding from undue and unexpected overstresses. For three-phase transformers, the connections to the transformer are numerous. If the impulsed winding is Y-connected, and the neutral is grounded, then the impulse voltage is applied between the line terminal of the winding and the ground. The other two terminals of the Y-connected winding are shorted to ground either directly or through a resistor. All the terminals of other windings in the transformer are similarly grounded (Fig. 12.15). The test configuration for isolated neutral is shown in Fig. 12.16. The connections for impulsing a delta-connected winding are shown in Fig. 12.17.

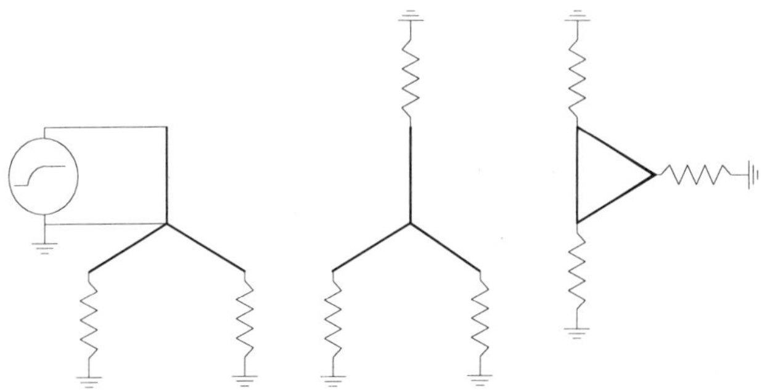

Fig. 12.15 Test configuration for impulse testing of Y-connected winding with grounded neutral.

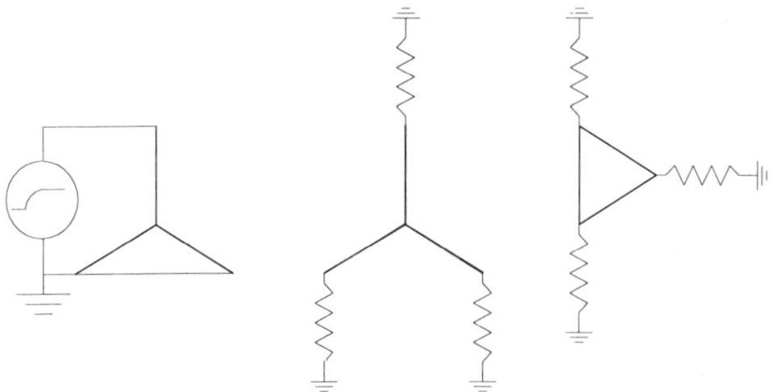

Fig. 12.16 Test configuration for impulse testing of Y-connected winding with isolated neutral.

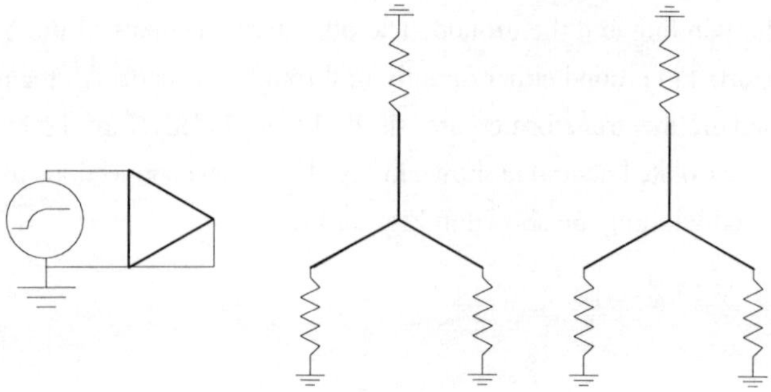

Fig. 12.17 Test configuration for impulse testing of delta-connected winding.

The objective of this test is to determine if the transformer can withstand the specified BIL - both full wave (FW) and chopped wave (CW). The waveshape of the applied voltage is universally monitored. The winding is initially impulsed at reduced voltage levels (50 and 80 percents of full BIL). Any change in the voltage waveshape will indicate a possible fault in the winding. It is difficult to detect a turn-to-turn fault across a few turns of a winding which contains many turns by monitoring the change in the waveshape of the applied voltage. A more sensitive method of fault detection is to monitor the ground or neutral current [55].

Three main components can be distinguished in the ground current [54]:

1. The capacitive component that provides the charging current of the series capacitance of the winding.

2. The period of small oscillations, caused by mutual inductance and capacitance between turns or coils in a winding.

3. The inductive component that flows through the inductance of the winding.

The relative importance of these three components varies widely, depending upon the type of the winding tested.

The type of current transducer to monitor the ground current depends upon what component of the ground current is considered to be the most sensitive for fault detection. A small non-inductive resistor-type transducer is most usually employed. However, this type of transducer is not suitable for detection of turn-to-turn faults in a winding which has many turns (i.e., high impedance). In such cases, a capacitive transducer (a capacitor shunted by a bleeder resistor) is highly sensitive.

12.4.2 Measurement of natural frequencies of coils

There are at least two methods by which the resonant (natural) frequencies of a coil can be experimentally determined [56]:

1. A cathode-ray oscillogram of free oscillations in a winding excited by a steep-front voltage wave generated by an impulse generator.

2. Sustained oscillations excited from a variable-frequency voltage source.

To determine the natural frequencies with an impulse generator, the coil is impulsed with a steep-front transient voltage. The mid-point voltage of the coil to the ground, monitored on an oscilloscope, will predominantly contain the fundamental space harmonic. The voltage across the 25 percent and the 75 percent points will contain the second harmonic. It will not be possible to determine the higher harmonics.

To determine the natural frequencies with the second method, a low-voltage (5 to 10 volts) signal generator is connected across the coil. The mid-point voltage is monitored either with an oscilloscope or a high-frequency voltmeter. The frequency of the signal generator is uniformly increased while the magnitude of the applied voltage is kept constant. The mid-point voltage will increase with the increasing frequency of the source until a critical frequency is reached. The mid-point voltage will decrease past the critical frequency. This critical frequency is the first natural frequency of the coil. To determine the second harmonic, the voltmeter or the oscilloscope is connected across the 25- and 75-percent points of the coil, and the frequency of the signal generator is varied until a maximum voltage is read across these two points. This is the second harmonic (Fig. 12.18a).

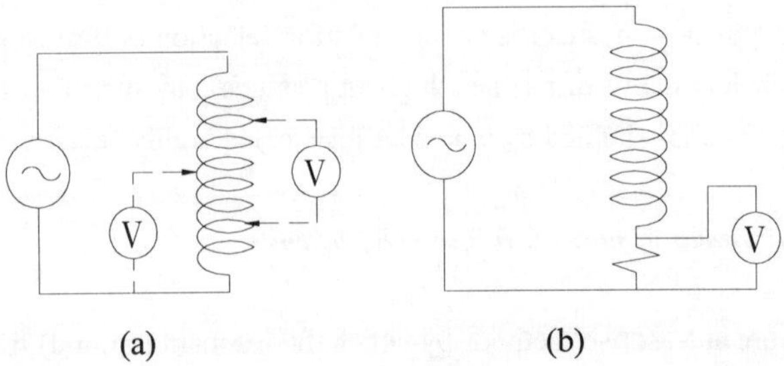

Fig. 12.18 Measurement of natural frequencies by sustained oscillations. [56]

The disadvantages of the above two methods are (i) only the first two natural frequencies can be measured, and (ii) three points (25-, 50- and 75-percent) of the coil must be accessible. The insulation of the coil must be stripped to have direct contact with the conductor of the coil. A calibrated capacitive probe can be used to avoid stripping the coil insulation required for direct contact.

Higher harmonics can be measured with a variable-frequency signal generator by measuring the current through the coil (Fig. 12.18b). The current will go through successive maxima as the frequency of the signal generator is uniformly increased. The first current peak will be at the first natural frequency, the second peak will be at the second natural frequency, etc. The natural frequencies can also be measured with a variable-frequency impedance meter, instead of the signal generator. The maxima given by the impedance bridge will give the condition of series resonance, and the minima the parallel resonance.

It must be emphasized that the natural frequencies measured according to Fig. 12.18a are caused by the ***internal*** resonances, and the resonance conditions obtained by Fig. 12.18b and the impedance bridge method are the ***terminal*** resonances. The two resonances may not always coincide.

The calculated and the measured natural frequencies of three large air-core magnets are shown in Table 12.1 [41,57,58].

Table 12.1: Resonant Frequencies of Three Large Air-Core Magnets *(in kHz)*

	Computed Natural Frequencies		Measured Natural Frequencies	
	1st Harmonic	2nd Harmonic	1st Harmonic	2nd Harmonic
300-kJ Magnet	532	821	577	970
20-MJ Magnet	139.5	335.9	125.3/143.8[1)]	352.5/338.6[1)]
30-MJ Magnet	8.8	15.9	7.3	15.3

[1)] Measured by Fig. 12.18a/12.18b

12.4.3 Surge testing of rotating machines

Rotating machines, in general, do not have specified basic impulse insulation levels (BIL). However, it should be recognized that a rotating machine will have a lower transient-voltage withstand capability than a transformer of the same power-frequency steady-state voltage rating, because (i) transformers are immersed in oil or gas of high dielectric strength, whereas a rotating machine operates in atmospheric air, and (ii) space in a rotating machine is more limited than in a transformer.

Small machines, particularly fractional horse-power motors, are regularly surge tested, the primary objective being to detect damaged insulation during manufacturing. The ground (or major) insulation of a machine can be tested by the power-frequency hipot test. However, surge testing is necessary to test the turns insulation. Foust and Rohats developed a test technique which has been used universally in various modifications [59]. One modification is shown in Fig. 12.19. Two identical stator windings - one master winding, and the other the test winding - are simultaneously impulsed from an impulse generator, and the difference voltage is monitored. For no fault in the winding, the monitored voltage will be zero. One disadvantage of this technique is that an unintentional fault in the master winding would give a false reading. For a sensitive fault detecting technique, even a slight dissymmetry between the two stators may give false indications.

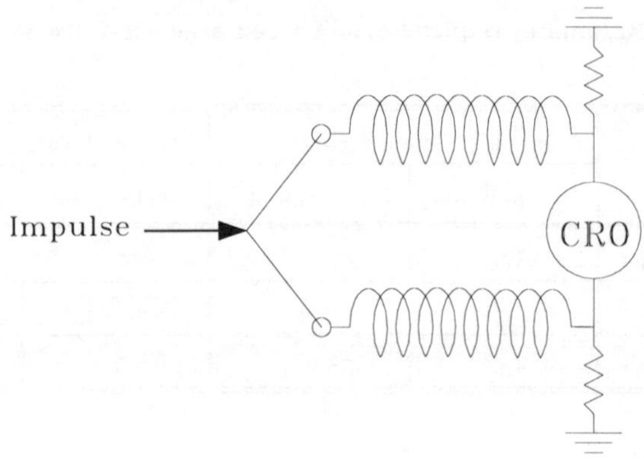

Fig. 12.19 Schematic of surge testing of small motors.

Large rotating machines are seldom surge tested. These machines are somewhat protected by the step-up transformer and the interconnecting cable which is generally several hundred meters long. Special rotating-machine type surge protectors are used to limit the magnitude of the transients which manage to appear across the machine terminals. Moreover, capacitors (usually 0.25 μF) are also connected across the machine terminals to elongate the wavefront of the oncoming voltage transients. However, faults in the machine windings do occur. A fault in a winding can be classified into two categories:

1. A solid fault where solid copper to copper contact is made through very little ohmic resistance; and

2. An incipient fault where the insulation between two conductors has been nicked or partially removed so that it may break down during the application of the voltage stress.

A solid fault is easier to detect than an incipient fault. However, a solid fault does not occur frequently. To detect an incipient fault, the applied voltage should be such that the voltage along the winding is sufficiently high to break down any damaged dielectric path.

The conventional surge testing techniques (section 12.4.1) fail to detect turn faults in large rotating machines which have many parallel paths per phase. The exploring-coil and the inducing-coil methods are excellent to detect faults in large machines [60-62]. In these methods, an exploring coil of about the size and shape of the stator coil is successively placed over stator slot pairs while the stator winding is being impulsed. The induced voltage across the exploring coil gives an indication if the stator coil directly beneath is faulted. However, these methods are time consuming. Moreover, during the maintenance tests in the field, it would be convenient to perform the surge test with the rotor in place. These methods are unsuitable in that case because of the narrow air gap between the rotor and the stator. In fact, surge testing with the rotor in place is rather difficult because of the low impedance given to the impulse generator.

Three methods were proposed to detect turn faults in large three-phase rotating machines with the rotor in place [63]. The first two methods (Fig. 12.20) require impulse testing of the stator windings. In the first method, one phase of the stator winding is impulsed at a time, while the voltage difference across the terminals of the other two phases is monitored (Fig. 12.20a). In the second method, two phases of the stator winding are simultaneously impulsed from a dual surge generator while the voltage difference across the two impulsed terminals is monitored (Fig. 12.20b). The third method is similar to the test method I except that a sinusoidal voltage at a selected frequency, instead of a surge voltage, is applied to one phase. Figures 12.21 to 12.23 show the oscillograms of the difference voltage for a one-turn solid fault for the three methods.

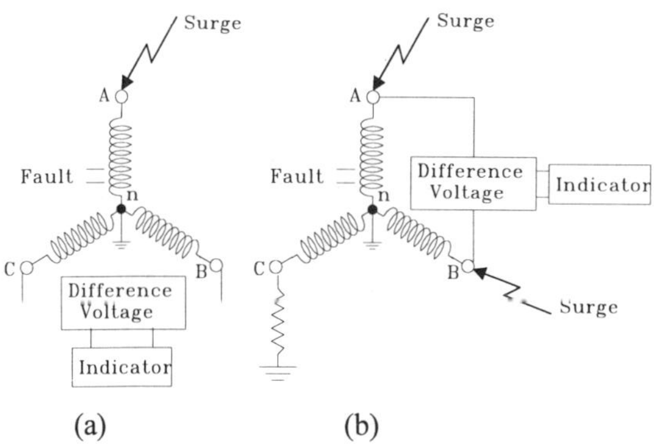

Fig. 12.20 Schematic diagrams of fault detection in three-phase machines. [63]

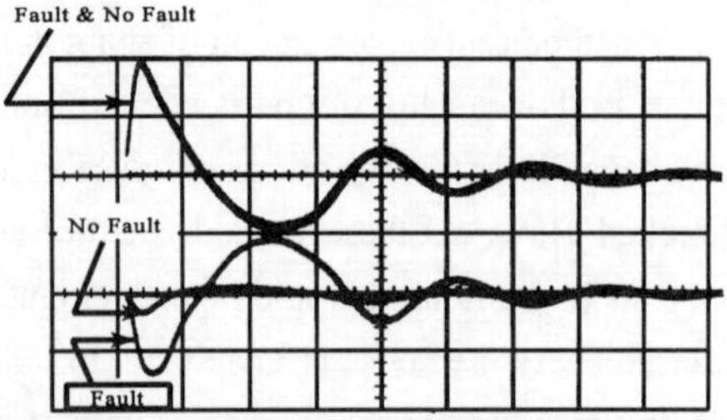

Fig. 12.21 Oscillograms of difference voltage with method I. [63]
One-turn solid fault; rotor balanced in place.
Upper trace: Applied voltage; 200 V/div: 10 μs/div.
Lower trace: Voltage across terminals B and C (Fig. 12.20a); 10 V/div: 10 μs/div.

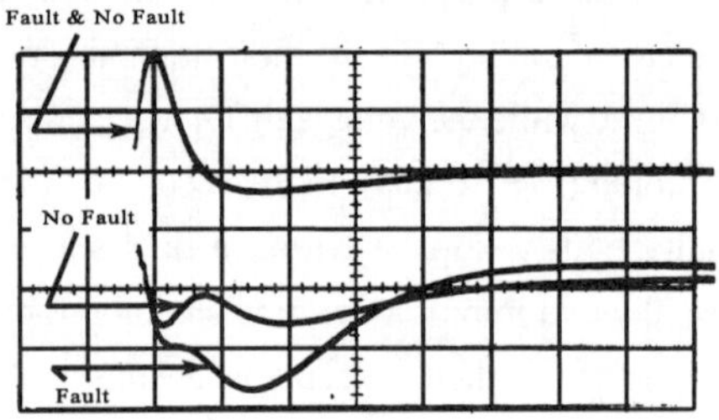

Fig. 12.22 Oscillograms of difference voltage with method II. [63]
One-turn solid fault; rotor balanced in place.
Upper trace: Applied voltage; 200 V/div: 5 μs/div.
Lower trace: Voltage across terminals A and B (Fig. 12.20b); 10 V/div: 5 μs/div.

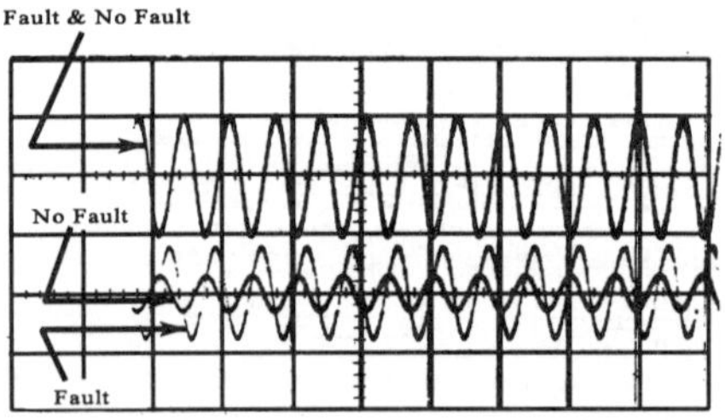

Fig. 12.23 Oscillograms of difference voltage with method III. [63]
One-turn solid fault; rotor balanced in place.
Upper trace: Applied voltage at 158 kHz; 10 V/div: 10 μs/div.
Lower trace: Voltage across terminals B and C (Fig. 12.20a); 0.5 V/div: 10 μs/div.

12.5 References to Chapter 12

1. W.J. McNutt, T.J. Blalock and R.A. Hinton, "Response of Transformer Windings to System Transient Voltages," ***IEEE Trans. on Power Apparatus and Systems***," Vol. PAS-93, pp. 457-467, 1974.

2. H.B. Margolis, et al, "Experience with Part-Winding Resonance in EHV Auto-Transformers: Diagnosis and Corrective Measures," ***IEEE Trans. on Power Apparatus and Systems***, Vol. PAS-94, pp.1294-1300, 1975.

3. A.J. McElroy, "On the Significance of Recent EHV Transformer Failures Involving Winding Resonance," ***IEEE Trans. on Power Apparatus and Systems***, Vol. PAS-94, pp. 1301-1316, 1975.

4. "Surge Characteristics and Protection of Distribution Transformers," EPRI Report No. EL-3385, Electric Power Research Institute, Palo Alto, Ca., January, 1984.

5. D.J. Allan, "Special Report for Group 12 (transformers)," ***Proc. CIGRE***, Paper No.12-00, 1984.

6. W. Stein, et al., "Oscillatory Switching Voltages and Relevant Response of HV Power Transformers," ***Proc. CIGRE***, Paper No. 12-03, 1984.

7. A. Schei, et al, "Resonant Overvoltages in Power Station Transformers Initiated by Switching Transients in the Connected Cable Network," ***Proc. CIGRE***, Paper No. 12-07, 1984.

8. R.E. Pretorius and P.V. Goosen, "Practical Investigation into Repeated Failures of 400/220 kV Auto Transformers in the ESCOM Network - Results and Solutions," ***Proc. CIGRE***, Paper No. 12-10, 1984.

9. H. D'Heure and A. Even, "Contribution to Studies on an Unusual Overvoltage Phenomenon in the Tapchanger on the HV Side of a 400 kV Station Transformer," ***Proc. CIGRE***, Paper No. 12-11, 1984.

10. Working Group 12.07 Report: "Resonance Behaviour of High-Voltage Transformers," ***Proc. CIGRE***, Paper No. 12-14, 1984.

11. J.L. Puri, et al., "Lightning Induced Failures in Distribution Transformers," ***IEEE Trans. on Power Delivery***, Vol. 3, pp. 1784-1801, 1988.

12. P.A. Abetti, "Survey and Classification of Published Data on the Surge Performance of Transformers and Rotating Machines", ***AIEE Power Apparatus and Systems***, No. 40, pp. 1-11, February 1959.

13. P.A. Abetti, "Bibliography on the Surge Performance of Transformers and Rotating Machines", ***AIEE Power Apparatus and Systems***, No. 39, pp.1150-1168, December 1958.

14. R. Rudenberg, "Performance of Traveling Waves in Coils and Windings", ***AIEE Trans.***, Vol. 59, pp. 1031-1040, 1257-1262, 1940.

15. L.F. Blume, A. Boyajian, G. Camilli, T.C. Lennox, S. Minneci and V.M. Montsinger, ***Transformer Engineering***, Chapter 17, Second Edition. John Wiley & Sons, 1951.

16. R.C. Degeneff, "A General Method for Determining Resonances in Transformer Windings", ***IEEE Trans. on Power Apparatus and Systems***, Vol. PAS-96, pp. 423-430, 1977.

17. K.W. Wagner, "Penetration of an Electromagnetic Wave in a Coil with Winding Capacitance", (in German), ***Elektrotechn. u. Masch.-Bau***, Vol. 33, pp. 89-92, 105-108, 1915.

18. L.F. Blume and A. Boyajian, "Abnormal Voltages Within Transformers", ***AIEE Trans.***, Vol. 33, Pt. 1, pp. 577-620, 1919.

19. N. Balabanian and T.A. Bickart, ***Electrical Network Theory***, Chapter 4, First Edition. John Wiley & Sons, 1969.

20. J.A. Edminister, ***Electric Circuits***, Chapter 15, Second Edition. Schaum's Outline Series, McGraw-Hill Book Co., 1983.

21. P.I. Fergestad, ***Transient Oscillations in Transformer Windings***, Doctoral Dissertation, Norwegian Institute of Technology, Oslo, 1972.

22. P.I. Fergestad and T. Henriksen, "Transient Oscillations in Multiwinding Transformers", ***IEEE Trans. on Power Apparatus and Systems***, Vol. PAS-93, pp. 500-509, 1974.

23. S. Hayashi, ***Surges on Transmission Systems***, First Edition. Denki-Shoin, Kyoto, 1955.

24. ***Surge Phenomena***, Chapters 12-15, Ref.S/T35, The British Electrical and Allied Industries Research Association, London, 1941.

25. L.V. Bewley, ***Traveling Waves on Transmission Systems***, Second Edition. New York: John Wiley, 1951.

26. H.B. Dwight and L.S. Dzung, "A Formula for the Reactance of the Interleaved Component of Transformers", ***Electrical Engineering***, pp. 1368-1371, 1937.

27. F.W. Grover, ***Inductance Calculations: Working Formulas and Tables***, Dover Publications, New York, 1946.

28. L. Rabins, "Transformer Reactance Calculations with Digital Computers", ***AIEE Trans.***, Vol. 75, Pt. I, pp. 261-267, 1956.

29. P.I. Fergestad and T. Henriksen, "Inductances for the Calculation of Transient Oscillations in Transformers", ***IEEE Trans. on Power Apparatus and Systems***, Vol. PAS-93, pp. 510-517, 1974.

30. K.A. Wirgau, "Inductance Calculation on an Air-Core Disk Winding", ***IEEE Trans. on Power Apparatus and Systems***, Vol. PAS-95, pp. 394-400, 1976.

31. T.H. Fawzi and P.E. Burke, "The Accurate Computation of Self and Mutual Inductances of Circular Coils", ***IEEE Trans. on Power Apparatus and Systems***, Vol. PAS-97, pp. 464-468, 1978.

32. P. Sarma-Maruvada and N. Hylten-Cavallius, "Capacitance Calculations for Some Basic High Voltage Electrode Configurations", ***IEEE Trans. on Power Apparatus and Systems***, Vol. PAS-94, pp. 1708-1713, 1975.

33. V.A. Karasev, ***Overvoltages in Windings of Transformers with Grounded Neutral***, (in Russian), ONTI, 1934.

34. Bulgakov, ***Computation in Transformers***. Moscow, 1950. This book was mentioned by G. Karady in "A Simple Method for Determining the Stresses Caused by Impulse Voltages in a Transformer of Dissimilar Coils", (in German), ***Periodica Polytechnica***, Budapest, Vol. 2, pp. 15-38, 1958.

35. E.R. Hartill, "The Initial Impulse Voltage Distribution on a Transformer Winding", ***Bulletin Electrical Engineering Education***, Vol. 10, pp. 45-52, 1953.

36. P. Waldvogel and M. Rouxel, "Predetermination by Calculation of the Electric Stresses in Winding Subjected to a Surge Voltage", ***Proc. CIGRE***, Paper No. 125, 1956.

37. H. Brechna, "Impulse-Proof Transformer Windings", (in German), ***Bulletin Oerlikon***, Zurich, Vol. 328/329, pp. 89-101, 1958.

38. B.N. Jayaram, "Determination of Impulse Voltage Distribution in Transformers with a Digital Computer", (in German), ***Elektrotechnische Zeitschrift - A***, Vol. 82, pp. 1-9, 1961.

39. G.M. Stein, "A Study of the Initial Surge Distribution in Concentric Transformer Windings", ***IEEE Trans. on Power Apparatus and Systems*** , Vol. 83, pp. 877-893, 1964.

40. A.K. Lokhanin and V.M. Pogostin, "Longitudinal Capacitance of Transformer Coil Windings", ***Elektrotekhnika (Soviet Electrical Engineering)***, Vol. 36, pp. 33-35, 1965.

41. P. Chowdhuri, "Calculation of Series Capacitance for Transient Analysis of Windings", ***IEEE Trans. on Power Delivery***, Vol. PWRD-2, pp. 133-139, 1987.

42. H.S. Carslaw and J.C. Jaeger, ***Operational Methods in Applied Mathematics***, pp. 42-43, Dover Publications, 1963.

43. K.K. Palueff and J.H. Hagenguth, "Effect of Transient Voltages on Power Transformer Design - IV Transition of Lightning Waves from One Circuit to Another through Transformers", ***AIEE Trans.***, Vol. 51, pp. 601-620, 1932.

44. P.A. Abetti, I.B. Johnson and A.J. Schultz, 'Surge Phenomena in Large Unit-Connected Steam Turbine Generators", ***AIEE Trans.***, Vol. 71, Part III, pp. 1035-1047, 1952.

45. P.A. Abetti and H.F. Davis, "Surge Transfer in 3-Winding Transformers", ***AIEE Trans.***, Vol. 73, Part III-B, pp. 1395-1407, 1954.

46. A.P. Hayward, J.K. Dillard and A.R. Hileman, "Lightning Protection of Unit-Connected Turbine Generators - Field and Laboratory Studies", ***AIEE Trans..***, Vol. 75, Part III, pp. 1370-1381, 1956.

47. J.K. Dillard and A.R. Hileman, "Lightning Protective Requirements of Generators Connected to the System through Y-Grounded-Delta Transformers", ***AIEE Trans.***, Vol. 77, Part III, pp. 1426-1434, 1958.

48. A.R. Hileman, "Surge Transfer through 3-Phase transformers", ***AIEE Trans.***, Vol. 77, Part III, pp. 1543-1554, 1958.

49. B. Heller and A. Veverka, ***Surge Phenomena in Electrical Machines***, Chapter 4. Translated by J.S. Vosper, Iliffe Books, Ltd., London, 1968.

50. R. Rudenberg, ***Electrical Shock Waves in Power Systems***, Chapter 22. Translated by H.J. Wetzstein, Harvard University Press, 1968.

51. E.L. White, "Surge-Transference Characteristics of Generator-Transformer Installations", ***Proc.IEE*** (G.B.), Vol. 116, pp. 575-587, 1969.

52. A. Miki, T. Hosoya and K. Okuyama, "A Calculation Method for Impulse Voltage Distribution and Transferred Voltage in Transformer Windings", ***IEEE Trans. on Power Apparatus and Systems***, Vol. PAS-97, pp. 930-939, 1978.

53. ***General Requirements for Liquid-Immersed Distribution, Power, and Regulating Transformers***, ANSI/IEEE Standard No. C57.12.00-1987.

54. ***Guide for Transformer Impulse Tests***, ANSI/IEEE Standard No. C57.98-1986.

55. J.H. Hagenguth, "Progress in Impulse Testing of Transformers", ***AIEE Trans.***, Vol. 63, pp. 999-1005, 1944.

56. L.V. Bewley, J.H. Hagenguth and F.R. Jackson, Jr., "Methods of Determining Natural Frequencies in Coils and Windings", ***AIEE Trans.***, Vol. 60, pp. 1145-1150, 1941.

57. P. Chowdhuri and M. Anderson, "Performance of Large Magnets under Transient Voltages", ***Proc. Ninth Symposium on Engineering Problems of Fusion Research***, IEEE Publ. No. 81CH175-2 NPS, pp. 448-451, 1981.

58. P. Chowdhuri, "Transient Voltage Oscillations in Coils," ***Fusion Technology***, Vol. 8, pp. 1708-1713, 1985.

59. C.M. Foust and N. Rohats, "Insulation Testing of Electric Windings", ***AIEE Trans.***, Vol. 62, pp. 203-206, 1943.

60. G.L. Moses and E.F. Harter, "Winding-Fault Detection and Location by Surge-Comparison Testing", ***AIEE Trans.***, Vol. 64, pp. 499-503, 1945.

61. R.M. Sexton and R.J. Alke, "Detection of Turn-to-Turn Faults in Large High-Voltage Turbine Generators", ***AIEE Trans.***, Vol. 70, Part I, pp. 270-274, 1951.

62. J.A. Oliver, H.H. Woodson and J.S. Johnson, "A Turn Insulation Test for Stator Coils", ***IEEE Trans. on Power Apparatus and Systems***, Vol. PAS-87, pp. 669-678, 1968.

63. P. Chowdhuri, "Fault Detection in Three-Phase Rotating Machines", ***IEEE Trans. on Power Apparatus and Systems***, Vol. PAS-91, pp. 160-167, 1972.

CHAPTER 13
LIGHTNING ARRESTERS

13.1 Introduction

It has been estimated that the annual cost of lightning damage to power lines, transformers and other utility equipment in the United States is in the order of $50 million [1]. In the U.S. eastern seaboard, lightning may be the cause for 45 percent of all power outages. This cost estimate does not include the loss of revenue for the electric utilities, nor the cost of production loss for the industrial and commercial sectors of the nation. In today's society, many of the industrial and commercial functions are controlled by microprocessors, which are intolerant of power failures even for a few cycles. These microprocessors and other solid state devices are also being used to control household appliances. Even without a power outage, these devices can be permanently damaged if a transient voltage gets through to these devices via the power lines. Therefore, the total cost of damages caused by thunderstorms will be many times the estimated value of $50 million per year. Hence, the importance of lightning protection of the power system is obvious. Lightning arresters are the most effective means of protecting an electrical apparatus against traveling voltage waves caused by lightning and switching.

Lightning arresters are connected across an apparatus to provide a low-resistance path to ground, thus limiting the transient voltages below the BIL of the apparatus. The functions of a lightning arrester are:

1. To act like an open circuit during the normal operation of the system i.e., to hold off the system voltage,

2. To limit the transient voltage to a safe level (i.e., below the apparatus BIL) with the minimum delay and jitter, and

3. To bring the system back to its normal operational mode as soon as the transient voltage is suppressed, i.e., to interrupt the power-follow current and to reseal itself.

The normal operation or operational mode includes the system under faulted condition. Under several types of system faults, such as the single line-to-ground faults, the voltage to ground across the unfaulted phases will rise above the normal voltage level. The arrester must not go into conduction during this faulted condition. It should also be able to interrupt the power-follow current and reseal itself under system fault conditions when the power-frequency voltage across it rises.

An ordinary air gap, such as a rod gap, a needle gap or a sphere gap is not a lightning arrester because it does not fulfill the third function. There are two main types of lightning arresters: the expulsion type and the valve type. Although the expulsion type of arresters are no longer manufactured, they are still in use in many power systems. Therefore, the expulsion type arrester is reviewed below.

13.2 Expulsion Arresters

The expulsion arrester, also known as a protector tube, is like an expulsion fuse with the fuse link removed [2]. The schematic of an expulsion type arrester is shown in Fig. 13.1.

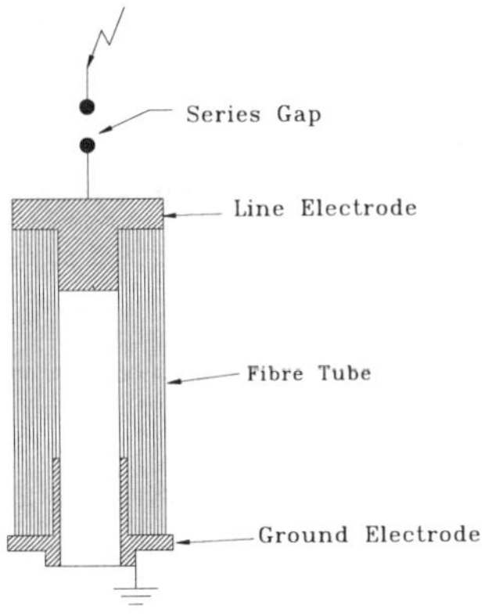

Fig. 13.1 Expulsion type arrester. [2]

The expulsion arrester essentially consists of a hollow tube made of a fibrous material which is capable of producing gases when heated. This hollow fibrous tube is fitted with a metal electrode at each end; the one at the grounded end is hollow. External to the tube is a series gap at the line end. This series gap keeps the fibrous tube isolated from the power-frequency line voltage during normal operation to prevent carbonization of the tube wall under long-duration voltage stress. The series-connected external gap and the internal gap spark over when a transient voltage is applied across the arrester. The heat of the arc produces gas inside the hollow fibrous tube, which forces the arc outside the tube through the lower hollow electrode. Once the energy of the transient voltage is dissipated, power-follow current flows through the arc with very little resistance offered by the arc to the current flow. The power-follow current is virtually the short circuit current of the system, limited by the system impedance. The arc is finally quenched at the natural zero of the follow current, and the system is restored to its normal operating mode. The expulsion arrester has several disadvantages:

1. Its current-carrying duty is entirely system dependent; therefore, its application must be carefully planned so that its maximum power-frequency current rating is not exceeded;

2. Its ability to quench the power-follow arc depends upon the system recovery voltage. If the system recovery voltage rises faster than the dielectric recovery rate of the expulsion tube, the arc will restrike and the arrester would destroy itself;

3. Because the voltage across the expulsion tube completely collapses once the two series-connected gaps spark over, it applies a steep-fronted voltage across the apparatus it protects. This may excite a transformer winding to internal oscillations;

4. The sparkover voltage level of the external gap is dependent upon the environmental conditions; it will vary with atmospheric temperature, pressure and humidity. Moreover, dirt, insects etc. may partially bridge the gap. Thus the protective level of this type of arrester cannot be precisely controlled. However, because of their cost effectiveness, the expulsion arresters have been extensively

used to protect line insulators where shield wires were uneconomical because of high grounding resistance.

13.3 Valve Arresters

The valve arresters limit the magnitude of the power-follow current regardless of the system MVA capacity once the energy in the transient overvoltage is dissipated. This is in contrast to the expulsion arresters where the magnitude of the power-follow current depends entirely upon the system short-circuit characteristics. The valve arresters got their name from the nonlinear resistors which are incorporated in such arresters. A nonlinear resistor acts like a valve when a voltage is applied to it. Ideally, the voltage drop across a nonlinear resistor should remain constant irrespective of the magnitude of the current flowing through it. In reality, the voltage drop across the valve elements increases somewhat with current. There are two types of valve arresters: (i) silicon carbide, and (ii) zinc oxide.

13.3.1 Silicon-carbide arresters

The nonlinear resistors in a silicon-carbide arrester are built from powdered silicon carbide mixed with a binding material (clay), molded into a circular disc and baked. The disc diameter depends upon its energy rating, and the thickness upon its voltage rating. The volt-ampere characteristic of a SiC (silicon carbide) nonlinear valve block is shown in Fig. 13.2. For comparison, the volt-ampere characteristic of a linear resistor is also shown. The volt-ampere characteristic of SiC has a hysteresis-type loop; the resistance is higher during the rising part of the impulse current than that during the tail of the current wave. This is because of the temperature difference in the valve block during these two parts of the current wave.

If the SiC valve-type arrester were directly connected to the power apparatus, then a power-frequency leakage current would be constantly flowing through the valve block. This would overheat the block, and would eventually lead to a thermal runaway condition. A set of air gaps is connected in series with the SiC valve block to prevent such a runaway condition.

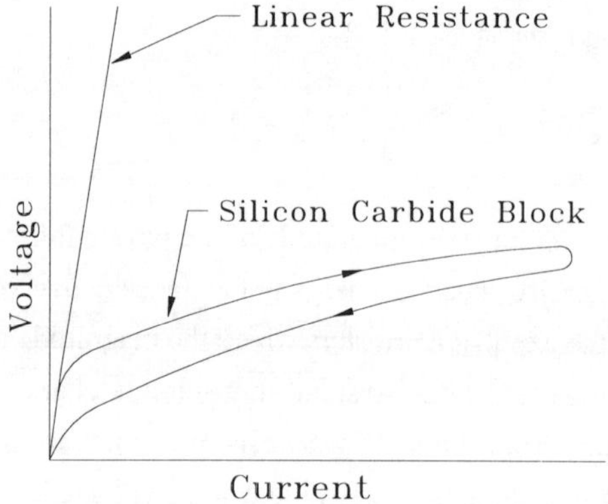

Fig. 13.2 Volt-ampere characteristics of linear resistor and of silicon carbide block.

The series gap elements will spark over if the magnitude of the applied transient voltage exceeds a preset level; this will result in a heavy current flow through the nonlinear valve elements. The voltage drop across the valve elements, called the discharge voltage of the arrester, will be held to the design value. Once the energy in the transient voltage is dissipated, power-follow current will flow through the valve elements. However, because of the nonlinearity of the valve blocks, the flow of current will be restricted to such a low value that the arc across the series gap elements will become unstable and be quenched, thus isolating the power-frequency voltage from the arrester. The sparkover and the discharge voltages of the arrester are not equal (Fig. 13.3).

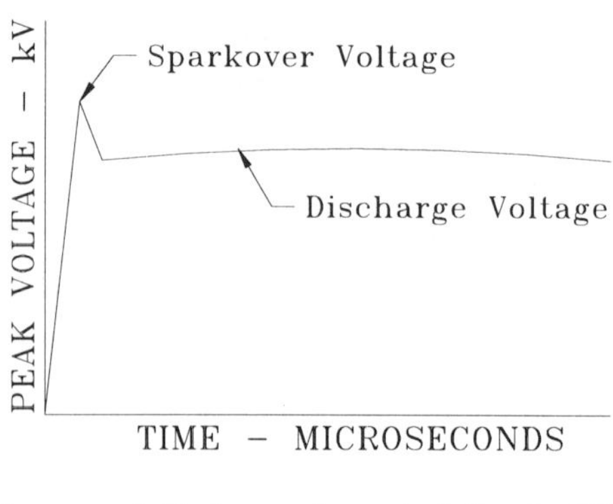

Fig. 13.3 Sparkover and discharge characteristic of a SiC valve arrester.

The sparkover voltage depends upon the voltage waveshape: the steeper the voltage rise, the higher the sparkover voltage. A good design of the gap elements will make the sparkover voltage less dependent on the voltage waveshape. The discharge voltage level rises with the discharge current through the valve elements. The dependence on the current amplitude becomes less as the nonlinearity increases. The proper application of the arrester will require that the sparkover voltage and the discharge voltage of the arrester must be lower than the BIL of the apparatus to be protected with a comfortable margin. This will be discussed in Chapter 15: (Insulation Coordination).

The design of the series gap elements is delicate. They must carry the heavy lightning discharge current; voltage equalization among the gap elements for high voltage applications is sometimes difficult; and the gap elements have some time delay to break down. The schematic of a SiC arrester is shown in Fig. 13.4.

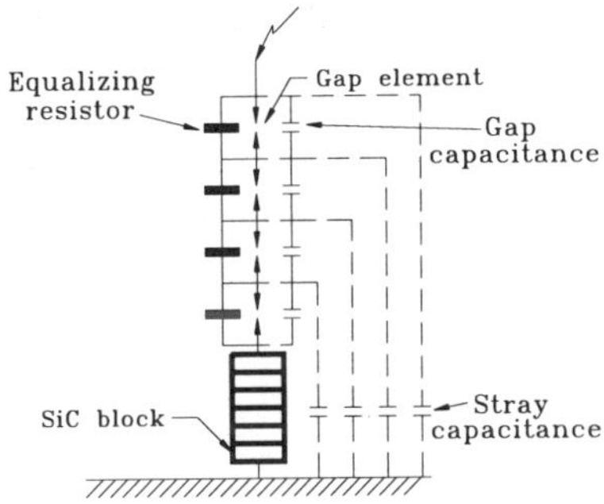

Fig. 13.4 Schematic of a SiC arrester.

The string of gap elements acts like a series-connected capacitive network. As all the gap elements are similar, their capacitances are also equal. Therefore, the applied voltage should be equally divided among them. However, because of the proximity of the ground, stray capacitance exists between each gap element and the ground plane, which diverts part of the capacitive current in the gap to the ground. As a result, the gap capacitive current progressively decreases from the line end to the ground end. Thus, the applied voltage across the gap is unequally divided among the gap elements, putting more stress across the line-end gap elements than that across the ground-end elements. This unequal voltage stress across the gap may precipitate a premature sparkover of the gap even under the normal power-frequency voltage, especially under adverse weather conditions when the external weather shed is contaminated and wet. This situation is corrected by connecting a resistor (fixed or nonlinear) across each gap element, as shown in Fig. 13.4. These resistors carry more current than the stray capacitances to the ground, forcing voltage equalization across the gap elements. These voltage equalizing resistors are in the form of rings which also serve as support for the gap electrodes.

The time lag of breakdown of the gap elements under transient voltages is minimized by proper shaping of the gap electrodes so that the electric field distribution across each gap element is uniform.To maintain consistent sparkover of the gaps under an impulse voltage by providing an ample supply of electrons, the gap elements are preionized [2]. In the early days of the valve arresters, radium salts were placed inside the gap. In modern arresters, gap preionization is achieved by using a ceramic material of high dielectric constant as part of the support structure for the gaps. Microgaps are provided between the edge of the ceramic material and the gap electrode. When voltage is applied across a space filled with two different dielectric materials, the electric stresses across the boundary between the two dielectric materials are inversely proportional to their dielectric constants, so that the material with lower dielectric constant is stressed more than the one with the higher dielectric constant. In the case of arresters, the microgaps are intensely ionized on the application of the transient voltage, illuminating the main gap elements.

The function of the series gap in a SiC arrester is to isolate the SiC valve blocks from the power-frequency voltage during normal operation. The gap sparks over on the arrival of a transient voltage, dissipating the transient energy in the nonlinear resistance of the SiC blocks. Once the transient enery is dissipated, the power follow current is limited by the valve blocks, and the gap interrupts the follow current at its normal current zero. To alleviate the dual duty of the valve blocks of dissipating the transient energy and also limiting the follow current, current-limiting gaps were developed where the high arc voltage not only shared part of the transient energy dissipation but also limited the follow current, thus reducing the arcing time. The reduced energy dissipation in the valve blocks also made it possible to select valve blocks of lower discharge voltage. Consequently, the insulation level of the protected apparatus could also be reduced.

There are four different classes of arresters [3]: (i) station, (ii) intermediate, (iii) distribution, and (iv) secondary. The station arresters have the greatest range of ratings, and are designed for heavy duty applications; they have the best protective characteristics and the highest durability. The intermediate arresters can be installed only in 138-kV systems and below; they are designed for moderate duty. The distribution arresters are applied to protect low-voltage (i.e., distribution) transformers and lines where the duty is light. The secondary arresters are applied to systems operating at less than 1000 volts.

The voltage rating of an arrester is an index of the highest permissible temporary overvoltage (TOV) at the arrester terminals [4]. An arrester must be able to reseal itself against the maximum anticipated TOV. The most common source of TOV is the voltage to ground of the unfaulted phases during a single-line-to-ground fault. Analysis of the system under fault conditions must be performed to estimate the TOV. Curves are also available [4].

13.3.2 Metal-oxide valve (MOV) arresters

Since the 1970's, a new type of valve block has been developed which is much more nonlinear than the SiC block. This is called the metal-oxide block. It is mostly made up of ZnO (zinc oxide) powder and traces of oxides of other metals bound in a ceramic mold. The power-frequency leakage current through the MOV arrester is so small that there is no danger of a thermal runaway. Therefore, many of the MOV lightning arresters are designed gapless. The volt-ampere characteristics of both the SiC and ZnO valve elements are shown in Fig. 13.5.

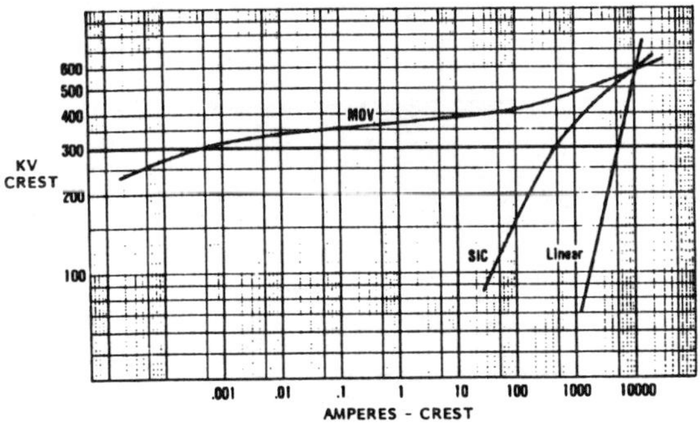

Fig. 13.5 Comparison between the nonlinear characteristics of ZnO and SiC valve blocks. (Courtesy of Ohio Brass Company)

The nonlinear volt-ampere (V-I) characteristic of an arrester valve (SiC or MO) is given by:

$$I=kV^{\alpha}. \tag{13.1}$$

The parameter k depends upon the dimensions of the valve block, i.e., its transient rating, and α , which imparts the nonlinear characteristic, depends upon the valve-block material. For a SiC block, α is typically 5, whereas for a MO block α is greater than 30. In Fig. 13.5, it should be noted that the current through this particular MO block at 300 kV is 1 mA, whereas in a comparable SiC block it is 500 A. Therefore, series gaps are essential for a SiC valve arrester to prevent thermal runaway during normal operation. Gapless operation is possible with MOV arresters because of the low leakage current during normal operation.

In a SiC block, the SiC particles which are typically 200 microns in diameter are held together by the bonding agent which is usually clay. In a MO block, the ZnO particles are 5 to 10 microns in diameter, and are separated by a highly resistive layer bonding the ZnO particles (Fig. 13.6). This bonding layer, which gives the highly nonlinear V-I characteristic of the MO block, acts like a Schottky barrier; it gives high resistance to the flow of current when the applied voltage is below a threshold level, but gives low resistance caused by tunneling of electrons through the layer when the applied voltage exceeds the threshold. The gapped SiC arresters cannot be applied to dc systems, because the series gaps can interrupt the power-follow current only when it goes through its periodic current zero. As the MOV arresters do not need series gaps, they can be used in dc systems. The photo of MO powder and valve blocks is shown in Fig. 13.7. Several references [5-12] are given for further information on the MO valve arrester.

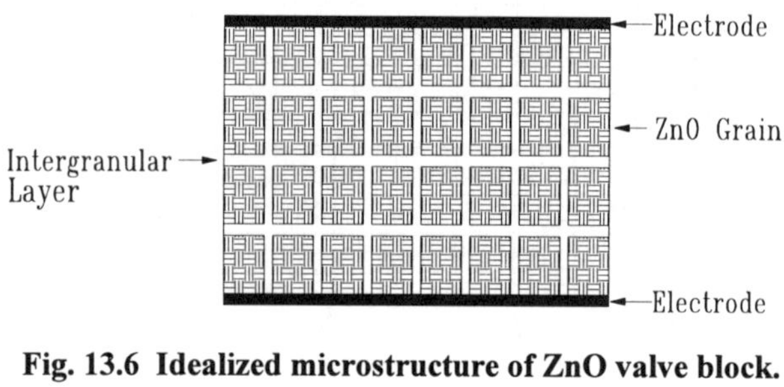

Fig. 13.6 Idealized microstructure of ZnO valve block.

Fig. 13.7 Photo of metal-oxide powder and valve blocks.
(Courtsey of ABB Power T&D Company, Inc.)

The early gapless MOV arresters showed thermal instability caused by continuous power loss in the valve blocks during normal operation. The resistance of a valve block (both SiC and MO) has a negative temperature coefficient, meaning that its resistance decreases with an increase in temperature. If the steady-state voltage across a gapless MOV arrester is too high then high leakage current will flow and the block temperature will rise, causing the block resistance to decrease and further increase of the block temperature. This will lead to a thermal runaway condition when the block will not be able to withstand any voltage. This situation will be aggravated by repetitive application of high surge currents.

The maximum continuous overvoltage (MCOV) is a very important parameter of a MO valve block. The MCOV is the maximum continuous voltage that can be applied to a valve block without exceeding its thermal and electrical stability. This is the maximum system line-to-ground voltage across a MO arrester. Temporary overvoltage (TOV) is another important parameter of MOV arresters, since the MOV block cannot tolerate voltages above its turn-on voltage for long periods of time. The TOV limit of a MOV arrester is a function of the duration of the overvoltage.

Although most of the MOV arresters are designed without series gap structures, some distribution-class arresters use series gaps with advantage. In this design, one or more of the MOV disks of the standard gapless design are replaced by resistance-graded gap structures consisting of silicon-carbide rings and electrodes (Fig. 13.8). The preionizing gaps produce intense ionization during the application of a transient voltage, thus illuminating the main gaps with ultraviolet rays. As a result, the time lag of breakdown is shortened during the application of a steep-front voltage surge. The silicon-carbide grading rings are designed so that the gap structure and the MOV blocks equally share the steady-state voltage. Thus, fewer MOV blocks can be used in series, resulting in lower discharge voltage and hence lower protective level. Incorporation of the gap assembly significantly increases the arrester's ability to survive 60-Hz overvoltage conditions. Figure 13.9 shows the cutaway view of a distribution-class gapped MOV arrester.

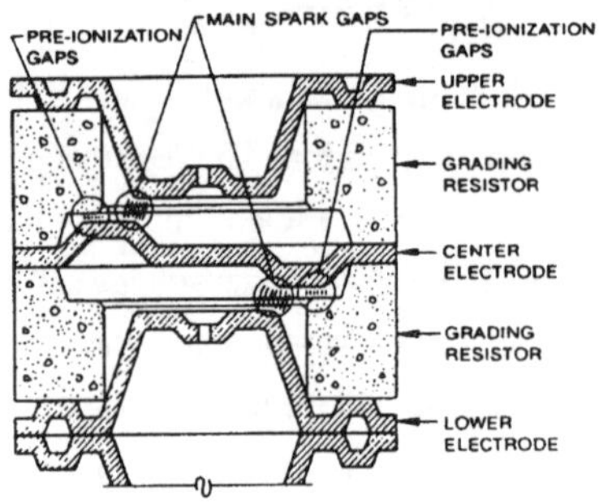

Fig. 13.8 (above) Gap structure of a distribution-class gapped MOV arrester.
(Courtesy of Cooper Power Systems)

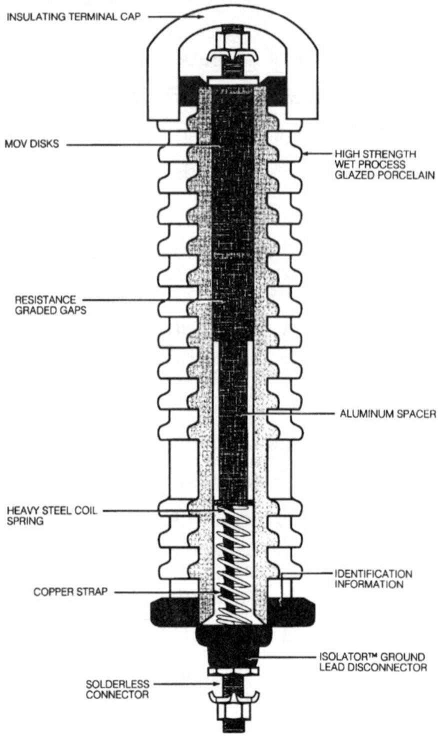

Fig. 13.9 (right) Cutaway of a distribution-class gapped MOV arrester.
(Courtesy of Cooper Power Systems)

13.3.3 Rate of current rise on discharge voltage

The operation of a lightning arrester is known to be sensitive to the rate of rise of the incoming surge current; the higher the rate of rise of the current, the more the arrester limiting voltage rises [13-18]. For the SiC arrester, the time lag of breakdown voltage of the series gap was known to be the cause of the higher protective level of the arrester. However, this effect has also been detected in the gapless MOV arrester. Sakshaug [13] proposed that the higher discharge voltage of the MO block for higher rates of current rise is caused by the negative temperature coefficient of the valve resistance. The instantaneous temperature and resistance of a valve block is a function of the energy dissipated in the block up to that instant. The dissipated energy in the block, in turn, is the integral of the VxI product with time. The energy dissipated in a valve block for any specific current level on the front of the current wave is smaller for a faster rising current than that for a slower rising current at the same current level. Then, the instantaneous resistance of a valve block will be higher for a faster rising current than that for a slower rising current. Hence, the discharge voltage for a faster rising current will be higher. Schmidt and colleagues [16] have proposed a model as shown in Fig. 13.10.

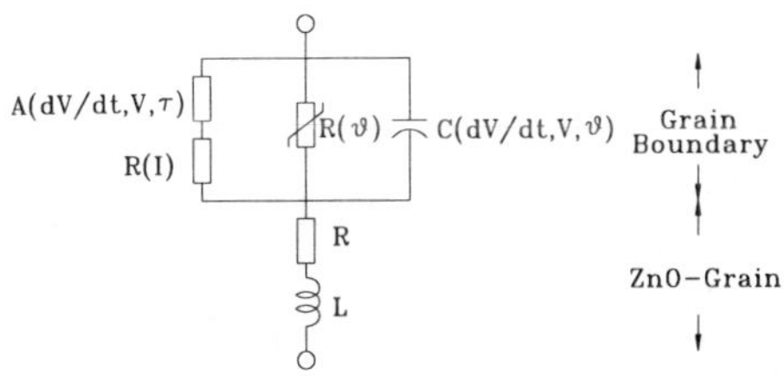

Fig. 13.10 Model of the valve block of an arrester [16].

In Fig. 13.10, the turn-on element, A, reflects the dynamic charge distribution at the grain boundary. Its dynamic behavior depends on the voltage waveshape (dV/dt, V) and the time constant, τ, for reaching equilibrium of electrons and holes at the grain boundary. The capacitance, C, depends on the steepness of the applied voltage and the temperature, θ. The inductance, L, represents the internal inductance of the valve block.

Figure 13.11 shows a station-class MOV arrester of 230 kV voltage rating. This arrester is of special interest. The lower half of the arrester is connected between one phase and the ground. The top of the arrester is connected to another phase for phase-to-phase protection. Three such arresters are installed for the protection of a three-phase system. Figure 13.12 shows MO arresters of various voltage ratings, and Fig. 13.13 shows the installation of arresters for protection of overhead power lines. Tables 13.1 to 13.3 show the lightning arrester characteristics. Application of lightning arresters will be discussed in Chapter 15: Insulation Coordination.

Fig. 13.11 Station class MOV lightning arrester.
(Courtesy of General Electric Company)

Fig. 13.12 MOV lightning arresters of various voltage ratings.
(Courtesy of ABB Power T&D Co., Inc.)

Fig. 13.13 Installation of MOV arresters for protection of overhead power lines.
(Courtesy of Ohio Brass Company)

Table 13.1: Protective Characteristics of Distribution-Class Gapped MOV Arresters (Courtesy of Cooper Power Systems)

Arrester Rating	MCOV	Min. 60-Hz Sparkover	FOW (V_1/V_2)	Max. Discharge Voltage (kV crest) under 8/20-μs Impulse Current @				
kV rms	kV rms	kV crest/√2	kV crest	1.5 kA	5.0 kA	10.0 kA	20.0 kA	40.0 kA
3	2.55	7	8/13	8.6	9.8	10.9	12.7	16
9	7.65	16.5	21/26	17.2	19.6	21.7	25.5	31.9
15	12.7	25	37/42	28.4	32.4	35.9	42.2	52.7
21	17	34	51/57	37.9	43.2	47.8	56.2	70.3
27	22	45	64/77	51.5	58.8	65.1	76.6	95.6
30	24.4	50	71/85	56.8	64.8	71.8	84.4	105.4
36	29	60	85/99	66.3	75.6	83.7	98.4	123

V_1=sparkover of gap assembly based on a wave rising 100 kV per 12 kV of arrester rating
V_2=discharge voltage for a 10-kA impulse current which produces a voltage cresting in 0.5 μs

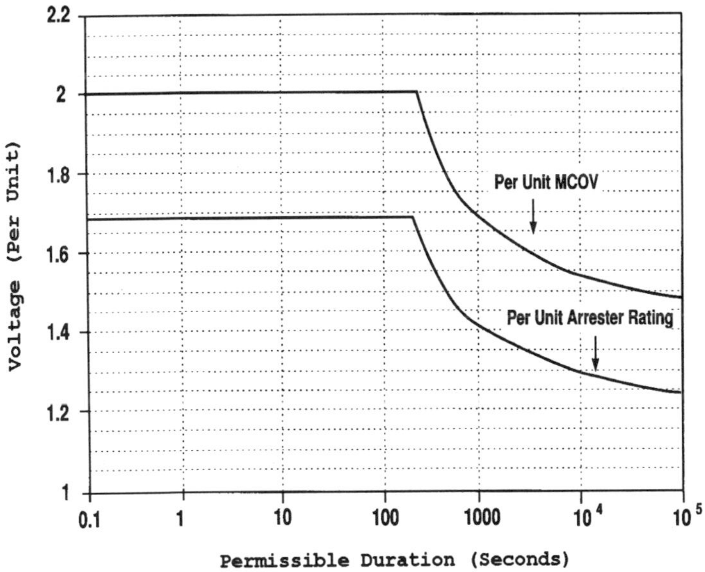

Temporary Overvoltage Withstand Capability
(Courtesy of Cooper Power Systems)

Table 13.2: Protective Characteristics of Intermediate-Class Gapless MOV Arresters
(Courtesy of Ohio Brass Company)

Arrester Rating	MCOV	Max. 0.5-μs Discharge Voltage[1)]	Max. Discharge Voltage (kV crest) under 8/20-μs Impulse Current @					Max. Switching-Surge Protective Level[2)]
kV rms	kV rms	kV crest	1.5 kA	5.0 kA	10.0 kA	20.0 kA	40.0 kA	kV crest
3	2.55	7.9/8.7	6.6	7.4	8.2	9.2	10.7	6
9	7.65	23.2/25.5	19.4	21.9	24.2	27	31.5	17.7
15	12.7	38.6/42.4	32.2	36.4	40.2	44.9	52.3	29.3
21	17	53.8/59.1	44.8	50.7	56	62.5	72.8	40.9
30	24.4	76.8/84.4	64	72.5	80	89.3	104	58.4
45	36.5	115/127	96	109	120	134	156	87.6
60	48	154/169	128	145	160	179	208	117
72	57	184/203	154	174	192	214	250	140
90	70	230/253	192	217	240	268	312	175
108	84	276/304	230	261	288	321	374	210

1):Maximum discharge voltages for 5- and 10-kA impulse currents which produce voltages cresting in 0.5 μs
2): Based on a 500-A surge of 45-μs time to crest

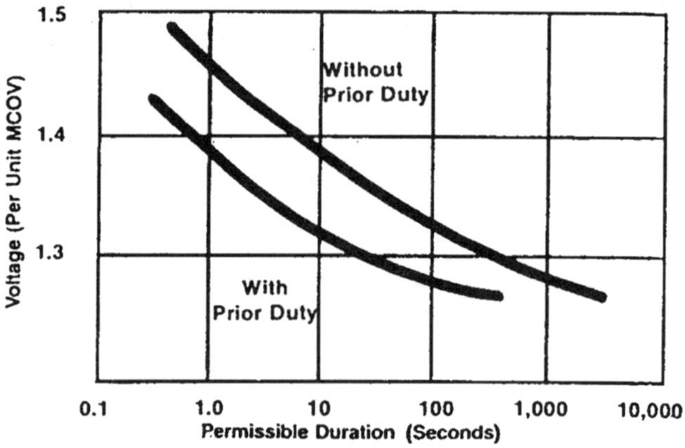

Temporary Overvoltage Withstand Capability
(Courtesy of Ohio Brass Company)

Table 13.3: Protective Characteristics of Station-Class Gapless MOV Arresters
(Courtesy of General Electric Company)

Arrester Rating	MCOV	One-Sec. TOV	FOW	Max. Discharge Voltage (kV crest) under 8/20-μs Impulse Current @					Max. Switching-Surge Protective Level
kV rms	kV rms	kV rms	kV crest	1.5 kA	5.0 kA	10.0 kA	20.0 kA	40.0 kA	kV crest
3	2.55	3.6	9.1	6.9	7.5	8.0	9.0	10.3	6.3 @ 0.5 kA
9	7.65	11.0	26.6	20.2	22.0	23.5	26.4	30.2	18.4 @ 0.5 kA
15	12.7	18.3	44.2	33.5	36.6	39.1	43.9	50.3	30.6 @ 0.5 kA
30	24.4	35.2	84.9	64.3	70.3	75.1	84.2	96.5	58.7 @ 0.5 kA
60	49	70	160	124	134	141	157	177	113 @ 0.5 kA
90	73	106	239	185	200	211	234	264	169 @0.5 kA
108	87	127	287	222	240	254	282	318	204 @ 0.5 kA
180	146	212	477	369	399	422	468	528	349 @ 1.0 kA
240	194	283	635	491	531	562	623	703	465 @ 1.0 kA
300	243	354	795	615	665	703	779	879	603 @ 2.0 kA
360	292	424	953	737	797	843	934	1054	723 @ 2.0 kA

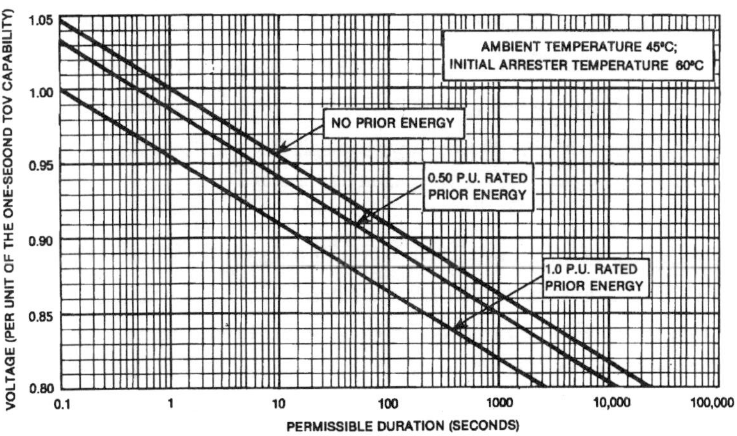

Temporary Overvoltage Withstand Capability
(Courtesy of General Electric Company)

In Tables 13.1 to 13.3, the arrester rating (column 1) is the nominal power-frequency steady-state voltage across the arrester terminals. This is the same as the duty-cycle voltage rating designating the maximum permissible voltage between the arrester's terminals at which the arrester is designed to perform its duty cycle. MCOV (maximum continuous operating voltage) in column 2 sets the voltage limit to prevent thermal runaway condition under steady-state operation. Even for the design with series gap assembly, the steady-state voltage is divided nearly equally between the gap assembly and the metal-oxide valve blocks. Therefore, there is a continuous power loss in the valve blocks during the normal operation. In column 3 of Table 13.1, the minimum 60-Hz sparkover voltage levels of the gapped MOV arresters are tabulated. A sparkover under power-frequency voltage will destroy the arrester.

MOV arresters are able to operate for limited periods of time at voltages in excess of their MCOV rating. This excess voltage defines the TOV (temporary overvoltage) limit of the arrester. The most common source of TOV is the voltage rise on the unfaulted phases during a single-line-to-ground fault. TOV can be generated also by other sources, such as load rejection, ferroresonance and others. In column 3 of Table 13.3, the one-second TOVs are tabulated. The effects of TOV are thermal in nature, so TOV is a function of the duration of the overvoltage - the longer the duration the lower the TOV limit. This is shown in the plots of TOV ***vs.*** time accompanying the three tables. TOV is designated as per unit of either MCOV or arrester voltage rating or one-second TOV (Table 13.3). Under many circumstances, temporary overvoltages occur after switching operations. Switching operations generate switching surges, which cause the arrester to discharge the energy of the switching surges, raising the temperature of the metal-oxide valve blocks. Such prior energy dissipation will lower the TOV limits. This is shown in the plots of TOV accompanying the tables.

FOW (front-of-wave) protective levels of arresters are shown in column 4 of Tables 13.1 and 13.3. Originally, this test was devised to check the time lag of sparkover of the series gap assembly of the gapped silicon-carbide valve arresters. Linearly rising impulse voltages of a specified steepness (kV/μs) were applied to the arrester, and the sparkover voltage of the gap assembly on the front of the applied voltage wave was recorded. With the advent of the gapless metal-oxide valve arresters, the same terminology has been kept; however, an additional test is performed where an impulse current is passed through the arrester which produces a discharge voltage cresting at 0.5 μs. The magnitude of the impulse current is selected so that the prospective peak of the discharge voltage at 0.5 μs is at least 1.2 times the arrester's discharge voltage at a classifying current resulting in a voltage cresting in 0.5 μs. The impulse classifying current varies from 1500 A for secondary-class arresters to 20 kA for 800-kV station-class arresters. The entries in column 3 of Table 13.2 are the same as the FOW protective levels, given for two levels of impulse currents. The five columns of discharge voltage levels for discharge currents of five peak amplitudes show the nonlinear characteristics of the valve blocks.

Distribution-class arresters do not have specifications on switching surges (Table 13.1). The switching-surge protective levels for the various intermediate- and station-class arresters are shown in the last column of Tables 13.2 and 13.3. Detailed specifications and guidance for application of metal-oxide arresters are given in references 19 and 20. Tables 13.1 to 13.3 are shown for the purpose of illustration. These were abbreviated from extensive data which are available from the arrester manufacturers.

13.4 References to Chapter 13

1. J. Mitsche, "Reducing Lightning Damages and Outages," ***EPRI Journal***, pp. 40-42, December 1988.

2. E. Beck, ***Lightning Protection for Electric Systems***. New York: McGraw-Hill, 1954.

3. **ANSI/IEEE Standard C62.1-1989: IEEE Standard for Gapped Silicon-Carbide Surge Arresters for AC Power Circuits.** The Institute of Electrical and Electronics Engineers, Inc., New York, N.Y., 1989.

4. **ANSI/IEEE Standard C62.2-1987: IEEE Guide for the Application of Gapped Silicon-Carbide Surge Arresters for Alternating Current Systems.** The Institute of Electrical and Electronics Engineers, Inc., New York, N.Y., 1988.

5. E.C. Sakshaug, J.S. Kresge and S.A. Miske, Jr., "A New Concept in Station Arrester Design", ***IEEE Trans. on Power Apparatus and Systems***, Vol. PAS-96, pp. 647-656, 1977.

6. M. Kobayashi, M. Mizuno, T. Aizawa, M. Hayashi and K. Mitani, "Development of Zinc-Oxide Non-Linear Resistors and Their Applications to Gapless Surge Arresters", ***IEEE Trans. on Power Apparatus and Systems***, Vol. PAS-97, pp. 1149-1158, 1978.

7. S. Tominaga, K. Azumi, Y. Shibuya, M. Imataki, Y. Fujiwara and S. Nishida, "Protective Performance of Metal Oxide Surge Arrester Based on the Dynamic V-I Characteristics", ***IEEE Trans. on Power Apparatus and Systems***, Vol. PAS-98, pp. 1860-1871, 1979.

8. M. Kobayashi, M. Mizuno, M. Hayashi and Y. Sughita, "Metal Oxide Surge Arrester", ***IEEE Trans. on Electrical Insulation***, Vol. EI-21, pp. 989-996, 1986.

9. **Surge Arrester Seminar Book**. The Ohio Brass Company, Wadsworth, OH, 1988.

10. E.C. Sakshaug, J.J. Burke and J.S. Kresge, "Metal Oxide Arresters on Distribution Systems Fundamental Considerations", ***IEEE Trans. on Power Delivery***, Vol. 4, pp. 2076-2089, 1989.

11. ***Fundamental Research on Metal Oxide Varistor Technology***, prepared by General Electric Company. Report EPRI EL-6960, Electric Power Research Institute, Palo Alto, CA. August 1990.

12. ***High-Energy Metal Oxide Valve Blocks***, prepared by Westinghouse Electric Corporation. Report EPRI EL-7155, Electric Power Research Institute, Palo Alto, CA. February 1991.

13. E.C. Sakshaug, "Influence of Rate-Of-Rise on Distribution Arrester Protective Characteristics", ***IEEE Trans. on Power Apparatus and Systems***, Vol. PAS-98, pp. 519-526, 1979.

14. R.E. Clayton, I.S. Grant, D.E. Hedman and D.D. Wilson, "Surge Arrester Protection and Very Fast Surges", ***IEEE Trans. on Power Apparatus and Systems***, Vol. PAS-102, pp. 2400-2412, 1983.

15. C. Dang, T.M. Parnell and P.J. Price, "The Response of Metal Oxide Surge Arresters to Steep Fronted Current Impulses", ***IEEE Trans. on Power Delivery***, Vol. PWRD-1, pp. 157-163, 1986.

16. W. Schmidt, J. Meppelink, B. Richter, K. Feser, L. Kehl and D. Qiu, "Behaviour of MO-Surge-Arrester Blocks to Fast Transients", ***IEEE Trans. on Power Delivery***, Vol. 4, pp. 292-300, 1989.

17. D.B. Miller, H.B. Fan and P.R. Barnes, "The Response of MOV and SiC Arresters to Steep-Front Longer Duration Current Pulses", ***IEEE Trans. on Power Delivery***, Vol. 6, pp. 666-671, 1991.

18. D.B. Miller and H. Fan, "Step Response of Distribution Arresters at Lower Currents", ***IEEE Trans. on Power Delivery***, Vol. 9, pp. 285-292, 1994.

19. **ANSI/IEEE Standard C62.11-1993: IEEE Standard for Metal-Oxide Surge Arresters for Alternating Current Power Circuits.** The Institute of Electrical and Electronics Engineers, Inc., New York, N.Y., 1993.

20. **ANSI/IEEE Standard C62.22-1991: IEEE Guide for the Application of Metal-Oxide Surge Arresters for Alternating-Current Systems.** The Institute of Electrical and Electronics Engineers, Inc., New York, N.Y., 1991.

CHAPTER 14
PROTECTION OF OUTDOOR SUBSTATIONS

14.1 Introduction

Broadly speaking, a power system consists of the generating stations, the substations, and the power lines. The generating stations, the indoor and the totally enclosed substations are immune from direct strokes of lightning. They require protection against traveling voltage waves coming from the open overhead lines. Outdoor substations and switchyards as well as the overhead power lines need protection against direct hits and against traveling voltage waves coming from another part of the system which is directly hit. The protection against a direct stroke is best accomplished by shielding, so that lightning strikes the shield and is led safely to the ground. For complete protection, the apparatus must be completely enclosed by a metallic shield.The generating station is an example of a completely shielded system. However, it will be prohibitively expensive to enclose the entire power system. Many substations and most power lines are the open air type. The overhead lines are shielded by shield (ground) wires, and the substations and the switchyards are shielded by shield wires and masts. We have discussed the shielding of overhead lines in Chapter 7. We shall discuss shielding of substations in this chapter. Later in this chapter, we will discuss the protection of power apparatus against traveling waves by lightning arresters. In Chapter 15, we will discuss insulation coordination among various power apparatus with protective devices.

14.2 Shielding of Substations: A Critique

Because of the limited surface area of the station, the frequency of direct lightning strokes to the station will be low; however, damages caused by a direct strike to the station will be permanent and hence costly. Therefore, a station should be shielded against direct lightning strokes.

Outdoor substations and switchyards are shielded from direct lightning strokes by shield wires and masts. Three different methods are used to design the shielding of substations against direct lightning strokes:

1. Fixed angles;
2. Empirical curves; and
3. Electrogeometric models.

For the fixed-angle method, it is assumed that a horizontal shield wire will protect all objects against direct lightning strokes if they are placed within a fixed angle to the vertical plane containing the shield wire. Similarly, for vertical masts, all objects placed within a fixed solid angle from the mast tip will be protected against direct lightning strokes. It is assumed in this method of shielding design that the shielding will be perfect, i.e., there will be no shielding failure. The rule of thumb for the protection of substations has been to use a 30° or 45° shielding angle [1]. For single masts or ground wires, a 30° shielding angle is used. For multiple masts or ground wires, a 45° shielding angle is used for the space between these protective structures, and a 30° shielding angle is used for the outside zones. A similar rule of thumb has been used for overhead lines. However, it has been realized that the shielding angle is dependent upon the height of the protected structure. Although the fixed-angle method of shielding design for the overhead lines has long been discarded, substation shielding in many cases is still being used because of its simplicity.

Empirical curves, based on field experience and model tests, are also extensively used for the design of shielding of substations against direct lightning strokes [2]. This method recognizes the effects of the height of the protected object on the design of shields. This method, unlike the fixed-angle method, allows the designer to specify an allowable shielding failure rate. Then, knowing the allowable clearance between the live parts and grounded objects (i.e., the shield), and the height of the shielded object, the height of the shield is determined from the empirical curves for a specified shielding failure rate. This method of shielding by empirical curves is also being widely used for the protection of substations against direct lightning strokes.

The excessive outage rates of the newly designed 345-kV transmission lines in the 1950's showed that neither the fixed-angle nor the empirical method of shielding was adequate for the design of shielding against direct lightning strokes [3]. The electrogeometric model of shielding against direct lightning strokes was then developed for the design of shielding of transmission lines [4].

According to the electrogeometric model, the striking distance (i.e., the distance over which a stepped leader of lightning is drawn to strike an object) of a lightning stroke is proportional to the charge density of the stepped leader. The current in the return stroke, however, is proportioanal to the charge density of the preceding stepped leader. As the current in the return stroke is a widely measured parameter, the striking distance, r_s, of a lightning stroke is expressed as a function of the return-stroke current. Several expressions have been suggested; the most frequently used one is given by

$$r_s = 8 I^{0.65} \quad (m), \tag{14.1}$$

where I is the return-stroke current in kA. It should be noticed in (14.1) that the striking distance is not a function of the height of the struck object, although it is generally believed that r_s should also be a function of the height of the struck object. Several variations of (14.1), some including height dependency, have been offered by various workers. However, none has been completely accepted by the industry.

For a specified return-stroke current, I_s, a cylinder with a radius, r_s, is drawn around a horizontal overhead line. The overhead line will be hit if the stroke with the prospective return-stroke current, I_s kA, arrives at the periphery of this cylinder. For a stroke to earth, a parallel plane at a distance, r_s, above ground specifies a boundary which when crossed by a stepped leader of the prospective return-stroke current, I_s, will signal the final jump of the stroke to earth. If there are two overhead conductors, and conductor 1 is to serve as a shield for conductor 2 for the return-stroke current, I_s, then the cylinder of radius r_s around conductor 2 must be entirely surrounded by the corresponding cylinder of conductor 1 and by the plane distant r_s from the earth (Fig. 14.1). It can be shown that conductor 2, if shielded for I_s, will be automatically shielded for any currents greater than I_s. Currents smaller than I_s will be able to penetrate the shielding and strike conductor 2.

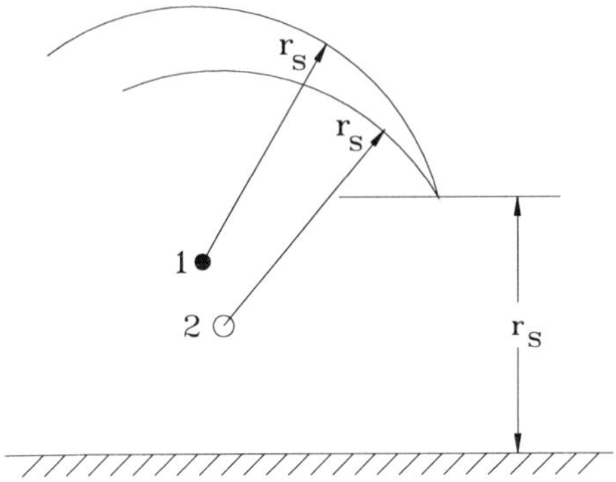

Fig. 14.1 Electrogeometric model of shielding of overhead line by shield wire.

For a vertical mast, the locus of the attractive radius will be a sphere surrounding the tip of the mast. An object will be protected against strokes exceeding I_s if its locus of attractive radius is completely covered by the spherical locus of the mast tip and the attractive plane above the earth. A horizontal overhead line will lose its protection against strokes greater than I_s as soon as the stroke pierces through the protective cone provided by the vertical mast. This will be discussed later.

For the design of shielding of overhead lines, the critical current, I_c, is defined as:

$$I_c = \frac{2\,BIL}{Z_p}, \tag{14.2}$$

where Z_p is the line surge impedance. This means that no flashover will occur if the stroke current is equal to or less than I_c. The objective of shielding against lightning is to place the shield wires so that all lightning strokes carrying currents higher than the critical current of the line will be intercepted by the shield wire(s). In the design of practical line shielding, the shielding current is generally higher than the critical current. Therefore, a few shielding failures for the overhead lines can be expected.

The heart of the electrogeometric model is the striking distance of the stepped leader to a grounded object. It has been suggested that this striking distance is a function of the polarity of the lightning stroke, and of the shape and height of the grounded object which is struck [5]. The diversity of suggestions on the striking distance has created several versions of the electrogeometric model [6]. As a result, the application of the electrogeometric model to shielding against lightning has been complex. For overhead power lines, some deviation from the predicted shielding effectiveness can be tolerated, bearing in mind that most of the outages on overhead lines by direct lightning strokes are caused by backflashover due to the tower-footing impedance.

The electrogeometric model has been proposed in the past for substation protection [7-14]. Sargent was the first to apply this model to the design of substation shielding [7]. He considered the possibility of nonvertical strokes, and applied the Monte Carlo technique to take into consideration the appropriate probabilistic distributions of several variables, such as the direction of approach of the stroke, the stroke current amplitude, etc. His Monte Carlo approach included up to 200 years of operation. Sargent's equation for the striking distance was somewhat different from (14.1). In addition, Sargent used the probability distribution of return-stroke current which was recommended by an AIEE Committee in 1950, and which is no longer in use [15]. Linck extended the electrogeometric model to the shielding of substations in a very practical manner by subdividing the protected area into several sections, and installing a shield wire at the center of each section [8]. He introduced the concept of the shielding failure risk, defined as the number of years in which one shielding failure event is predicted. He assumed that the protected area will be equal to the attractive area which will attract lightning strokes to the protected area. However, it is well known that the attractive area is larger than the protected area because the shields cast an 'electrical' shadow over the protective area. Another possible error in Linck's analysis may lie in the calculation of the attractive width. This attractive width is a function of the height of the bus, h_b. According to Linck, the attractive width is a function of the effective height of the shield, which is defined as the difference in height between the lightning conductor and the protected equipment.

However, another work - also the byproduct of the electrogeometric model - is worth mentioning. Lee developed the 'rolling sphere' method of designing substation shielding against direct lightning strokes [10]. According to this method, a sphere is rolled over the contour of the substation; the radius of the sphere is the striking distance that corresponds to the critical current of the return stroke. Those live parts which touch the surface of the sphere will be vulnerable to lightning strikes. However, if the surface of the rolling sphere touches only the shielding structures, i.e., the shield wires and masts, then the station is completely protected.

Several designers have applied the rolling sphere method to design substation shielding [12,13]. Like Linck, these papers assume the protective area to be equal to the attractive area. It will be shown later that the attractive area could be significantly larger than the protective area. The assumption of equality of the two areas will lead to optimistic designs.

Mousa has performed extensive work both on the electrogeometric model as well as its application to substation shielding [9,14]. Mousa's work closely follows Whitehead's electrogeometric model, except that a few of the 'intuitive' factors were modified. Mousa also assumes that no outage (i.e., no shielding failure) will be allowed. According to Mousa's method, the critical current is calculated from (14.2), where the radius of the conductor (or bus) is replaced by its equivalent corona radius. This is done to take into account the virtual increase of the conductor radius caused by impulse corona. An empirical relationship between the BIL and the critical flashover voltage of an insulator is assumed; finally, the critical flashover strength of the insulator is related to its length. In brief, Mousa follows the same procedure as the shielding of overhead lines.

Except for very fast-front return-stroke currents, traveling voltage waves are not developed by direct strokes due to the relatively short lengths of the buswork. This is particularly true if the stroke hits the terminals of the equipment, such as a transformer, an open circuit breaker or an open disconnect switch. Moreover, (14.2) is valid for transmission lines where no reflections are expected to be encountered. For the bus systems in a substation, multiple reflections will be encountered because of the short lengths, if traveling waves do come into play. Using the corona radius for the bus structure will reduce its surge impedance, thus increasing the computed critical current. This may give results that are too optimistic. Moreover, relating the critical flashover voltage of an apparatus to its insulator length or even its BIL is erroneous. It is widely known that the critical flashover strength of an apparatus (such as insulators) depends upon the shape of the impinging traveling voltage wave. These procedures are acceptable for line protection, because the flashovers are over air (insulator string) where the dielectric can be recovered. In a substation where nonrecoverable dielectrics (transformers, circuit breakers, etc.) coexist, there is no guaranty that an insulator string will flash over instead of a transformer blowing up. Moreover, because of the relatively short spacing between insulators, many insulators will be stressed simultaneously by the lightning stroke. The critical flashover strength of several insulators simultaneously stressed will be lower than that of a single insulator, as discussed in Chapter 11. Finally, lightning arresters are almost always installed in the substations. In that case, the effects of lightning arresters should be taken into account in the computation of the critical current. Therefore, for a direct stroke to a substation, computation of the critical current using the bus surge impedance will not be appropriate.

Because of the low ground-mat impedance of a substation, backflashovers by direct strokes to a substation will be minimal. Therefore, most of the substation outages by direct strokes will be caused by shielding failure. For substation shielding, deviations from the predicted shielding effectiveness would be very costly and unacceptable. The electrogeometric model has not been universally applied to substation shielding because of these complexities. The 30°/45° shielding angle for the protection of substations has generally provided a conservative design. However, such a simplified protection scheme has also caused expensive failures of power apparatus in the substations. A better method of shielding design against direct lightning strokes will not only avoid costly shutdowns but also may result in a more economical design.

14.3 Analysis of Substation Shielding [16]

The basic assumption of the method of substation shielding against direct lightning strokes is that a direct hit to any substation power apparatus may cause an outage. The substation is set up as a rectangular plot of land ($\ell_s x \omega_s$) with the live parts within the periphery of the rectangle. The substation can also be set up as several rectangular plots depending upon the construction of the substation. Shielding design by shield wires and masts are considered separately. The designer decides which shielding design is economical after both alternative designs are completed.

It is also assumed that the distance over which a lightning stroke strikes an object (i.e., the striking distance) is independent of the shape or height of the target object. In other words, the striking distances to a level ground, a shield wire or a mast are all equal. It is further assumed that the striking distance is solely a function of the peak value of the return-stroke current, given by the Whitehead equation (14.1). The acceptable shielding failure rate will be specified by the designer. If the outage rate is specified as SOR failures per year, the following equality should hold to protect an area of A km^2 which has a ground flash density of n_g strikes per km^2 per year:

$$SOR = n_g A P_\ell(I_s)\,, \qquad (14.3)$$

where $P_\ell(I_s)$ = the probability of occurrence of a lightning stroke which has a return-stroke current equal to or less than I_s kA, and I_s is the shielding current. All strokes with return-stroke current greater than I_s will hit the shield. The shielding of the substation must be designed to intercept all strokes with currents greater than I_s. Lightning strokes with currents equal to or less than I_s will hit the substation and result in the SOR per year.

Assuming the statistical variation of the peak return-stroke current, I, to follow log-normal distribution and recalling (4.9), the probability, $P_\ell(I_s)$, is given by [17]:

$$P_\ell(I_s) = 0.5\,[1 + \mathrm{erf}(u)]\,, \qquad (14.4)$$

where $u = \dfrac{\ell n I_s - \ell n I_m}{\sqrt{2}\,\sigma_{\ell n I_s}}$,

I_m = median value of return-stroke current, and
$\sigma_{\ell nI}$ = standard deviation of ℓnI.

The statistical parameters of the return-stroke current are as follows [18,19]:

For peak current ***less than or equal to*** 20 kA:

Median peak current, I_m = 61.1 kA (Fig. 4.10)
Log (to base e) of standard deviation, $\sigma_{\ell nI}$ = 1.33.

For $I_s \leq I_m$, $\mathrm{erf}(u) = 2P_\ell(I_s) - 1$. As $P_\ell(I_s)$ is known from (14.3), the value of u can be evaluated. Then,

$$I_s = \exp(1.8809u + 4.1125)\,. \qquad (14.5)$$

For peak current ***greater than*** 20 kA:

$I_m = 33.3$ kA (Fig. 4.10), and $\sigma_{\ell nI} = 0.605$.

$$\text{Then,}\quad I_s = \exp(0.8556u + 3.5056)\,. \tag{14.6}$$

Alternatively, the probability, $P(I_s)$, that a lightning stroke will have a current exceeding I_s kA can be estimated [20]:

$$P(I_s) = \frac{1}{1 + \left(\frac{I_s}{31}\right)^{2.6}}\,. \tag{14.7}$$

$$\text{Then,}\quad P_\ell(I_s) = 1 - P(I_s)\,. \tag{14.8}$$

I_s can be estimated from (14.3), (14.7) and (14.8), giving:

$$I_s = 31\left(\frac{\frac{SOR}{n_g A}}{1 - \frac{SOR}{n_g A}}\right)^{\frac{1}{2.6}} \text{ kA}\,. \tag{14.9}$$

The estimation of the shielding current, I_s, from (14.9) is simpler than that from (14.5) or (14.6). However, at low values of the shielding currents, such as for substation shielding, the results would vary. Equations 14.5 and 14.6 would give more conservative results.

Once I_s is found, the corresponding striking distance, r_s, can be computed from (14.1). The electrogeometric model of lightning strike is then applied to design the shielding. Different procedures will be required to design shielding by shield wires or by masts. These procedures will be described next.

14.4 Design of Substation Shielding by Shield Wires

Knowing the maximum height, h_b, of the bus to be protected and the minimum required clearance, S_b, between a live part and the ground, the shield wires must be positioned so that all strokes exceeding the estimated shielding current, I_s, will be intercepted by the shield wires. The geometrical process is shown in Fig. 14.2.

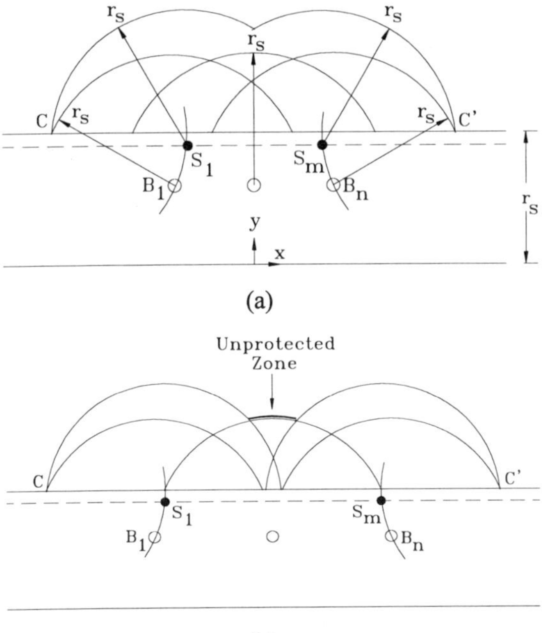

Fig. 14.2 Geometric construction for the shielding of a substation.
(a) middle bus protected; (b) middle bus unprotected.

In Fig. 14.2a, B_1 and B_n are the highest points in the substation to be protected. The distance between the two horizontal lines S_1S_m and B_1B_n is S_b, which is the minimum allowable clearance between a live part and a grounded object. The horizontal line CC' is drawn parallel to the ground plane, separated by r_s, which is the striking distance corresponding to I_s. An arc is drawn with B_n as the center and r_s as the radius, intersecting CC' at C'. Another arc is drawn with C' as the center and r_s as the radius which intercepts the line S_1S_m at S_m and obviously passes through B_n. S_m is the position of the outermost shield wire on the right-hand side.

Shifting the origin to B_n, the coordinates of C' are x_c,r_s ,where:

$$x_c = \sqrt{r_s^2 - (r_s - h_b)^2}\,. \tag{14.10}$$

The x-coordinate, x_m, of S_m, with respect to that of B_n, is found from the equation of the circle with C' as center and r_s as radius:

$$(x - x_c)^2 + (y - r_s)^2 = r_s^2 . \qquad (14.11)$$

For the point, S_m, this reduces to

$$(x_m - x_c)^2 + (h_s - r_s)^2 = r_s^2 . \qquad (14.12)$$

$$\text{Whence, } x_m = x_c - \sqrt{r_s^2 - (h_s - r_s)^2} . \qquad (14.13)$$

And, the height of the shield wire, h_s, is obviously given by

$$h_s = h_b + S_b . \qquad (14.14)$$

If $r_s \leq h_s$, then the attractive distance beyond the outer bus will be (r_s+x_m). Defining the attractive distance beyond the outer bus as x_a:

$$\text{For } r_s > h_s,\ x_a = x_c . \qquad (14.15)$$

$$\text{For } r_s \leq h_s,\ x_a = r_s + x_m . \qquad (14.16)$$

The coordinates (x_m, h_s) pertain to the coordinates of the shield wire, with respect to B_n, protecting one (right-hand) of the two outermost busses. The coordinates of the shield wire for the other (left-hand) outermost bus can be similarly found. However, if the two outermost busses are of the same height, the positions of the two shield wires relative to the outer busses will be the same. In addition to protecting the outer busses, the shield wires must also protect the area between them. To check if the two shield wires would also protect the area between them, circles with r_s as radius and the two shield wires as centers are drawn as shown in Fig.14.2. A third circle is drawn with r_s as the radius and the highest point between the two shield wires as the center. If the circle around the highest point is contained within the two intersecting circles around the two shield wires, then the entire area is protected (Fig. 14.2a). If any part of the third circle is outside the first two circles, then that part is not protected (Fig. 14.2b). Lightning strokes passing through this unprotected zone would strike the substation and cause outages. Two conditions must be met for the complete protection of the midsection:

1. the attractive circles of the two shield wires must intersect at x=0; and
2. the point of intersection must lie above the attractive circle of the middle bus.

If (x_{s1},h_{s1}) and (x_{s2},h_{s2}) are the coordinates of the right and left shield wires, respectively, and $x_{s2} = - x_{s1} = x_s$, and $h_{s1} = h_{s2} = h_s$, then the first condition is met if:

$$r_s \geq |x_s| . \tag{14.17}$$

If condition 1 is met, i.e., if (14.17) is valid, then the ordinate (y_o) of the point of intersection of the attractive circles of the two shield wires needs to be determined. This is done by writing the equations of these two circles.

$$(x - x_{s1})^2 + (y - h_{s1})^2 = r_s^2 , \text{ and} \tag{14.18}$$

$$(x - x_{s2})^2 + (y - h_{s2})^2 = r_s^2 . \tag{14.19}$$

For $|x_{s1}| = |x_{s2}| = x_s$, and $h_{s1} = h_{s2} = h_s$, the ordinates of the points of intersection of the two shield-wire attractive circles are given by:

$$y_o = h_s \pm \sqrt{r_s^2 - x_s^2} . \tag{14.20}$$

Taking the higher value of y_o, the second condition is met if:

$$h_s + \sqrt{r_s^2 - x_s^2} \geq h_b + r_s . \tag{14.21}$$

If the condition of (14.21) is met, then the actual width of the protected area will be $\omega_c = \omega_s + 2x_a$. The total protected area, $A_{tp} = \omega_c x \ell_s$. The new area, A_{tp}, is then substituted in the right-hand side of (14.3), to compute the actual outage rate, AOR. If the AOR differs from the SOR (specified outage rate) by more than a given value, the shielding current is recalculated, and the whole procedure is iterated until the AOR and SOR are within the limits set forth by the designer in (14.3).

If the conditions of (14.17) and/or (14.21) are not met, then the middle part will remain unprotected unless the number of shield wires is increased. The number of shield wires needed for complete protection can be calculated as follows.

From Fig. 14.2a, the separation distance, Δx_s, between two contiguous shield wires for complete protection is:

$$\Delta x_s = 2\sqrt{r_s^2 - (h_b + r_s - h_s)^2}\,. \tag{14.22}$$

The separation distance between the two outer shield wires, ω_{swo}, is given by:

$$\omega_{swo} = \omega_s + 2\,x_s\,. \tag{14.23}$$

Then the required number of shield wires, n_{sw}, is:

$$n_{sw} = \text{integer}(\frac{\omega_{swo}}{\Delta x_s}) + 2\,. \tag{14.24}$$

The actual separation distance between two contiguous shield wires, D_s, is given by:

$$D_s = \frac{\omega_{swo}}{n_{sw} - 1}\,. \tag{14.25}$$

The following variables are needed as input:

ℓ_s = length of the substation in meters,
ω_s = width of the substation in meters,
h_b = height of the highest energized object in meters,
S_b = minimum required clearance between a live part and ground in meters,
n_g = ground flash density of lightning strike for the region in number of strokes/km^2/year, and
SOR = specified outage rate in number of outages per year.

The computed outputs are the following:

n_{sw} = number of shield wires,
D_s = distance between adjacent shield wires,
h_s = height of the shield wires in meters,
I_s = shielding current in kA,
θ_s = shielding angle of the shield wires in degrees, and
AOR = actual outage rate.

The actual outage rate (AOR) is a check for the specified outage rate (SOR).

14.5 Design of Substation Shielding by Masts

The principle of protection of a horizontal bus by a vertical mast is shown in Fig. 14.3. On the elevation view, the line GG' is the ground, M is the tip of the vertical mast and B is the end view of the horizontal bus, situated at a height h_b. The line HH' is at a height r_s from the ground, where r_s is the striking distance corresponding to the shielding current, I_s. An arc is drawn of radius r_s with the center at M, cutting the line HH' at A. With A as center, an arc, MDF, of radius r_s is drawn. Objects lying within MDF will be protected from direct lightning strokes for return-stroke currents equal to or greater than I_s. However, this object must be contained within the cone of revolution formed by the arc MDF. For the horizontal bus, situated at a lateral distance x_b from the mast, the protected length, l_p, of the bus is shown in the plan view.

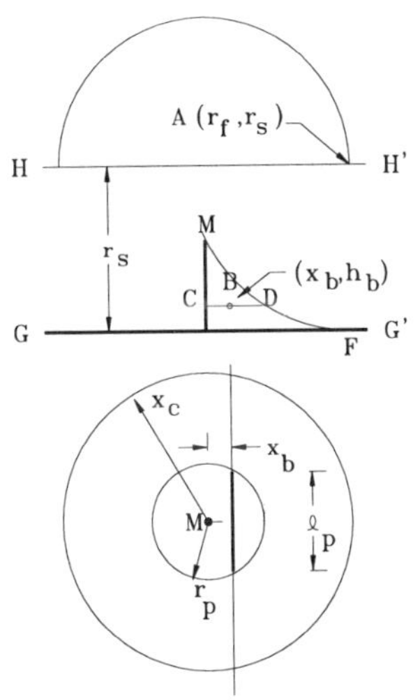

Fig. 14.3 Simple illustration of shielding by mast.

For shielding of bus structures by masts in a substation, rows of masts on either side of a parallel run of busses are generally used. Fig. 14.4 shows the plan view of protection by masts of a $\omega_s\ell_s$ rectangular plot. Here, ω_m = width between two rows of masts = width of the live bus structure + 2 times the minimum specified clearance between grounded object and live parts = $\omega_s + 2m_b$, and ℓ_m = distance between two adjacent masts in the same row. In Fig. 14.4, four masts are used for the protection of the area.

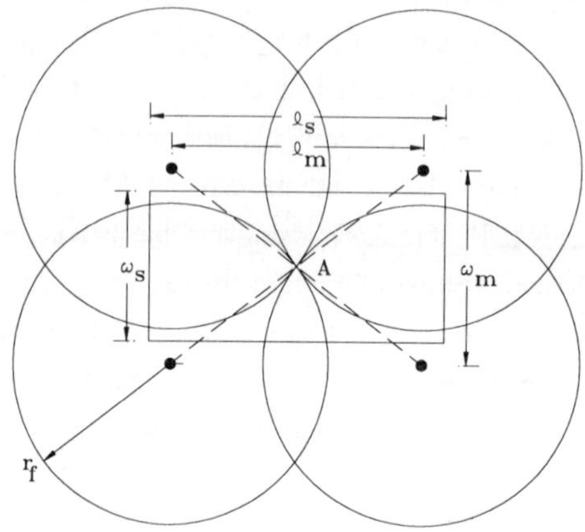

Fig. 14.4 Illustration of protection of horizontal bus by masts.

Each mast has a spherical dome above it, casting a protective shadow below. Each sphere has a radius, r_s, (striking distance) that corresponds to the shielding current, I_s. The circles around the masts in Fig. 14.4 are the circles of intersection of the r_s-spheres around the masts with the r_s-plane above the earth's surface. The radius, r_f, of these circles is:

$$r_f = \sqrt{r_s^2 - (r_s - h_m)^2}, \tag{14.26}$$

where h_m is the mast height. For $h_m = r_s$, $r_f = r_s$. For complete shielding of the rectangular plot ($\ell_s x \omega_s$), the r_s-spheres must cover that area completely; in other words, the r_f-circles must overlap each other so that there is no 'hole' through them. This condition is expressed as:

$$d_m \leq 2r_f, \tag{14.27}$$

where d_m is the length between two diagonally opposite masts. This, however, is not the only condition for complete shielding. The bus section below point A in Fig. 14.4 where the two diagonals of the rectangle ($\ell_m x \omega_m$) intersect is the farthest from the masts, and the height of the point on the r_s-sphere above A is the lowest above the ground plane.

Its height is given by:

$$y_s = h_m + \sqrt{r_s^2 - (0.5d_m)^2}\,. \tag{14.28}$$

Therefore, for protection against return-stroke currents equal to or greater than the shielding current, I_s:

$$y_s - h_b \geq r_s\,. \tag{14.29}$$

$$\text{Or, } h_m \geq r_s + h_b - \sqrt{r_s^2 - (0.5d_m)^2}\,, \tag{14.30}$$

$$\text{where } d_m = \sqrt{\omega_m^2 + \ell_m^2}\,.$$

Equation 14.30 holds if $r_s \geq 0.5d_m$. If $r_s < 0.5d_m$, then obviously more masts must be placed in a row. Assuming the limiting value of $d_m = 2r_s$, ℓ_m is determined as follows:

$$d_{mo}^2 = \ell_{mo}^2 + \omega_m^2 = 4r_s^2\,. \tag{14.31}$$

$$\text{or, } \ell_{mo}^2 = 4r_s^2 - \omega_m^2\,. \tag{14.32}$$

In (14.32), if $\omega_m \geq 2r_s$, then the separation between adjacent rows is too large; therefore, another row of masts will be required. If, however, $\omega_m < 2r_s$, then choose the integer n_{smo} = integer of (ℓ_s/ℓ_{mo}), and let $n_{sm} = n_{smo} + 1$. The total number of masts on each side of the bus will then be $n_{mr} = n_{sm} + 1$, and the distance between adjacent masts in a row, $\ell_m = \ell_s/n_{sm}$.

14.5.1 Bus-end protection

The bus ends may be struck by lightning if they are flush with the outermost masts. The bus section which is situated midway between the two adjacent outermost masts is the most vulnerable spot. This is illustrated by Fig. 14.5.

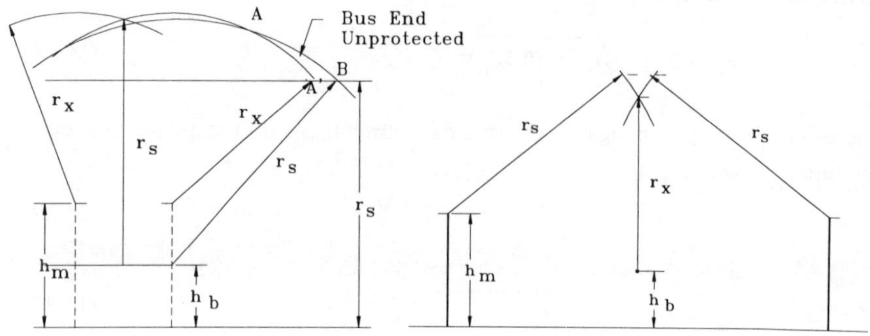

a. Side view of mid-section bus bay　　　　b. End view of bus bay

Fig. 14.5 Bus-end protection.

The r_s-spheres around the masts on two adjacent rows will meet midway between the two masts. The locus of this intersection will be a circle of radius, r_x, given by:

$$r_x = \sqrt{r_s^2 - 0.25\omega_m^2}\,. \tag{14.33}$$

If a circle drawn with the midpoint of the bus end as the center and r_s as the radius (Fig. 14.5a) intersects the r_x-circle at A and the r_s-plane at B, then the midsection of the bus end is exposed to lightning in the AB region. In that case, the outermost masts must be moved so that the points A' and B coincide on the r_s-plane. The horizontal projections of A' and B are given by:

$$x_a^{\prime} = \sqrt{r_x^2 - (r_s - h_m)^2}, \quad \text{and} \tag{14.34}$$

$$x_b = \sqrt{r_s^2 - (r_s - h_b)^2}\,. \tag{14.35}$$

If x_b is greater than x'_a, then the outermost masts must be laterally moved by an amount $(x_b - x'_a)$ on either end. The total length of the substation is thus increased by $2(x_b - x'_a)$. The new area of the substation is then put in (14.3), and the computations repeated until $x'_a \geq x_b$.

14.5.2 Attractive area

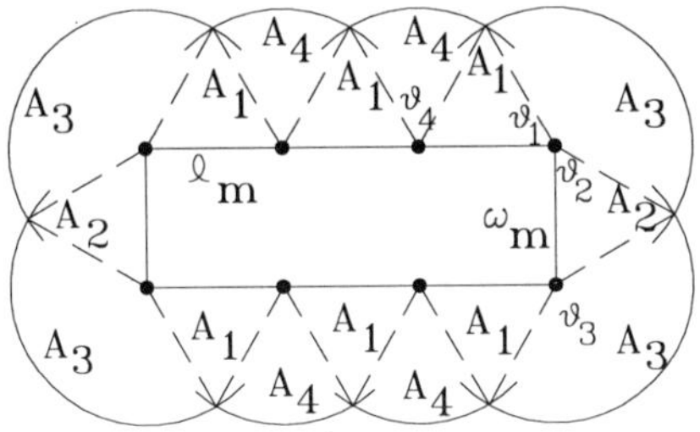

Fig. 14.6 Attractive area for shielding of a substation by vertical masts.

The total area which would attract lightning to the substation is the area covered by the 'electrical shadow' of the masts (Fig. 14.6). This area, A_{ta}, is much larger than the area encompassing the live parts of the substation. This area can be calculated as follows with reference to Fig. 14.6:

$$A_{ta} = A_1 + A_2 + A_3 + A_4 + A_5 + A_6 , \quad (14.36)$$

where

$$A_1 = (n_{mr} - 1)\ell_m r_f \sin\theta_1 , \quad (14.37)$$

$$A_2 = (n_r - 1)\omega_m r_f \sin\theta_2 , \quad (14.38)$$

$$A_3 = 2\theta_3 r_f^2 , \quad (14.39)$$

$$A_4 = (n_{mr} - 1)\theta_4 r_f^2 , \quad (14.40)$$

$$A_5 = (n_r - 2)\theta_5 r_f^2 , \quad (14.41)$$

$$A_6 = (n_{mr} - 1)\ell_m (n_r - 1)\omega_m , \quad (14.42)$$

$$\theta_1 = \arccos\left(\frac{0.5\,\ell_m}{r_f}\right), \tag{14.43}$$

$$\theta_2 = \arccos\left(\frac{0.5\,\omega_m}{r_f}\right), \tag{14.44}$$

$$\theta_3 = 1.5\pi - (\theta_1 + \theta_2), \tag{14.45}$$

$$\theta_4 = \pi - 2\theta_1, \text{ and} \tag{14.46}$$

$$\theta_5 = \pi - 2\theta_2. \tag{14.47}$$

n_{mr}=number of masts per row (4 in Fig. 14.6); n_r=number of rows of masts (2 in Fig. 14.6). A_5 =0 for n_r=2.

The new area, A_{ta}, is then substituted in the right-hand side of (14.3), to compute the actual outage rate, AOR. If the AOR differs from the SOR (specified outage rate) by more than a specified value, the shielding current is recalculated by adjusting the SOR, and the whole procedure is iterated until the AOR and the original SOR are within the limits set forth by the designer.

It should be pointed out that the following condition may sometimes arise initially in (14.3):

$$\frac{SOR}{n_g A} \geq 1\,. \tag{14.48}$$

In that case, the initial value of the shielding current is assumed to be a large number (e.g., 300 kA). The shielding current is automatically adjusted during the iteration process to achieve the specified outage rate (SOR).

The input variables for shielding by masts are the same as in the case for the shielding by shield wires, except that the minimum clearance between live parts and the ground is now the distance between a mast and a live part (M_b). The outputs from the computation are the following:

I_s = shielding current, kA,
h_m = height of each mast, meters, and
n_{mr} = number of masts on each side of bus.

During computation, the number of masts and h_m must be adjusted to provide complete protection. The designer has the option to specify the number of masts per row to optimize the design. If the width of the substation (w_s) is too large, then the height of the masts will be too high to be practical or cost effective. In that case, the width of the substation should be subdivided, and the number of masts and the mast height, h_m, for the subdivided area must be determined.

14.6 Shielding Design by the Method of Rolling Sphere

Lee developed the 'rolling sphere' method of designing substation shielding against lightning strokes [10,11]. According to this method, an imaginary sphere is rolled over the contour of the substation; the radius of the sphere is the striking distance corresponding to the critical current of the return stroke. If the surface of the rolling sphere touches only the shielding structures, i.e., the shield wires and masts, then the station is completely protected. Any live parts if touched by the sphere will be vulnerable to lightning strikes. The principle of rolling sphere is described below.

Figure 14.7 shows a substation protected by four masts, each of height, h_m. The bus height is h_b. Figure 14.8 shows a cross section of Fig. 14.7 cut across the plane connecting the diagonally situated masts M_1 and M_4. The cross section of the r_s-sphere is shown as a circle. The r_s-circle is held by the masts M_1 and M_4, and just touches the bus, B. The radius, r_s, is given by:

$$r_s = \sqrt{(M_4A)^2 - OA^2}, \tag{14.49}$$

where $M_4A = 0.5d_m = 0.5\sqrt{\ell_m^2 + \omega_m^2}$, and (14.50)

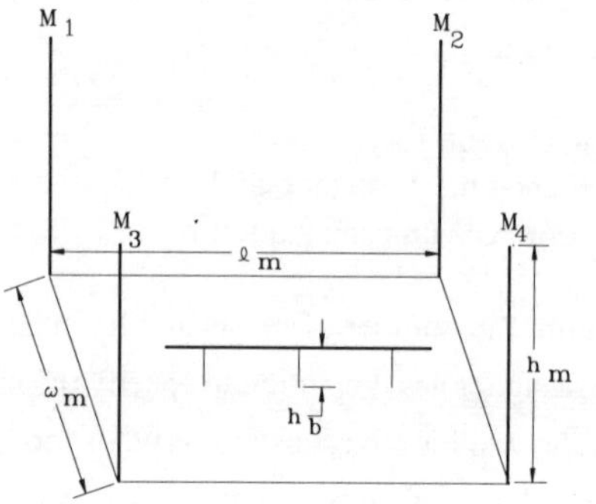

Fig. 14.7 Substation shielded by four vertical masts.

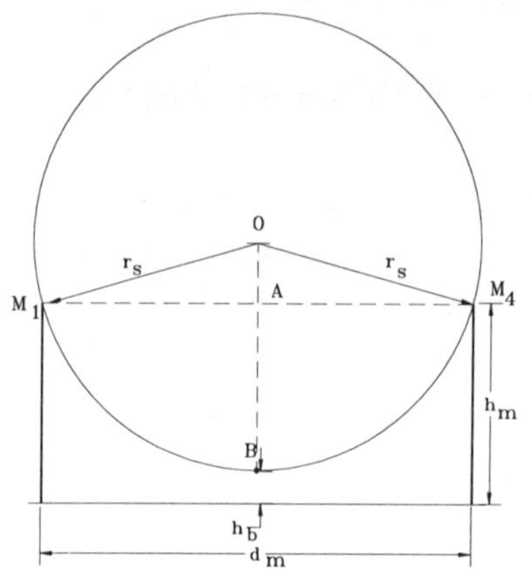

Fig. 14.8 Cross section of Fig. 14.7 across M_1 and M_4 supporting rolling sphere.

$$OA = OB - AB = r_s - (h_m - h_b)\,. \tag{14.51}$$

From (14.49):

$$r_s = \frac{0.25\,d_m^2 + (h_m - h_b)^2}{2\,(h_m - h_b)}\,. \tag{14.52}$$

Once r_s is computed, the shielding current, I_s, can be derived from (14.1). The probability, $P_\ell(I_s)$ of a return-stroke current, I, less than I_s, can be estimated from (14.4).

The attracting area, A_{ta}, is computed from (14.36), and the outage rate of the substation of the given design is estimated from (14.3), with A_s replaced by A_{ta}.

If shielding consists of shield wires, then d_m in the above analysis should be replaced by the lateral distance between the contiguous shield wires.

It should be borne in mind that the rolling sphere method can be applied only when the shielding design is known. In other words, the mast (or shield wire) height and the diagonal distance, d_m, between masts (or lateral distance between shield wires) must be known to estimate the outage rate. However, the rolling sphere method can be used for shielding design by trial and error.

14.7 An Example of Substation Shielding

An analysis was performed for shielding of a substation against direct lightning strokes. The dimensions of the substation were arbitrarily chosen as follows:

Substation length, $\ell_s = 100.0$ m; Substation width, $\omega_s = 100.0$ m
Bus height, $h_b = 10.0$ m; Minimum clearance, M_b or $S_b = 1.0$ m
Ground flash density, $n_g = 10$ per km^2 per year
Specified outage rate, SOR = 1 per 100 years

Table 14.1: Design with Shield Wires

Shield-wire height, $h_s = h_b + S_b = 11.0$ m

n_{sw}	D_{sw}, m	I_{sh}, kA	AOR
8	14.03	8.46	1.004

n_{sw} = number of shield wires; D_{sw} = distance between shield wires
I_{sh} = shielding current; AOR = actual outage rate per 100 years

Table 14.2: Design with Masts

Bus-end protection	n_r	n_{mr}	h_m, m	I_{sh}, kA	AOR	A_{ta}, km^2
No	3	8	32.61	6.43	1.005	0.0222
No	3	12	29.88	6.43	1.006	0.0223
No	3	16	29.23	6.43	1.007	0.0223
No	4	4	24.61	6.39	1.003	0.0224
Yes	4	4	24.89	6.36	0.999	0.0225
Yes	4	6	18.93	6.37	1.004	0.0225
Yes	4	8	17.55	6.36	1.001	0.0225

n_r = number of rows of masts; n_{mr} = number of masts per row
h_m = height of masts; A_{ta} = total attractive area

Comparing Tables 14.1 and 14.2, it can be seen that the number and the height of the shield wires are smaller than those for the masts. However, supports such as towers will be needed for stringing the shield wires. Increasing the number of masts per row allows decrease of the mast height. However, after a certain number, the decrease in the mast height is not that significant. It should also be noticed in Table 12.2 that the total attractive area (A_{ta}) is more than double the area (0.01 km^2) to be protected. The shielding of horizontal components, such as bus or overhead lines, is more effective with a horizontal shield, such as the shield wires, than the vertical shield, i.e., mast. However, masts will be very effective for shielding objects whose vertical dimension is comparatively larger than the horizontal dimension. Ultimately, cost, aesthetics and other non-electrical engineering problems must be considered before a final choice of design is made.

It should also be borne in mind that substation shielding is not designed using solely either shield wires or masts. Shield wires and masts can both be used for protection against direct lightning strokes to a substation.

14.8 Protection of Substations against Traveling Waves [23-38]

Lightning overvoltages can impinge on a substation by (i) voltages induced by lightning strokes to nearby ground, (ii) shielding failure of the overhead lines leading to the substation, and (iii) backflashover from a stroke to a tower. Apart from lightning, substation apparatus can be subjected to transient overvoltages by switching operations either in the substation or in another substation and eventually traveling via the transmission lines. Lightning-induced voltages in substations have not been of concern in the past. However, two recent studies [21,22] have shown that these induced voltages can indeed be high. Failure of the shield wires to intercept lightning will result in a direct stroke to the phase conductor. If the amplitude of the stroke current is lower than the critical current of the line, then no insulator flashover will occur, and the voltage will impinge on the substation. A direct stroke to a tower may cause an insulator string of the tower to flash over, thus applying the tower crossarm voltage to the phase conductor. This backflashover voltage will travel towards the substation and eventually impinge on the substation apparatus.

These transient voltages can have a wavefront in the submicrosecond regime at the source. The transient voltages are attenuated and their wavefront flattened as they travel, caused by (i) conductor losses, (ii) ground losses due to finite soil electrical conductivity, and (iii) impulse corona on the line. The effect of the conductor losses is the least of the three effects, and is generally neglected. The losses in the ground attenuate the traveling wave, and the velocity of travel is slower than that in free space. For a three-phase line, three modes of propagation exist - two line modes and one ground mode (Chapter 3). The two line-mode components are not affected by the soil resistivity. Because of the attenuation and slower velocity of the ground mode, the composite traveling wave is sheared (i.e., elongated) as it travels along the line. All three modes of propagation are affected by impulse corona due to ionization surrounding the conductors. They become distorted and attenuated as they travel. The attenuation and distortion of traveling waves have been discussed in Chapter 3.

Ideally, a lightning arrester should be installed directly across the apparatus it protects. However, this is not always possible because of the minimum clearance required between a live conductor and a grounded object. Sometimes, one single lightning arrester is installed to protect two apparatus for cost effectiveness (Fig. 14.9). The effects of the lightning-arrester location and the steepness of the wavefront of the oncoming traveling voltage on the actual voltage level across the apparatus is shown by an example in Fig.14.10.

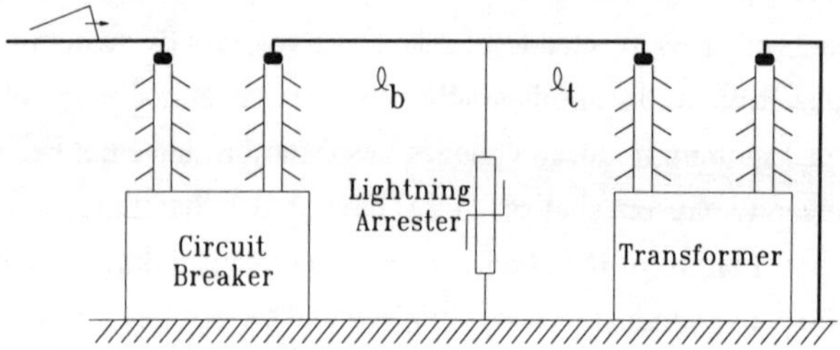

Fig. 14.9 Voltages across substation equipment from an oncoming traveling voltage wave.

In Fig. 14.10a, a traveling voltage wave, with a rate of rise of 1000 kV/μs, arrives at the location of the circuit breaker (Fig. 14.9) at t=0. At t=0.1 μs it arrives at the lightning arrester, and rises at the same rate (1000 kV/μs). It arrives at the transformer at t=0.2 μs, but rises at 2000 kV/μs because of the high impedance of the transformer. A reflected voltage wave, rising at 1000 kV/μs, travels back towards the arrester and the circuit breaker. The reflected voltage arrives at the arrester and at the circuit breaker at t=0.3 μs and t=0.4 μs, respectively, when the total voltage at each of these two locations starts to rise at 2000 kV/μs. The voltage across the arrester reaches the arrester discharge voltage level of 400 kV at t=0.4 μs. The arrester voltage is held at this level after this instant. To hold this voltage at 400 kV against the oncoming voltage rising at 2000 kV/μs rate, a -2000 kV/μs voltage is transmitted from the arrester location in both directions. It reaches the circuit breaker at t=0.5 μs where it is added to the existing voltage which is rising at 2000 kV/μs. As a result, the total voltage at the circuit breaker remains constant at 600 kV after t=0.5 μs. The voltage transmitted from the arrester towards the transformer (at -2000 kV/μs rate) also reaches the transformer at t=0.5 μs, rising at -4000 kV/μs at the transformer terminal because of the high transformer impedance. The total voltage at the transformer falls at 2000 kV/μs rate at this instant. It should be noticed that the circuit breaker and the transformer voltages are 600 kV although the arrester discharge voltage is only 400 kV. This rise in voltage is caused by the separation distance between the arrester and the protected apparatus. In Fig. 14.10b, the rate of rise of the oncoming voltage wave is 3000 kV/μs. The arrester

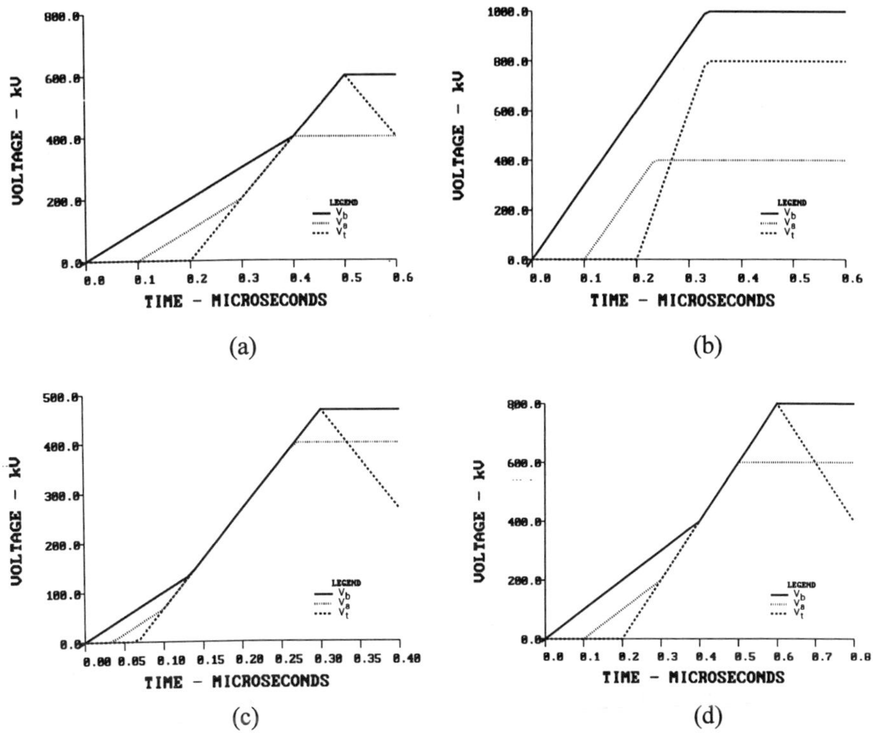

Fig. 14.10 Voltages across substation apparatus caused by traveling voltage waves on incoming lines.

(a) α=1000 kV/μs, $\ell_b=\ell_t$=30 m, V_{ad}=400 kV;
(b) α=3000 kV/μs, $\ell_b=\ell_t$=30 m, V_{ad}=400 kV;
(c) α=1000 kV/μs, $\ell_b=\ell_t$=10 m, V_{ad}=400 kV;
(d) α=1000 kV/μs; $\ell_b=\ell_t$=30 m; V_{ad}=600 kV.

Note: V_a=arrester voltage; V_b=breaker voltage; V_t=transformer voltage; α= voltage rate of rise; V_{ad}=lightning arrester discharge voltage.

discharge voltage level is reached sooner, but the voltages at the circuit breaker and the transformer are higher than those in Fig. 14.10a because of the higher rate of rise of the oncoming voltage. Shorter separation distance between the arrester and the protected apparatus lowers the voltage at the protected apparatus (Fig. 14.10c). Higher arrester discharge voltage increases the apparatus voltage (Fig. 14.10d).

A simple relationship can be obtained between the rate of rise, α, of the oncoming voltage, the arrester discharge voltage, V_{ad}, the peak voltages of the circuit breaker, V_{bp}, and the transformer, V_{tp}, and the separation distances of the apparatus and the arrester, ℓ_b, and ℓ_t. This is shown below by redrawing Fig. 14.10a.

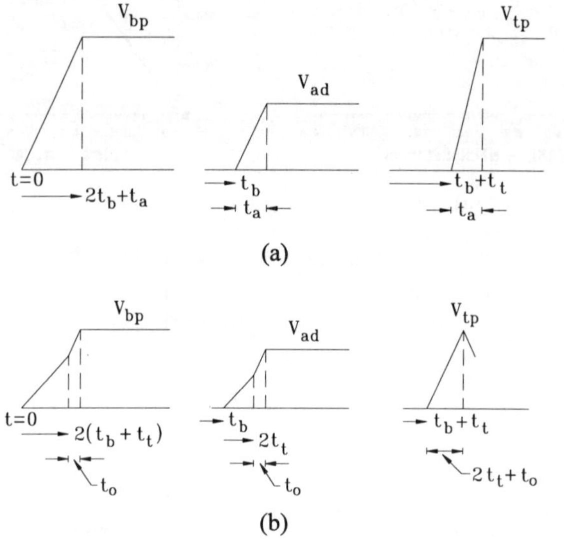

Fig. 14.11 Voltage profiles of substation apparatus caused by traveling voltage wave.

Two cases are shown in Fig. 14.11. In the first case, the arrester discharges before the reflected voltage arrives from the transformer (Fig. 14.11a); in the second case, the arrester discharges after the reflected voltage has arrived from the transformer. In both cases, $t_b=\ell_b/300$ (μs) and $t_t=\ell_t/300$ (μs). In the first case, $t_a=V_{ad}/\alpha$, where V_{ad} is the arrester discharge voltage in kV and α is the rate of rise of the oncoming traveling voltage in kV/μs. In the second case, $t_o=(V_{ad}/2\alpha) - t_t$.

For case I:

$$V_{bp}=2\,\alpha\,t_b+V_{ad}=\frac{2\,\alpha\,\ell_b}{300}+V_{ad}\,,\ \text{and} \tag{14.53}$$

$$V_{tp}=2\,\alpha\,t_a=2\,V_{ad}\,. \tag{14.54}$$

For case II:

$$V_{bp}=2\alpha t_b+V_{ad}=\frac{2\alpha \ell_b}{300}+V_{ad}\text{, and} \tag{14.55}$$

$$V_{tp}=2\alpha t_t+V_{ad}=\frac{2\alpha \ell_t}{300}+V_{ad}\,. \tag{14.56}$$

The arrester discharge voltage, V_{ad}, is not a constant parameter; it is a function of the discharge current and the rate of rise of the oncoming wave. The permissible steepness of the oncoming voltage wave can be estimated once the withstand voltage levels of the station apparatus are known, the proper arrester is selected and the separation distance of the arrester from the protected equipment has been specified from other design constraints. If the expected steepness of the voltage at its source is higher than the permissible steepness, then action must be taken to reduce the steepness of the oncoming voltage wave as it enters the station. This is discussed below.

Modal propagation over imperfect earth and impulse corona (Chapter 3) will also reduce the steepness of the oncoming wave. Generally, the effect of the modal propagation is small over a short distance, and hence, neglected. The effect of the impulse corona is predominant. The flattening of the wavefront can be estimated as shown in Chapter 3, or the effects of the impulse corona on a specific conductor arrangement can be obtained from field data. Eriksson and Weck [34] have suggested simple corona damping constants for negative-polarity lightning overvoltages as shown in Table 14.3. The damping constant, K, is expressed in terms of microseconds of front elongation per kV of applied voltage per one meter of travel. For instance, the front of a 1000 kV voltage wave will be elongated by 1.5 μs for every 1 km of travel along a single-conductor overhead line. In terms of front steepness, a voltage wave of α=1000 kV/μs will have an α=667 kV/μs after 1 km of travel.

If the permissible steepness of a traveling voltage wave entering a substation is 500 kV/μs, and the expected steepness at the source is 1000 kV/μs, then this voltage must travel 1.33 km over a single-conductor line before it enters the substation. This means that the line entering the station must be shielded and the tower-footing resistance minimized for a distance of 1.33 km beyond the station so that no shielding

Table 14.3: Damping Constant, K, for Impulse Corona [34]
Negative-polarity voltage wave

Conductor	K (μs/kV.m)
Single conductor	1.5×10^{-6}
2-bundle conductor	1.0×10^{-6}
3- and 4-bundle conductor	0.6×10^{-6}
6- and 8-bundle conductor	0.4×10^{-6}

failure nor backflashover occurs within this length of the line. This is the limiting distance. If the line design outage rate is one outage per 100 km of line per year, then there will be 0.0133 strikes per year within this stretch of line length or about one outage in the substation in about 75 years. If this outage rate of the station is not permissible, then better shielding and lower tower-footing resistance will be required for this length of the line. As an alternative, the separation distance between the arrester and the protected apparatus must be shortened.

Frequently, multiple lines enter a substation. If the number of incoming lines is n, then the steepness of the wavefront at the station entrance will be reduced by n. Consequently, the α's in (14.53) to (14.56) should be replaced by α/n. The limiting distance over which the wave must travel to reduce its steepness to the permissible value will also be reduced by n. Although the number of strikes within the limiting distance will also be reduced by n of that for a single line, the total number of traveling waves originated within the limiting distance and entering the station will remain the same because of n number of entry paths. Therefore, the station outage rate will remain the same. This is for the case when all n lines have separate rights of way.

Further gains will be made if the lines have the same right of way. For example, the limiting distance for a double-circuit line will be half that for a single-circuit line. The outage rate will also be half that of a single-circuit line because of the shorter limiting distance. For the same outage rate, then, the limiting distance for the double-circuit line will be further reduced.

An underground cable, connected between the overhead line and the station, will significantly reduce the steepness of the entering traveling wave. The front of the wave will rise exponentially with a time constant CZ_0, where C is the capacitance of the cable and Z_0 is the surge impedance of the overhead line (Problem 3 of Chapter 2). Because of the short tail (generally, less than 10 μs to half value) for the voltages due to backflashover, eventual overvoltage will be avoided. A discrete capacitance, such as a voltage transformer, at the station entrance will also reduce the front steepness of the oncoming wave. The apparatus capacitance, such as the bushing capacitance of the circuit breaker and transformer, and transformer winding capacitance, will also reduce the front steepness. However, their effect is not that significant.

14.9 References to Chapter 14

1. C.F. Wagner, G.D. McCann and G.L. MacLane, "Shielding of Transmission Lines", ***AIEE Trans.***, Vol. 60, pp. 313-328, 1941.

2. C.F. Wagner, G.D. McCann and C.M. Lear, "Shielding of Substations", ***AIEE Trans.***, Vol. 61, pp. 96-100, 1942.

3. R.H. Schlomann, I.B. Johnson, W.S. Price and J.G. Anderson, "1956 Lightning Field Investigation on the OVEC 345-kV System", ***AIEE Trans.***, Vol. 76, Part III, pp. 1447-1459, 1958.

4. G.W. Brown and E.R. Whitehead, "Field and Analytical Studies of Transmission Line Shielding: Part II", ***IEEE Trans. on Power Apparatus and Systems***,Vol. PAS-88, pp. 617-626, 1969.

5. P. Chowdhuri and A.K. Kotapalli, "Significant Parameters in Estimating the Striking Distance of Lightning Strokes to Overhead Lines", ***IEEE Trans. on Power Delivery***, Vol. 4, pp. 1970-1981, 1989.

6. A.M. Mousa and K.D. Srivastava, "A Revised Electrogeometric Model for the Termination of Lightning Strokes on Ground Objects", ***Proc. International Aerospace and Ground Conference on Lightning and Static Electricity***, pp. 342-352, 1988.

7. M.A. Sargent, "Monte Carlo Simulation of the Lightning Performance of Overhead Shielding Networks of High Voltage Stations", ***IEEE Trans. on Power Apparatus and Systems***, Vol. PAS-91, pp. 1651-1656, 1972.

8. H. Linck, "Shielding of Modern Substations against Lightning Strokes", ***IEEE Trans. on Power Apparatus and Systems***, Vol. PAS-94, pp. 1674-1679, 1975.

9. A.M. Mousa, "Shielding of High-Voltage and Extra-High-Voltage Substations", ***IEEE Trans. on Power Apparatus and Systems***, Vol. PAS-95, pp. 1303-1310, 1976.

10. R.H. Lee, "Lightning Protection of Buildings", ***IEEE Trans. on Industry Applications***, Vol. IA-15, pp. 236-240, 1979.

11. R.H. Lee, "Protect Your Plant Against Lightning", ***Instruments & Control Systems***, pp. 31-34, 1982.

12. F. Hofbauer, "The Protection of High Voltage Substations Against Lightning", ***Proc. CIGRE***, Paper No. 33-02, 1988.

13. S. Sadovic, M. Rascic and M. Cirkovic, "An Iterative Program for the High Voltage Substation Shielding Analysis Against Lightning", ***Proc. 20th International Conf. on Lightning Protection***, Paper No. 2.1, Interlaken, Switzerland, 1990.

14. A.M. Mousa, "A Computer Program for Designing the Lightning Shielding Systems of Substations", ***IEEE Trans. on Power Delivery***, Vol. 6, pp. 143-149, 1991.

15. AIEE Committee Report, "A Method of Estimating the Lightning Performance of Transmission Lines", ***AIEE Trans.***, Vol. 69, pp. 1187-1196, 1950.

16. P. Chowdhuri, "Shielding of Substations Against Direct Lightning Strokes", ***IEEE Trans. on Power Delivery***, Vol. 9, pp. 314-323, 1994.

17. M.R. Spiegel, ***Probability and Statistics***, Schaum's Outline Series in Mathematics. New York: McGraw-Hill, 1975.

18. R.B. Anderson and A.J. Eriksson, "Lightning Parameters for Engineering Applications", ***Electra***, No. 69, pp. 65-102, 1980.

19. A.J. Eriksson, "Notes on Lightning Parameters for System Performance Estimation", CIGRE Note 33-86 (WG33-01) IWD, July 15, 1986.

20. J.G. Anderson, "Lightning Performance of Transmission Lines", Chapter 12 of ***Transmission Line Reference Book 345 kV and Above***, Second Edition, Revised. Electric Power Research Institute Publication No. EL-2500, 1987.

21. K.H. Lee and J.M. Schneider, "Rockport Transient Voltage Monitoring System: Analysis and Simulation of Recorded Waveforms", ***IEEE Trans. on Power Delivery***, Vol. 4, pp. 1794-1805, 1989.

22. S. Okabe, M. Kan and T. Kouno, "Analysis of Surges Measured at 550 kV Substations", ***IEEE Trans. on Power Delivery***, Vol. 6, pp. 1462-1468, 1991.

23. E.R. Whitehead, "Development in Lightning Protection of Stations", ***AIEE Trans.***, Vol. 60, pp. 898-900, 1941.

24. I.W. Gross, T.J. Bliss and J.K. Dillard, "Lightning Protection in Extra-High-Voltage Stations - Analysis, Anacom Study, and Results", ***AIEE Trans.***, Vol. 71, Part. III, pp. 482-492, 1952.

25. I.W. Gross, L.B. Levesconte and J.K. Dillard, "Lightning Protection in Extra-High-Voltage Stations - Influence of Multiple Circuits", ***AIEE Trans.***, Vol. 72, Part. III, pp. 882-890, 1953.

26. I.W. Gross, S.B. Griscom, J.M. Clayton and W.S. Price, "High-Voltage Impulse Tests in Substations", ***AIEE Trans.***, Vol. 73, Part. III, pp. 210-222, 1954.

27. S.B. Griscom, J.K. Dillard and A.R. Hileman, "Direct-Stroke Protection of High-Voltage Switching Stations and Transformers", ***AIEE Trans.***, Vol. 74, Part. III, pp. 354-368, 1955.

28. J.M. Clayton and R.W. Powell, "Application of Arresters for Complete Lightning Protection of Substations", ***AIEE Trans. on Power Apparatus and Systems***, No. 44, pp. 1608-1614, 1959.

29. J.M. Clayton and F.S. Young, "Application of Arresters for Lightning Protection of Multiline Substations", ***AIEE Trans. on Power Apparatus and Systems***, No. 49, pp. 566-575, 1960.

30. AIEE Working Group, "Simplified Method for Determining Permissible Separation between Arresters and Transformers", ***IEEE Trans. on Power Apparatus and Systems***, Special Supplement, Paper No. 63-229, pp. 35-57, 1963.

31. A.R. Hileman, W.M. Guyker, R.W. Powell, W.A. Richter and J.M. DeSalvo, "Insulation Coordination in APS 500-kV Stations", ***IEEE Trans. on Power Apparatus and Systems***, Vol. PAS-86, pp. 655-665, 1967.

32. J.P. Bickford and M.S. Savic, "Some Aspects of System Modelling for the High Voltage Substations", ***IEE Proc.***, Vol. 131, Part. C, pp. 204-209, 1984.

33. M.S. Savic, "Sensitivity Analysis of Lightning Performance Calculations for Transmission Lines and Substations", ***IEE Proc.***, Vol. 132, Part. C, pp. 217-223, 1985.

34. A.J. Eriksson and K.H. Weck, "Simplified Procedures for Determining Representative Substation Impinging Lightning Overvoltages", ***Proc. CIGRE***, Paper No. 33-16, 1988.

35. M.S. Savic, "Engineering Method for High-Voltage Substations Lightning Performance Estimation", ***IEE Proc.***, Vol. 136, Part. C, pp. 222-229, 1989.

36. J. Panek, P.F. Albrecht and H. Elahi, "Design Criteria for Lightning Protection of HV Substations", ***IEEE Trans. on Power Delivery***, Vol. 7, pp. 489-498, 1992.

37. M.S. Savic and Z. Stojkovic, "An Expert System for High-Voltage Substations Lightning Performance Estimation", ***IEEE Trans. on Power Delivery***, Vol. 7, pp. 1223-1231, 1992.

38. M. Savic and Z. Stojkovic, "High Voltage Substation Equivalent Circuits in the Lightning Performance Estimation", ***IEE Proc.-Generation, Transmission and Distribution***, Vol. 141, pp. 99-105, 1994.

CHAPTER 15
INSULATION COORDINATION

15.1 Introduction

By definition, insulation coordination is "the selection of the dielectric strength of equipment in relation to the voltages which can appear on the system for which the equipment is intended and taking into account the service environment and the characteristics of the available protective devices" [1]. Insulation coordination is the judicious correlation of the insulation of electrical equipment and circuits with the characteristics of the protective devices for obtaining the optimal margin of protection against overvoltages [2,3]. Yet it must be consistent with good economics and non-interference with other operating conditions. In other words, the insulation of a power apparatus should be selected in such a way that its withstand voltage level under transient voltages is higher than the level to which an oncoming transient voltage will be limited by the protective device. Naturally, insulation coordination does not give a unique solution. It depends heavily on the design engineer's judgment and previous experience.

The voltage withstand capability of an electrical insulation system is not unique. Its withstand capability varies with the rate of rise and the duration (i.e., waveshape) of the applied voltage. It is inconceivable to design an insulation system which can withstand all anticipated transient overvoltages. Therefore, surge protectors (i.e., lightning arresters) are widely applied in power systems to limit the transient overvoltages to safe levels. The protective level of the surge protector, like the system insulation, is also sensitive to the waveshape of the applied transient voltage. The basic principle of insulation coordination is to select surge protectors for an insulation system such that the protective level of the protector is always below the withstand level of the insulation with a safe margin. The concept of insulation coordination is illustrated in Fig. 15.1.

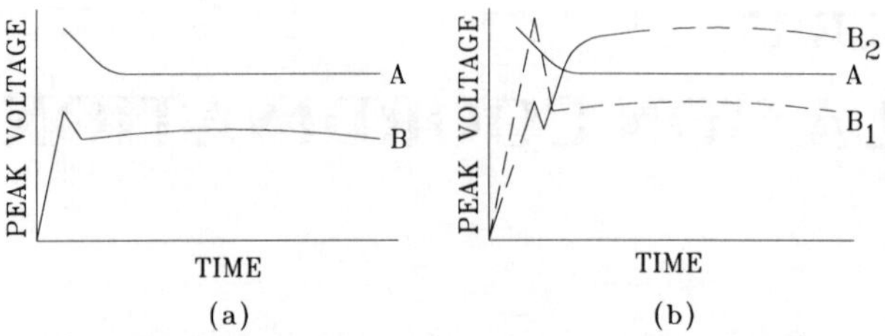

Fig. 15.1 Illustration of insulation coordination with a gapped valve-type arrester.
(a) proper coordination; (b) improper coordination.
Curve A: insulation strength of power apparatus;
Curve B: voltage-limiting characteristic of arrester.

Curve A is the demonstrated (rated) impulse strength of the insulation of a power apparatus. Curve B is the voltage developed across a valve-type lightning arrester when it conducts under a lightning voltage wave. Figure 15.1(a) illustrates a proper insulation coordination where the impulse strength of the apparatus insulation is higher than the voltage developed across the lightning arrester. The peak of curve B denotes the sparkover voltage of the gap elements connected in series with the valve blocks; the relatively flat part of curve B is the discharge voltage across the valve blocks due to current flow. Figure 15.1(b) illustrates improper insulation coordination, where the voltage across the protective device is higher than the impulse strength of the power apparatus for a certain period of time. For B_1, the sparkover voltage of the arrester is higher than the impulse strength of the power apparatus, even though its discharge voltage is below it; for B_2, although the sparkover voltage is below the impulse strength of the apparatus, its discharge voltage is higher. The power apparatus will not be completely protected in either case.

The objective of insulation coordination is not only to protect the power apparatus against transient overvoltages, but also to prevent damages to the protective device as well. It should be borne in mind that the protective device has a limited capability to dissipate energy. A system fault may raise the power-frequency line-to-ground voltage above its normal value. For instance, a ground fault on one phase of a three-phase system will raise the voltages of the two remaining unfaulted phases; the amount depends upon the type of neutral grounding of the system. For a solidly grounded neutral system, the voltage rise on the unfaulted phases will be minimum, and the voltage rise will be maximum for an isolated neutral system. The spark gaps in the gapped valve-type arrester must not spark over under these temporary power-frequency overvoltage conditions. If they do, then excessive energy will be dissipated in the arrester, and the arrester will be destroyed. Similarly, during recovery from discharging a lightning stroke, the arrester spark gaps may not be able to quench the power-frequency arc if the power-frequency voltage of the line is above that specified for the safe operation of the arrester, and consequently would burn themselves to destruction. For a gapless arrester, excessive power-frequency voltage will force large leakage current through the valve blocks, exceeding the energy-handling capability of the arrester. Thus the arrester must protect the power apparatus and also be capable of restoring itself as an insulator against whatever system voltage might exist across it. Proper insulation coordination thus boils down to: (i) selection of the basic impulse insulation level of the power apparatus, and (ii) application of protective devices that will provide a reliable margin of protection to the protected apparatus.

15.2 Overvoltage Classification for Insulation Coordination

The electrical insulation is continuously stressed by the system operating voltage, whereas it is occasionally stressed by overvoltages. The overvoltages are events whose statistical distribution can be assessed. An overvoltage is defined as a voltage whose peak value exceeds the peak of the maximum operating voltage. The waveshapes of the overvoltages are widely divergent. However, they can be broadly classified as (i) atmospheric overvoltages, (ii) switching overvoltages, and (iii) sustained overvoltages [4].

Atmospheric overvoltages are caused by lightning discharges, which are assumed to be current sources. The voltages developed across an apparatus by the atmospheric discharges are then dependent upon the impedance the apparatus gives to these discharges. These lightning-caused voltages are generally characterized by unidirectional impulse voltages. Although the waveshapes of the atmospheric overvoltages are numerous, a standard waveshape has been specified for the purpose of standardizing the impulse strength and testing of power apparatus [5,6]. This standard impulse voltage wave rises exponentially to its peak in 1.2 μs, and decays exponentially to its half value in 50 μs. It is designated as the 1.2/50-μs impulse voltage wave, and is shown in Fig. 15.2.

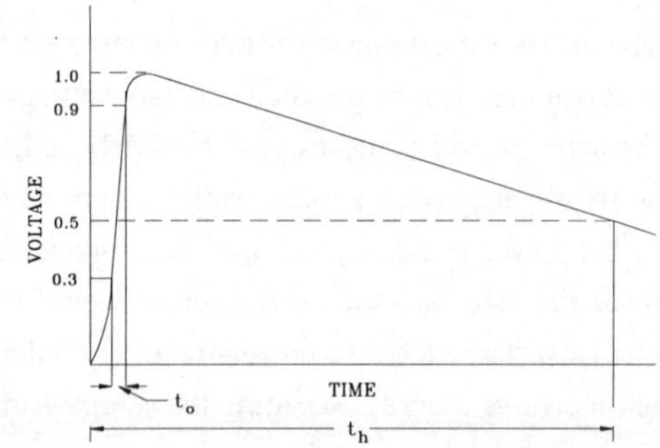

Fig. 15.2 Standard lightning impulse voltage.

If the time interval between the 30 percent and 90 percent points on the voltage scale is t_o, then the front time is defined as $t_f = t_o/0.6$. For the lightning impulse voltage, $t_f = 1.2$ μs±30 %, and time to half value, $t_h = 50$ μs±20 %.

Switching overvoltages occur during switching operations in a power system. Switching overvoltages are damped oscillatory voltages, oscillating at the system resonant frequencies. However, for ease of generation in the laboratories and testing of power apparatus, the standards [5,6] have specified the standard switching surge as an exponentially rising and exponentially falling unidirectional voltage wave with a 250/2500-μs waveshape. The front time is defined as the time to the actual peak of the voltage, unlike the standard lightning impulse voltage.

Sustained overvoltages are power-frequency dynamic voltages which may occur at the receiving end of a long high voltage unloaded or lightly loaded transmision line. Sustained overvoltages may also occur due to a sudden loss of load. High power-frequency voltage would occur across the unfaulted phases when a fault occurs on one or more phases of a three-phase system. The magnitudes of these sustained voltages will be maximum for isolated neutral systems.

15.3 Classification of Insulation for Insulation Coordination

For the purpose of insulation coordination, the electrical insulation system in a power system is broadly classified into two categories: (i) self-restoring insulation, and (ii) non-self-restoring insulation. A self-restoring insulation completely recovers its dielectric properties after a discharge (i.e., flashover). A non-self-restoring insulant is partially or completely damaged after a discharge. Gaseous and liquid insulants are self-restoring insulation, whereas solid insulants are non-self-restoring. In power systems, the external insulation in air, such as the line insulators and bushings, is considered as the self-restoring type. Although oil in transformers and circuit breakers, and gases, such as sulfur hexafluoride, in gas insulated apparatus are self-restoring, they are classified as internal or non-self-restoring insulants; an uncontrolled arc within an enclosure can cause damage to the enclosed apparatus due to enormous pressure built up during the discharge. For the coordination of insulation, the self-restoring insulation is subject to greater risk of failure than the non-self-restoring type.

The dielectric strength of an insulant is not a unique property; it is a function of the rate of change and duration of the applied voltage. The dielectric strength of an insulant is also a statistical quantity, meaning that the insulant may fail a few times and withstand other times the same voltage applied repeatedly under the same ambient conditions. Therefore, the voltage withstand capability of a power apparatus under a specific type of voltage is determined from statistically designed tests. For the external insulation in air, the withstand capability under a specified voltage type is gauged by its critical flashover voltage, V_{50}, defined as the voltage level of the specified voltage type where the insulant has the 50 percent probability of flashover, and 50 percent probability of withstand. The critical flashover voltage of an insulator is given for both the 1.2/50-μs lightning and the 250/2500-μs switching surge voltage waves. For the internal non-self-restoring insulants, the BIL (basic impulse insulation level) and the BSL (basic switching impulse insulation level) are specified to define their withstand capabilities under the standard lightning and switching surge voltage waves.

15.3.1 Lightning impulse characteristics of electrical insulation

The breakdown of an electrical insulation under transient voltages of short duration, such as those caused by lightning, may occur on the falling part of the voltage wave at a level lower than the peak. This is caused by the finite time needed to complete the breakdown process. The breakdown process is initiated by an avalanche process which may or may not lead to a complete breakdown. The avalanche process, in turn, is initiated by an electron which ionizes neutral atoms and molecules if its kinetic energy is raised to a critical level by the applied electric field. The electrons, liberated from the ionized atoms and molecules, take part in ionizing other neutral atoms and molecules, causing an ion space charge and an electron avalanche. The breakdown of the insulation occurs when the avalanche has traversed the space between the electrically stressed electrodes. The avalanche may start but may not result in a breakdown of the insulation if the applied electric field is not high enough for the avalanche to traverse the electrode space. The total time for the insulation breakdown, known as the time lag of breakdown, is comprised of the statistical time delay and the formative time delay. The statistical time delay results because the electron with the critical kinetic energy is not available at the instant of the application of the voltage. As the name implies, the statistical time delay is random. The formative time delay is the time required by the avalanche to span the whole interelectrode space. Longer interelectrode space causes longer formative time delay; higher applied voltage (i.e., electric field) causes shorter formative time delay.

An insulation system will withstand an applied transient voltage if the peak amplitude of the voltage is lower than the withstand voltage level of the insulation. If the applied transient voltage is increased, the insulation will break down on the tail (falling part) of the voltage wave (Fig. 15.3a). As the applied voltage is progressively increased, the time delay of breakdown on the tail of the voltage wave becomes shorter. At still higher applied voltage, the breakdown will occur on the front of the applied voltage wave (Fig. 15.3b).

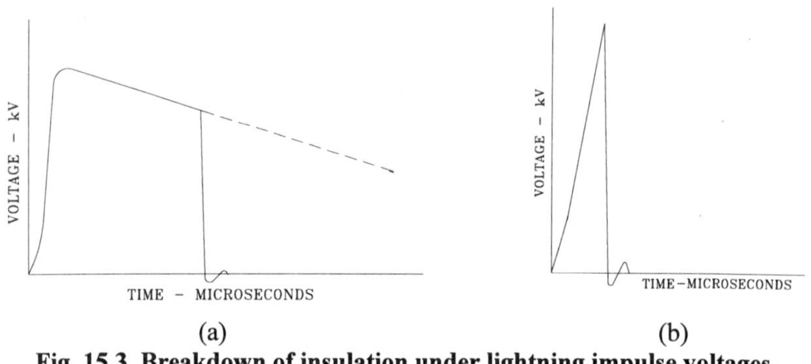

Fig. 15.3 Breakdown of insulation under lightning impulse voltages.
(a) Chop at tail; (b) Chop on front.

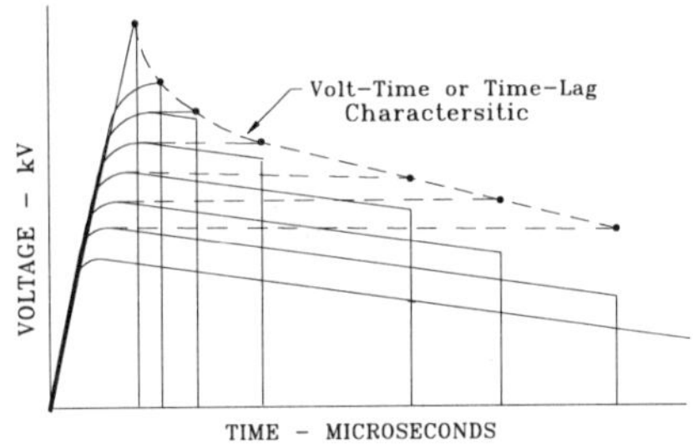

Fig. 15.4 Volt-time or time-lag characteristic of insulation.

The volt-time or the time-lag curve for an insulation is a plot of the breakdown voltage against the time at which breakdown occurs. If the breakdown occurs after the peak of the applied voltage, then the peak of the voltage, instead of the actual breakdown level, is plotted on the curve. The construction of the volt-time curve is shown in Fig. 15.4. The volt-time characteristic of an insulation system depends upon the waveshape as well as the polarity of the applied impulse voltage. The volt-time characteristic under the standard 1.2/50-μs impulse voltage is generally given by the manufacturer. The lower of the two (positive- or negative-polarity) volt-time curves is used for insulation coordination.

Apparatus insulation level must be equal to or greater than the specified BIL. The power transformer is a complex apparatus in regard to its withstand capability under transient voltages. We discussed in Chapter 12 that the transformer insulation experiences nonuniform voltage stresses under transient voltages, and that the internal voltage oscillations may be many times higher than the applied terminal voltage. Consequently, internal breakdown in a transformer may occur across the major insulation (insulation to ground), across the minor insulation (insulation between turns and coils), internally within the bushing, or external flashover may occur across the bushing live terminal and ground. The internal breakdown of transformer insulation is permanent, and therefore costly. Transformers are usually subjected to an additional impulse test, i.e., ***CW (chopped wave) test***. During the CW test, an external rod gap is sparked at the tail of the voltage wave (Fig. 15.3a). The liquid-immersed transformers may be subjected to one more test when called for, i.e., the ***FOW (front-of-wave) test***, where the external gap is sparked over at the front of the applied voltage wave (Fig. 15.3b). The CW test level is usually 15 to 20 percent higher than the FW (full-wave) test level, and the FOW test level is still higher. The various insulation levels of oil-immersed power transformers are shown in Table 15.1.

Table 15.1: Insulation Levels of Oil-Immersed Power Transformers [7]

Nominal System Voltage (kV r.m.s.)	BIL (kV crest)	C-W Level (kV crest)	BSL (kV crest)
15 kV and below	110	120	-
25	150	165	-
34.5	200	220	-
46	250	275	-
69	250	275	-
	350	385	-
115	350	385	280
	450	495	375
	550	605	460
138	450	495	375
	550	605	460
	650	715	540
161	550	605	460
	650	715	540
	750	825	620
230	650	715	540
	750	825	620
	825	905	685
	900	990	745
345	900	990	745
	1050	1155	870
	1175	1290	975
500	1300	1430	1080
	1425	1570	1180
	1550	1705	1290
	1675	1845	1390
765	1800	1980	1500
	1925	2120	1600
	2050	2255	1700

This table has been reprinted from *IEEE Std C57.12.00-1993, IEEE Standards General Requirements for Liquid-Immersed Distribution, Power and Regulating Transformers*, Copyright © 1993 by the Institute of Electrical and Electronics Engineers, Inc. The IEEE disclaims any responsibility or liability resulting from the placement and use in this publication. Information is printed with permission of the IEEE.

15.4 Characteristics of Protective Devices

The valve-type lightning arrester is the most reliable overvoltage protective device available today. Its characteristics were discussed in Chapter 13; it will again be discussed here, but as a system component with reference to insulation coordination.

Lightning arresters are classified in four categories. The different classes and their voltage ratings are specified as [8]:

Secondary class (175 and 650 volts),
Distribution class (1 to 30 kV),
Intermediate class (3 to 120 kV),
Station class (3 to 684 kV).

Arresters are classified according to the prescribed tests to be performed on them [6]. The station class arresters are suitable for protection of station equipment, and high-voltage line insulators. They can dissipate more transient energy, and they limit the transient voltages to lower limits (in per unit quantity) than the intermediate class.

The ***voltage rating*** of an arrester, as shown above, is the maximum permissible operating voltage across its terminals at which it is designed to perform its duty cycle tests successfully. Therefore, the actual operating voltage of the arrester should not exceed its rated voltage; otherwise, the risk of arrester failure will be great.

Lightning arresters with series-connected gaps cannot conduct the energy of the transient voltage unless the gap elements spark over. The level of the sparkover voltage is not constant, but varies with the front duration of the applied transient voltage; the shorter the front duration, the higher is the sparkover voltage level. Therefore, a ***FOW (front-of-wave) sparkover voltage rating*** is specified to a lightning arrester to ensure proper insulation coordination. Gapless lightning arresters do not encounter the sparkover problem; however, they are also assigned an FOW rating in conformity with the general arrester ratings. For gapless arresters, the FOW rating is the peak discharge voltage, rising to peak in 0.5 μs, across the arrester at a specified discharge current level. This specified discharge current varies from 20 kA for the highest station-class arrester to 1.5 kA for the secondary arresters.

Once a gapped lightning arrester sparks over, the transient current flows through the nonlinear valve elements and produces a voltage drop across these elements. This is the ***discharge voltage***. The discharge voltage increases somewhat with the magnitude of the discharge current. The discharge voltages of an arrester are determined by passing 8/20-μs impulse currents of various magnitudes.

The gapless arresters are continuously energized by the power-frequency voltage; consequently, there is constant energy dissipation in the valve blocks. For each arrester rating, there is a recommended limit of the magnitude of the power-frequency voltage which may be continuously applied without any ill effects to the arrester. This is called the ***MCOV (maximum continuous operating voltage) rating*** of an arrester. Examples of the protective characteristics of some of the gapped and gapless arresters were shown in Tables 13.3-13.5 in Chapter 13.

15.5 Application of Insulation Coordination

The degree of protection is measured by the protective ratio, PR, which is defined as [9]

$$PR = \frac{\text{insulation withstand level}}{\text{voltage at protected equipment}} \tag{15.1}$$

The following three protective ratios would ensure proper insulation coordination:

$$PR(1) = \frac{\text{chopped-wave withstand (CWW)}}{\text{front-of-wave sparkover (FOW)}} \geq 1.2 \tag{15.2}$$

$$PR(2) = \frac{\text{basic impulse insulation level (BIL)}}{\text{lightning protective level (LPL)}} \geq 1.2 \tag{15.3}$$

$$PR(3) = \frac{\text{switching impulse insulation level (BSL)}}{\text{switching surge protective level (SSP)}} \geq 1.15 \tag{15.4}$$

The lightning protective level is the higher of the arrester sparkover level and the arrester discharge voltage. As the arrester discharge voltage depends upon the discharge current level, it is necessary to estimate the magnitude of the expected arrester

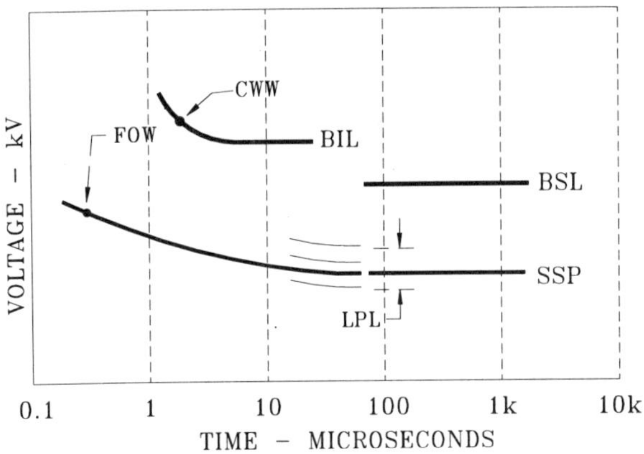

Fig. 15.5 Illustration of insulation coordination.

discharge current for each application. Figure 15.5 illustrates the insulation coordination process for an oil-immersed transformer.

15.5.1 Estimation of arrester discharge current

When the station and the incoming lines to the station are effectively shielded, then a conservative estimate of the discharge current through the station lightning arrester can be made [9].

1. For a lightning arrester situated at a dead-end station (Fig. 15.6):

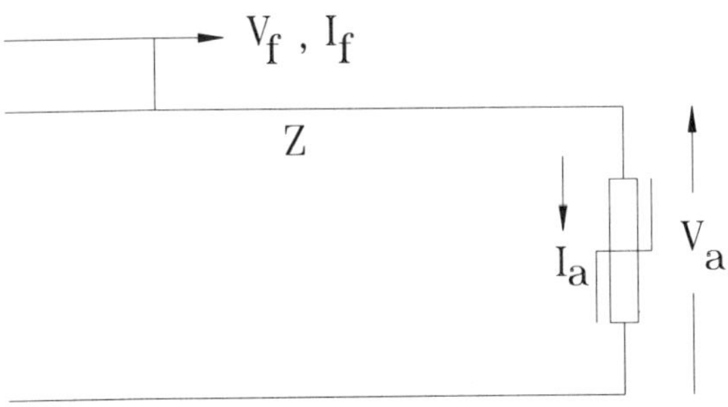

Fig. 15.6 Lightning arrester at dead-end station.

$$V_a = V_f + V_r \tag{15.5}$$

$$I_a = I_f + I_r = \frac{V_f}{Z} - \frac{V_r}{Z} = \frac{V_f}{Z} - \frac{V_a - V_f}{Z} = \frac{2V_f - V_a}{Z} \tag{15.6}$$

where I_a = estimated discharge current through the arrester,
V_f = 1.2 x line insulation level,
V_r = voltage reflected from the arrester terminal,
V_a = arrester discharge voltage, and
Z = line surge impedance.

2. ***For a lightning arrester at a tapped-off station (Fig. 15.7):***

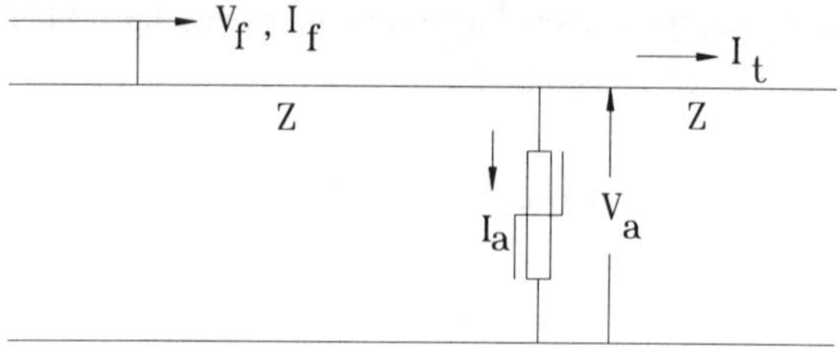

Fig. 15.7 Lightning arrester at a tapped-off station.

$$V_a = V_f + V_r \tag{15.7}$$

$$I_f + I_r = I_a + I_t \tag{15.8}$$

$$I_a = I_f + I_r - I_t = \frac{V_f}{Z} - \frac{V_r}{Z} - \frac{V_a}{Z} = \frac{2(V_f - V_a)}{Z}. \tag{15.9}$$

For unshielded lines, a discharge current level of 20 kA is recommended [9].

15.5.2 An example of insulation coordination

Let us consider a 115-kV system for two extreme cases:

1. neutral effectively grounded; and
2. neutral isolated.

The temporary power-frequency overvoltages (TOV) of the unfaulted phases will be different for the two cases. The arrester must be able to reseal itself against the TOV. Therefore, the arrester rating will have to be selected accordingly. The most frequent cause of TOV is the line-to-ground fault which raises the power-frequency voltage of the unfaulted phases. This rise in voltage is a function of the ratios R_0/X_1 and X_1/X_2, where R_0 is the zero-sequence resistance, and X_0, X_1 and X_2 are the zero-, positive- and negative-sequence inductive reactances, respectively. A system is designated as an effectively grounded system if its $R_0/X_1 \leq 1$, and $X_1/X_2 \leq 3$. For the effectively grounded neutral, the line-to-ground voltages of the unfaulted phases are generally contained below 75 percent of the normal line-to-line voltage. For an isolated-neutral system, the line-to-ground voltages will rise to 100 percent of the normal line-to-line voltage. Given a 5 percent margin of safety, the TOV will be 80 and 105 percent of the line-to-line voltage for effectively grounded and isolated-neutral systems.

Assuming ±5 percent fluctuation, the highest continuous power-frequency voltage across the arrester will be 69.72 ($=1.05 \times 115/\sqrt{3}$). The BIL of a 115-kV transformer can be either 450 or 350 kV (Table 15.1); the 350-kV level is the reduced BIL which can be used provided the neutral ground condition is appropriate to select an arrester of lower voltage rating. The corresponding CWW (chopped-wave withstand) and the BSL (switching surge withstand level) from Table 15.1 are shown below:

BIL = 450; 350 kV: CWW = 495; 385 kV; BSL = 375; 280 kV

For effectively grounded neutral:

Expected power-frequency phase-to-ground temporary overvoltage (TOV) = 0.8x115 = 92 kV

Nearest arrester ratings of higher voltage level (Table 13.3):

Power-frequency voltage rating = 90 kV
MCOV = 73 kV (>69.72 kV)
TOV = 106 kV (>92 kV)

FOW level = 239 kV
Discharge voltages: 185 kV @ 1.5 kA and 234 kV @ 20 kA
Switching surge protective level = 169 kV

From (15.2) to (15.4):

For BIL = 450 kV: PR(1) = 495/239 = 2.07
PR(2) = 450/234 = 1.92
PR(3) = 375/169 = 2.22

For BIL = 350 kV: PR(1) = 385/239 = 1.61
PR(2) = 350/234 = 1.50
PR(3) = 280/169 = 1.66

For isolated neutral:

Expected power-frequency phase-to-ground temporary overvoltage (TOV)
= 1.05x115 = 120.75 kV

Nearest arrester ratings of higher voltage level (Table 13.4):

Power-frequency voltage rating = 108 kV
MCOV = 87 kV (>69.72 kV)
TOV = 127 kV; FOW level = 287 kV
Discharge voltages = 222 kV @ 1.5kA and 282 kV @ 20 kA
Switching surge protective level = 204 kV

From (15.2) to (15.4):

For BIL = 450 kV: PR(1) = 495/287 = 1.72
PR(2) = 450/282 = 1.60
PR(3) = 375/204 = 1.84

For BIL = 350 kV: PR(1) = 385/287 = 1.34
PR(2) = 350/282 = 1.24
PR(3) = 280/204 = 1.37

It should be noticed from the above tabulations that the BIL of 350 kV will be a suitable choice. The above computations were made with the assumption that the arrester is located very close to the protected apparatus.

15.6 Location of Lightning Arresters

Lead lengths between the arrester and the protected apparatus would produce voltage oscillations, and naturally the voltage across the apparatus will be higher than that limited by the arrester. A simplistic calculation is made to show the adverse effects of lead length between an arrester and the protected apparatus. As an example, assume that a distribution transformer is connected to an arrester through a 3-m long lead which has an inductance of 1 μH/m. A lightning current, rising at a rate of 10 kA/μs, will produce an inductive voltage drop (LdI/dt) across the lead of 30 kV. The voltage across the transformer will be the sum of the arrester discharge voltage and the voltage drop across the lead. In fact, the stray capacitance of the lead, in conjunction with its inductance, will produce voltage oscillations whose effects could be even worse. Illustrations of arrester connections are shown in Fig. 15.8. In Fig. 15.8a, the leads, connecting the arrester to the power apparatus, carry the fast-rising lightning current. Therefore, the voltage across the power apparatus is the sum of the inductive voltage drops across the leads and the arrester discharge voltage. In Fig. 15.8b, the lightning current is diverted directly to ground through the arrester; therefore, the voltage impressed across the power apparatus is the discharge voltage of the arrester.

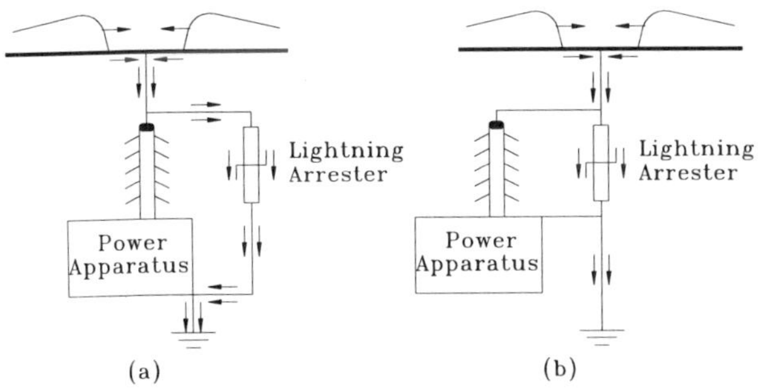

Fig. 15.8 Effect of lead length of arrester connected across power apparatus.

For the protection of station apparatus, a common arrester is often used to protect two or more apparatus. This has been discussed in Chapter 14. In addition to the reflected wave phenomena, it is possible that still higher peak voltages would exist at the apparatus as a result of oscillations caused by inductance of the line between the arrester and the apparatus, and the capacitance of the apparatus. Furthermore, the voltage drops in the lead from the line to the arrester and in the lead from the arrester to the ground, which are affected by the rate of rise of the surge current, add to the voltage drop across the arrester. Any difference in the ground potential between the apparatus ground and the arrester ground also adds to the voltage impressed across the apparatus insulation. In view of the above factors, it is important, particularly in stations where direct strokes may originate close to the station, to follow the following rules:

1. The protective devices should be located close to the apparatus they are to protect;

2. The leads to the devices should be kept as short and direct as possible; and

3. The arrester and the apparatus grounds should be interconnected and be as low in inductance as possible, preferably one μH or less.

The ultimate in this respect is reached when the protective device is mounted directly on the transformer. With the line side of the arrester connected to the transformer terminal, and the arrester ground connected directly to the transformer tank, the voltage between the winding and the core is definitely limited to the discharge voltage of the arrester.

To provide protection to an extended station, an arrester should be located directly ahead of the disconnect switch where the line enters the station, and another arrester located directly adjacent to or on the transformer.

15.7 Behavior of Insulation under Nonstandard Lightning Voltages

The international standard 1.2/50-μs waveshape of lightning voltages has helped the designers and application engineers to test, specify and evaluate the power apparatus and protective equipment. However, a lightning-caused transient voltage seldom has the standard waveshape. Researchers and designers of power apparatus and systems have endeavored over the past 50 years to understand how insulation systems behave under nonstandard lightning voltages. An extensive review of past research on the behavior of power apparatus and systems under nonstandard lightning voltages has been published by the IEEE Task Force [10,11].

15.7.1 Disruptive effect (DE) model of volt-time characteristics

In dealing with the effects of nonstandard voltage waves on transformers, Witzke and Bliss [12,13] introduced the term disruptive effect (DE), which was defined as:

$$DE = \int [V(t) - K_1]^{K_2} dt , \tag{15.10}$$

where, V(t)=applied impulse voltage, and
K1,K_2=fixed by the standard transformer test voltages.

This relationship was established on the basis that transformer insulation could withstand some constant voltage, K_1, for an extended period of time without suffering ill effects. It was assumed that the disruptive effect of a surge is a function of both voltage magnitude and time, but that these two factors are not of equal importance. Introducing the exponent, K_2, allows some variation in the relative weight given to magnitude and time. The constants, K1 and K_2, were evaluated from the standard impulse tests of transformers with the assumption that the full-wave, chopped-wave and the front-of-wave tests are of equal severity. K_1 and K_2 were obtained by solving the following equations:

$$\int [V_1(t) - K_1]^{K_2} dt = \int [V_2(t) - K_1]^{K_2} dt , \text{ and} \tag{15.11}$$

$$\int [V_2(t) - K_1]^{K_2} dt = \int [V_3(t) - K_1]^{K_2} dt , \tag{15.12}$$

where $V_1(t)$, V_2 (t) and $V_3(t)$ are the full-wave, chopped-wave and the front-of-wave test voltages, respectively.

Jones [14] simplified the Witzke-Bliss definition of the disruptive effect, and assumed K_1 in (15.10) to be zero, because it was impossible to obtain a set of positive constants which exactly represented the insulation strength of transformers in the voltage classes 138 kV and above.

Kind [15] suggested that once the applied voltage, V(t), exceeds a critical level, V_b, the breakdown process will start, and the breakdown process will be completed within the time, t_a. He defined the disruptive effect as follows:

$$DE = \int_{t_1}^{t_1+t_a} [V(t) - V_b]dt\,, \tag{15.13}$$

where t_1 is the time when the applied voltage exceeds the critical voltage level, V_b, i.e., $V(t_1)=V_b$. Kind's equation (15.13) is similar to the Witzke-Bliss equation (15.10), with $K_2=1$. Kind suggested that V_b is the breakdown voltage of the electrode configuration under static conditions provided no space charge alters the electric field profile. This would be true for electrode configurations with homogeneous or slightly inhomogeneous fields. In highly inhomogeneous fields, the formation of space charges will distort the electric field.

Rusck [16] proposed another variation of the Witzke-Bliss equation:

$$\int_{t_c}^{t_b} V(t)dt = \text{constant}\,, \tag{15.14}$$

where t_c is the time when the applied voltage reaches the critical voltage level, and t_b is the time to breakdown.

Darveniza and his associates [17-19] have done extensive research to extend the integration method, especially in the determination of the various constants, i.e., DE, K_1 and K_2 of (15.10). To determine these constants, they selected two points from the volt-time curve of a given air gap with the standard lightning voltage wave. To determine three constants from two test points, they first assumed $K_1=0$, and evaluated DE and K_2; they then assumed $K_2=1$, and determined DE and K_1. The values of these constants were found to be sensitive to the selected points of the standard volt-time curve. Darveniza [19] concluded that for lightning impulses applied to various insulants and electrode configurations, the integration method gives acceptable accuracy in reproducing standard volt-time curves and in predicting breakdown data for steep-front and for long-front lightning impulses with time-to-breakdown from 50 ns to 25 μs. He also concluded that the integration method is not accurate for switching impulses, with inaccuracies often greater than 25 percent or even 50 percent.

15.7.2 Leader progression model

It was stated in Section 15.3.1 that the time lag of breakdown of an air insulation is caused by statistical time delay and formative time delay. Neglecting the statistical time delay, the time lag of breakdown, t_b, can be expressed as the sum of the corona inception time, t_i, the streamer formation time, t_s, and the leader propagation time, t_l. For an air gap with asymmetrical nonuniform electric field (e.g., rod-plane gap), corona (i.e., electron avalanche) starts at the the surface of the electrode which has the highest electric field when the applied voltage exceeds a critical value. This time delay, t_i, is small compared with t_b. The avalanche travels toward the opposite electrode, ionizing more neutral molecules. The number of ionized particles is given by (11.6) in Chapter 11. Because of their high mobility, the electrons are rapidly swept out of the gap by the anode, leaving behind a positive space charge. The electric field of the space charge increases as the avalanche grows, thus interacting with the applied field. At a critical concentration of the electrons ($n \geq 10^8$), the avalanche grows into a streamer with a time delay, t_s. Current increases within the streamer, increasing its temperature, which eventually 'ignites' the streamer into a leader. Because of the leader's high conductivity, electrons flow freely to the leader tip, increasing the electric field at the tip, and thus enhancing the speed of progression of the streamer and the leader behind it toward the opposite electrode. Neutral molecules in the streamer are driven to higher energy states by collision with electrons, emitting photons when they fall back to their ground state. The photons travel at the speed of light and create secondary avalanches in mid gap, further increasing the speed of progression of the streamer. The cycle of avalanche, streamer and leader is repeated until the streamer from the leader tip reaches the opposite electrode, when the breakdown is completed. Leaders do not develop when an air gap is stressed by a lightning voltage, because of its short duration. Under lightning voltages, leaders develop once the air gap has been bridged by the streamers. For symmetrical gaps, such as rod-rod gaps, electron avalanche starts at both electrodes. The streamers from both electrodes meet at mid gap; the leaders start at mid gap, traveling in opposite directions, and eventually bridging the gap.

The propagation times of the streamers and the leaders are dependent on the waveshape and polarity of the applied voltage, as well as on the electrode configuration. Many researchers have contributed in the estimation of these propagation times, and hence the time lag of breakdown. The final analysis of the time lag of breakdown contains empirical constants which can be determined experimentally. No accord has been reached on their values because of the varying experimental conditions. A great deal of research needs to be performed before a practical method of estimating the time-lag characteristics of insulation systems can be established for the design engineers. A comprehensive review of the present state of art is given in [20,21].

15.7.3 Time-lag characteristics of non-self-restoring insulants

Although the DE model was originally developed to estimate the time lag characteristics of non-self-restoring insulation (e.g., transformer insulation) [12,13], most of the research in the DE model has been applied to self-restoring insulation (e.g., air gaps), perhaps because of the vast extent of utilization of air as an insulant for high-voltage power systems, and also because of the ease of experimentation with air in the laboratory. However, several attempts [22-26] have been made to model the time-lag characteristics of breakdown of non-self-restoring insulants.

A model for the time lag of breakdown for the oil impregnated paper insulation has been proposed [26] which postulates that the breakdown is the cumulative effect of charge injection into the dielectric from the cathode followed by structural fluctuation of the localized regions containing the charge carriers. The breakdown time lag consists of two components: (i) time lag, t_c, for the injection of charge carriers into the dielectric, and (ii) time lag, t_f, for the catastrophic events which follow the structural fluctuations. These time lag components are similar to the statistical and formative time lags of the gaseous dielectrics. The authors computed the cumulative probability function for the total time lag ($t_d=t_c+t_f$) from the probability functions of the individual time lags t_c and t_f. The various empirical constants were obtained from experiments.

15.8 Specific Applications of Insulation Coordination in AC Systems

For the protection of apparatus and systems against lightning, each case must be studied individually. However, some general discussion on the various types of apparatus is in order.

15.8.1 Station protection

The power apparatus in a substation may be exposed to lightning in two ways: (1) by direct lightning stroke to the station, and (2) by voltage and current waves traveling along the exposed lines leading to the station. Because of the limited surface area of the station, the frequency of direct lightning stroke to the station will be low; however, damages caused by a direct strike to the station will be permanent and hence costly. Therefore, a station should be completely shielded against direct lightning strokes. The station apparatus can be protected against traveling waves entering the station through the connected lines. This protection can be provided by lightning arresters with proper insulation coordination. Protection of substations was discussed in Chapter 14. However, if the lines entering the station are unshielded, lightning currents of high magnitude will be imposed on the arresters, and the apparatus may not be protected. It is generally a good practice to shield the lines entering the substation and lower the tower-footing resistance up to at least one kilometer from the station. In that case, lightning will strike the lines at least one kilometer from the station; if the lightning current is high, then the line insulators will flash over, limiting the voltage which will travel to the station. Moreover, the voltage waves will be somewhat attenuated by the impulse corona along the 1-km long shielded lines as well as by dissipation in the soil resistance before reaching the station. Sometimes, arresters are connected to the lines where the shielding begins. The discharge of these arresters will drain a large portion of the lightning current.

15.8.2 Protection of underground cables and cable-connected equipment

An insulation failure in an underground cable inflicts permanent damage to the cable. The faulted section of the cable must be located first, then removed and replaced by a new cable section. This is labor intensive, time consuming and costly. The voltage rating of the cable system varies from several hundred volts in the secondary circuits to the highest transmission voltages. The length of a cable system may vary from a few hundred meters to several hundred kilometers in the urban underground distribution and subtransmission systems.

Cables are almost always connected to overhead power lines. Lightning surges, therefore, reach the cable via the overhead lines. Traveling voltage waves on an overhead line are considerably reduced at the junction of the overhead line and the underground cable because of the lower surge impedance of the cable. However, multiple reflections at the cable terminals would build the voltage up to a dangerously high value. This was shown for a very simple case of a cable connected between an overhead line and a rotating machine in a problem in Chapter 2. Considerable work has been done to study the effects of traveling voltages on cable-connected distribution equipment [27,28]. An IEEE Committee concluded [27] that adequate protection could be obtained for system voltages 15 kV and below using distribution arresters at the overhead line-cable transition point (riser pole). Higher voltage distribution systems of 25-kV and 34.5-kV class were considered in [28]. It was concluded that the riser-pole arrester alone will not be able to protect the cable-connected equipment. It will be necessary to connect arresters at the equipment terminals as well. The U.S. Rural Electrification Administration specifies that a lightning arrester must be installed one span from the riser pole, in addition to the riser-pole arrester [29]. The arrester on the pole adjacent to the riser pole will be able to drain some of the lightning current of the oncoming transient to ground, thus lowering the burden on the riser-pole arrester. Lower discharge current through the riser-pole arrester will also lower the voltage transmitted to the cable. This arrangement will be particularly beneficial if gapless lightning arresters are used. If gapped arresters are used, then the sparkover voltage of the arrester at the adjacent pole should be higher than that of the riser-pole arrester.

The outer sheath of a coaxial cable is normally connected to the ground either at one terminal or at both terminals of the cable. For long cables, the distant points along the sheath do not remain at the ground potential under transient voltages. Moreover, the resistance of the grounding electrode is finite, and so is the soil resistivity. A study on a two-conductor coaxial cable showed that finite grounding resistances would reduce the potential gradient across the dielectric between the inner conductor and the sheath [30]; finite soil resistivity will tend to reduce such voltage stresses.

15.8.3 Protection of distribution systems

Overhead power distribution lines are normally not shielded by ground wires. Grounded neutral wires are strung generally below the phase conductors. Therefore, the distribution lines are exposed to direct lightning strokes. As the BIL of an overhead power distribution line is low (100-300 kV), a direct stroke to the line will almost always result in an outage [31,32]. However, because of their low heights, the distribution lines attract relatively few direct strokes. Reports from an IEEE Task Force showed the benefits of protection against direct strokes obtained by the application of lightning arresters on all phases [33,34]. However, the spacing between lightning arresters is an important design consideration; this spacing should not exceed 400 feet for a line of 300-kV BIL. For lower BIL, the spacing must be shorter.

Most of the lightning-caused damages to the distribution lines are caused by indirect strokes, i.e., voltages induced on the lines by strokes to nearby ground [35,36]. The underbuilt neutral wire will reduce the voltages induced on the phase conductors by nearby lightning strokes [37]. This, in turn, will reduce the outage frequency. However, the neutral wire must be grounded frequently and the grounding resistance must be low enough to increase the effectiveness of the neutral wire. The most effective overvoltage protection against indirect lightning strokes can be provided by lightning arresters, if they are installed on each pole. This is not cost effective. The combination of the neutral wire and the lightning arresters, even if not installed at each pole, will definitely improve the lightning performance of the distribution line. Unfortunately, no analysis exists yet to quantitatively assess the effectiveness of the various design options.

The standard design practice is to install lightning arresters at the terminals of the distribution transformers, and at the riser poles. Because of the relatively low BIL of the distribution apparatus, the lead lengths between the apparatus and the arrester must be as short as possible. This was discussed before, in Section 15.6. For transformer protection, the U.S. Rural Electrification Administration recommends that the arrester should be connected to the phase conductor ahead of the fuse [38]. This is recommended because the lightning current and the power-follow current will then flow through the arrester, bypassing the fuse; this will minimize the likelihood of fuse blowing during lightning storms.

It has been reported that in many cases the primary windings of single-phase pole-mounted transformers have failed in spite of lightning arresters on the primary side [39-41]. It was concluded that the layer-to-layer insulation failures of the high-voltage windings were caused by lightning voltages entering the transformers through the secondary side. Darveniza and Mercer [42] reported that a significant reduction of the transformer failure rate was achieved by relocating the primary-side lightning arresters close to the terminals, and by installing low-voltage side arresters.

15.8.4 Effects of wood poles [43-46]

Most overhead distribution lines and many subtransmission lines are strung on wood poles. Wood provides additional insulation, thus increasing the line BIL. It was stated before that most of the lightning-caused outages on the distribution lines are the result of indirect strokes. Voltages induced on a distribution line by indirect strokes seldom exceed 300 kV which is above the BIL of many distribution lines. Darveniza suggested that outages caused by indirect strokes to 11-kV lines can be effectively eliminated by a pin insulator in series with three feet of seasoned wet wood [45]. Lewis concluded that wood adds about 50 kV/ft of insulation strength under impulse conditions for conservative design [43]. The power-follow current after an insulator flashover activates the automatic reclosing mechanism of the circuit breaker, and causes a momentary power outage. However, wood has the property to quench the arc, and thus prevent the power-follow current. Darveniza recommends that the probability of a power-follow current after an insulator flashover will be very small if the ***peak*** power-frequency voltage gradient across the wooden arc path is kept below 1.13 kV(peak)/inch [45]. For example, about eight inches of wood will be required to quench the power-follow arc in an 11-kV line ($11\text{x}\sqrt{2}/1.13\text{x}\sqrt{3}=7.95$ inches).

The disadvantages of wood include the possibility of mechanical damages from the disruptive effects of the lightning currents, and leakage current from the power-frequency voltage. Darveniza reports [45] that the arc path is normally internal to the structure for green wood, and hence severe mechanical damage can result. This suggests that lines constructed with unseasoned wood will undergo an initial period of high risk of mechanical damage due to lightning. The resistivity and ignitability of wood are properties which determine the effects of power-frequency leakage currents. Leakage currents of sufficient magnitude to endanger human safety or damage the wood structure may result from: (1) complete or partial insulator failure; (2) bridging of insulator by a conductor (wire or live object); or (3) contamination of the insulator surface by dust, salt spray or industrial pollution.

15.9 Insulation Coordination in High-Voltage DC Systems

The basic principles of insulation coordination of high-voltage dc (hvdc) systems are the same as those of high-voltage ac (hvac) systems. However, the actual procedures for the implementation of insulation coordination are complex for the dc systems. The insulation coordination of a hvdc system comprises the insulation coordination of (i) the hvac lines leading to the converter stations, (ii) the converter stations, and (iii) the hvdc lines originating from the converter stations. The complexity of the insulation coordination of hvdc systems is caused by the complexity of the converter stations.

The protection of hvdc lines against lightning is similar to that of hvac lines, i.e., by (i) shield wires for protection against direct strikes, and (ii) lower tower-footing impedance for protection against backflashover. The effects of the power-frequency voltages are ignored for the protection of hvac lines against lightning. However, for the protection of hvdc lines, it should be borne in mind that the positive-polarity conductor will be hit by lightning more frequently because lightning strokes are mostly of negative polarity. A flashover path on a power line (ac or dc), caused by overvoltage, is air which is recoverable; consequently, no permanent damage occurs. However, reliability suffers because of power outage. The cost of complete protection of the line against reliability is weighed, and a few line outages per year are allowed in the design of the line protection scheme.

An outage in a converter station can cause permanent equipment damage in addition to revenue loss due to the outage. The per-unit cost of a converter station is significantly higher than that of an ac substation. The overvoltage protection of ac substations was discussed in Chapter 14. The overvoltage protection of a converter station is more stringent than that of an ac substation because of its per-unit cost as well as the converter station's electrical complexity.

A simplified schematic of a 12-pulse bipolar hvdc transmission system is shown in Fig. 15.9. No current flows through the neutral line under balanced conditions. However, during unbalanced operation or when one pole is out, large direct current will flow in the neutral conductor. As direct current flows far in the ground, the neutral conductor is not grounded near the converters to avoid corrosion

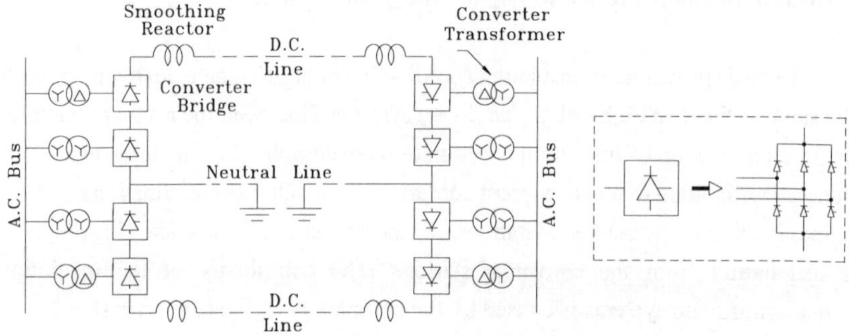

Fig. 15.9 Simplified schematic of a hvdc transmission system.

of buried metallic objects, such as underground cables and gas pipes. The neutral conductor is run isolated from the ground over a long distance and connected to ground electrodes at a remote location, preferably underwater. Two converter bridges per pole are shown in Fig. 15.9. Several bridges are connected in series, depending upon the voltage level of the dc line. For example, four bridges are connected in series per pole for a ±533-kV dc system. As shown in the insert of Fig. 15.9, a bridge consists of six 'valves', so called because they conduct in one direction. A valve, in turn, consists of numerous series-connected thyristors to withstand the forward and reverse voltage impressed upon a valve. Sometimes more than one string of thyristors is connected in parallel in a valve to satisfy the demand for current. Bridges have also been connected in parallel to meet the demand for higher currents. Obviously, the series-connected bridges and their associated transformers are at different voltage levels, referred to ground; therefore, surge protectors have to be selected accordingly. The series-connected thyristors in a valve do not share the impressed voltage equally because of the difference in their internal impedances and stray capacitances. An RC-snubber circuit is connected across each thyristor to force voltage sharing. These snubber circuits also limit the voltage 'spikes' during commutation when a thyristor ceases to conduct during the regular cycle.

To perform insulation coordination, one needs to have data on the following:

(a) Expected voltage stresses.

- (i) Maximum continuous operating voltage (MCOV);
- (ii) Temporary overvoltage (TOV);
- (iii) Lightning and switching surges.

(b) Voltage withstand level of each component.

(c) Characteristics of surge protectors.

(i) Discharge voltage level;
(ii) Reseal voltage;
(iii) Transient energy handling capability.

For the estimation of MCOV, the maximum ac bus voltage may be assumed to be 1.05 to 1.10 of the nominal ac bus voltage. The MCOV across the valves is the maximum converter transformer secondary line-to-line peak voltage. Superimposed on this voltage is the commutation spike which occurs during each cycle when one valve goes out of conduction. This becomes maximum when the firing angle, α, is 90° [48,49]. This commutation spike has been estimated to be below 1.3 p.u. for α=90° [49], but for typical firing angle (e.g., α=18°), the overshoot is about 1.15 [50]. On the dc line, the actual voltage will vary from the nominal voltage, depending upon the load current, firing angle and the commutating reactance. The dc line voltage is given by [53,54]:

$$V_d = V_{do}\cos\alpha - \frac{3X_c}{\pi}I_d = \left(\frac{n\sqrt{2}V_{\ell-\ell}}{\pi}\sin\frac{\pi}{n}\right)\cos\alpha - \frac{3X_c}{\pi}I_d, \tag{15.15}$$

where, V_{do} is the no-load dc voltage, X_c is the commutating reactance, n is the pulse number of the converter, and I_d is the load current. At no load, the dc bus voltage can be about 1.15 of the rated dc voltage [47].

Load rejection or large load changes lead to temporary overvoltage (TOV). Converter stations require large amounts of reactive power compensation in the form of static capacitors, synchronous condensers and filter capacitors. These may constitute as much as 40 to 50 percent of the dc load. When dc power transfer is blocked, this excessive capacitive reactive power still connected to the ac bus produces dynamic overvoltage as much as 1.10 p.u. [47,49].

Transient overvoltages can stress the components of the converter station, arriving from the dc and/or ac lines, as well as internally generated within the converter station. These transient voltages are caused by lightning, switching and faults.

Lightning-caused overvoltages can impinge on the converter station, arriving from the ac and/or dc lines. These transient voltages will be limited by the critical flashover voltages (CFO) of the respective lines. The lightning surges which finally penetrate the dc smoothing reactors and dc filters on the dc side, and the VAR compensators and ac filters on the ac side will be attenuated and the wavefronts flattened by these terminal equipments, so that these surges can be treated like the switching surges. However, as the terminal equipments (e.g., dc smoothing reactors, dc filters, var compensators and ac filters) are the first line of defense, their protection should be carefully planned.

Switching surges on ac systems have been discussed in Chapters 9, 10 and 11. The ac-side switching surges will appear across the valves by magnetic coupling through the converter transformer. As no dc circuit breaker exists, no switching surge will appear from the dc line.

A pole to ground fault of a bipolar dc line will induce a fast-rising transient voltage on the healthy pole. The worst case is when the fault occurs near the midpoint of the dc line. Fast-rising transient voltages would travel both ways from the location of the fault toward the converter stations, and then be reflected at the dc smoothing reactors, doubling the voltage at the line-end of the smoothing reactor. Analyses [55,56] on the Pacific hvdc intertie showed an overvoltage of 2 to 2.2 p.u. However, actual field data [57] showed overvoltages to be about 1.7 p.u. The higher analytical value may be attributed to lack of modeling of dissipation caused by impulse corona and finite earth resistivity.

The rigorous analysis of overvoltages in an ac/dc system is complex because of the difficulties encountered in modeling the various components in the system. Various methods have been applied in such analysis [50]. Transient network analyzer (TNA), which was originally developed for the analysis of ac systems, has been useful in estimating the ac-side overvoltages of the ac/dc system. The TNA incorporates scaled physical models of the power-system components. HVDC simulator is another analog device which incorporates scaled models of the system components for both the ac and dc sides of the system. The simulation of the ac/dc system by digital computers has also been widely used. It is prudent to check the analysis by more than one method. The analytical methods are generally more conservative. Field tests have been performed on many dc systems to estimate the expected overvoltages. The data from field tests have been very useful in refining the analytical tools.

The voltage withstand levels of the various components of a hvdc system have been suggested by Elahi and his associates [52], and are shown in Table 15.2.

Table 15.2: Typical Insulation Levels of a HVDC System [51]

Component	Base Voltage	BIL (p.u.)	BSL (p.u.)
AC System	$\sqrt{(2/3)}V_{\ell\text{-}\ell}$	2.8	2.4
AC-Side Converter Transformer	$\sqrt{(2/3)}V_{\ell\text{-}\ell}$	2.8	2.4
DC-Side Converter Transformer	$(\pi/3)V_{do}$	2.8 - 4.0	2.4 - 3.5
Valve	$(\pi/3)V_{do}$	1.8 - 2.2	1.8 - 2.2
DC Bus	V_{dc}	2.7	2.2
Midpoint Bus	$0.5V_{dc}$	2.7	2.2
DC Line	V_{dc}	2.9	2.0

Note: $V_{\ell\text{-}\ell}$=line-to-line rms voltage; V_{do}=no-load dc voltage

A surge protector in a hvdc system is required to have high discharge energy capability because of the presence of large energy storage devices (e.g., var compensators, filters, reactors and the dc lines) which discharge through the surge protector during its operation [48,58-61]. After the discharge of the transient voltages, the protector must reseal itself against the dynamic system voltage which exists across the protector terminals after the transient voltage is dissipated; otherwise, the protector will be damaged by thermal runaway. In other words, the maximum continuous operating voltage (MCOV) rating of the protector must be higher than the expected maximum continuous voltage across the protected apparatus, and its temporary overvoltage (TOV) rating must be higher than the expected maximum dynamic voltage of the apparatus. At the same time, the discharge voltage of the protector during the dissipation of the transient must be reasonable with a safety margin between the protective level and the withstand voltage level of the protected apparatus. The per-unit cost of a converter station is considerably higher than that of an ac substation; the cost of insulation is a significant part of the total cost. To reduce cost without reducing reliability, dc surge protectors are selected with lower protective levels and protective margin than the protectors installed in ac substations. An example of a Japanese application [59] is shown in Table 15.3.

Table 15.3: Specifications of Arresters for DC and AC High Voltage Systems

System	±250-kV DC System	500-kV AC System
Arrester Rated Voltage	392 kV dc	420 kV rms
Lightning Impulse Sparkover Voltage	(Rated Voltage)x1.9	(Rated Voltage)x2.9
Max. Switching Impulse Sparkover Voltage	(Rated Voltage)x1.8	(Rated Voltage)x2.6
Discharge Voltage @ 10 kA	(Rated Voltage)x2.0	(Rated Voltage)x2.9
Line Capacitance for Switching Surge Operating Duty Test	650 μF at 14 kV dc	78 μF at 14 kV rms

Although no standard exists yet for hvdc lightning arresters, significant commonality exists in testing of arresters for hvdc applications among manufacturers [61]. A schematic of the surge protection of a hvdc converter station is shown in Fig. 15.10.

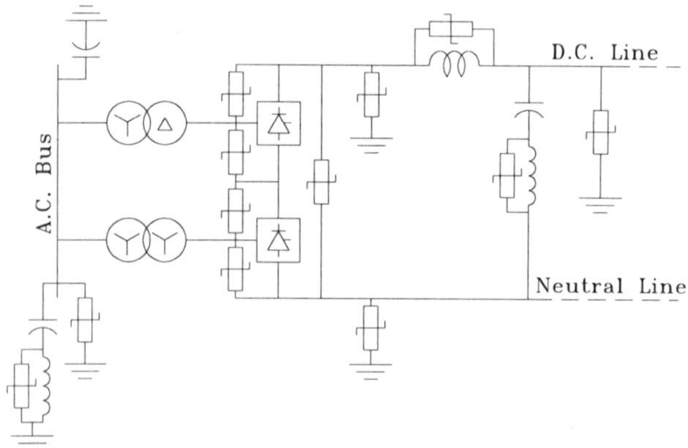

Fig. 15.10 Schematic of surge protection of a hvdc converter station.

15.10 References to Chapter 15

1. ***Insulation Co-ordination Part 1: Definitions, Principles and Rules.*** International Standard IEC 71-1, seventh edition. International Electrotechnical Commission, Geneva, Switzerland, 1993.

2. ***Electrical Transmission and Distribution Reference Book***. Westinghouse Electric Corp., East Pittsburgh, PA., 1964.

3. L.V. Bewley: ***Traveling Waves on Transmission Systems***, Second Edition. New York: John Wiley, 1951.

4. L. Paris, "Insulation Co-ordination", ***Elettrotecnica***, Vol. 57, pp. 566-580, 1970. Translated by the Electrical Research Association (G.B.), ERA Trans-2922, February 1972.

5. ***High-Voltage Test Techniques Part 1: General Definitions and Test Requirements***. International Standard IEC 60-1, second edition. International Electrotechnical Commission, Geneva, Switzerland, 1989.

6. ***IEEE Standard Techniques for High Voltage Testing***, IEEE Std 4-1995. Institute of Electrical and Electronics Engineers, New York, NY, 1995.

7. ***General Requirements for Liquid-Immersed Distribution, Power, and Regulating Transformers***, ANSI/IEEE C57.12.00-1993.

8. ***IEEE Standard for Surge Arresters for AC Power Circuits***, ANSI/IEEE C62.1-1984.

9. ***IEEE Guide for the Application of Gapped Silicon-Carbide Surge Arresters for Alternating-Current Systems***, ANSI/IEEE C62.2-1987.

10. IEEE Task Force 15.09, "Review of Research on Nonstandard Lightning Voltage Waves", ***IEEE Trans. on Power Delivery***, Vol. 9, pp. 1972-1981, 1994.

11. IEEE Task Force 15.09, "Bibliography of Research on Nonstandard Lightning Voltage Waves", ***IEEE Trans. on Power Delivery***, Vol. 9, pp. 1982-1999, 1994.

12. R.L. Witzke and T.J. Bliss, "Surge Protection of Cable-Connected Equipment", ***AIEE Trans.***, Vol. 69, Part I, pp. 527-542, 1950.

13. R.L. Witzke and T.J. Bliss, "Co-ordination of Lightning Arrester Location with Transformer Insulation Level", ***AIEE Trans.***, Vol. 69, Part I, pp. 964-975, 1950.

14. A.R. Jones, "Evaluation of the Integration Method for Analysis of Nonstandard Surge Voltages", ***AIEE Trans. on Power Apparatus and Systems***, No. 13, pp. 984-990, 1954.

15. D. Kind, "The Formative Area of Technical Electrode Configurations under Impulse Voltage Stress", (in German), ***Elektrotechnische Zeitschrift***, Vol. 79-A, pp. 65-69, 1958.

16. S. Rusck, "Effect of Non-Standard Surge Voltages on Insulation", ***ASEA Research***, Vol. 3, pp. 5-14, 1959.

17. R.O. Caldwell and M. Darveniza, "Experimental and Analytical Studies of the Effects of Non-Standard Waveshapes on the Impulse Strength of External Insulation", ***IEEE Trans. on Power Apparatus and Systems***, Vol. PAS-92, pp. 1420-1428, 1973.

18. M. Darveniza and A.E. Vlastos, "The Generalized Integration Method for Predicting Impulse Volt-Time Characteristics for Non-Standard Wave Shapes - A Theoretical Basis", ***IEEE Trans. on Electrical Insulation***, Vol. 23, pp. 373-381, 1988.

19. M. Darveniza, "Further Experience with the Integration Method of Estimating Non-Standard Impulse Insulation Strength", ***Proc. Sixth Intl. Symp. on High Voltage Eng.***, Paper No. 22.15, New Orleans, LA, 1989.

20. G. Baldo, A. Pigini and K.H. Weck, "Parameters Influencing the Lightning Performance of Systems Part IV: Non-Standard Lightning Impulse Strength", CIGRE 33.81(SC) 3.4 IWD Colloquium, Rio de Janeiro, 1981.

21. CIGRE Working Group 33.07 Report, "Guidelines for the Evaluation of the Dielectric Strength of External Insulation", 1992.

22. N. Klein, "A Theory of Localized Electrical Breakdown in Insulating Films", ***Advances in Physics***, Vol. 21, pp. 605-645, 1972.

23. R.M. Hill and L.A. Dissado, "Theoretical Basis for the Statistics of Dielectric Breakdown", ***Journal of Physics C - Solid State Physics***, Vol. 16, pp. 2145-2156, 1983.

24. R.M. Hill and L.A. Dissado, "Examination of the Statistics of Dielectric Breakdown", ***Journal of Physics C - Solid State Physics***, Vol. 16, pp. 4447-4468, 1983.

25. L.A. Dissado, J.C. Fothergill, S.V. Wolfe and R.M. Hill, "Weibull Statistics in Dielectric Breakdown; Theoretical Basis, Applications and Implications", ***IEEE Trans. on Electrical Insulation***, Vol. 19, pp. 227-233, 1984.

26. S.R. Naidu, J.B. Neilson and K.D. Srivastava, "The Volt-Time Characteristics of Oil-Impregnated Paper Insulation in the Submicrosecond and Microsecond Regime", ***IEEE Trans. on Electrical Insulation***, Vol. 24, pp. 39-46, 1989.

27. IEEE Committee Report, "Surge Protection of Cable-Connected Distribution Equipment on Underground Systems," ***IEEE Trans. on Power Apparatus and Systems***, Vol. PAS-89, pp. 263-267, 1970.

28. IEEE Committee Report, "Surge Protection of Cable-Connected Equipment on Higher-Voltage Distribution Systems," ***IEEE Trans. on Power Apparatus and Systems***, Vol. PAS-100, pp.154-157, 1981.

29. ***Underground Rural Distribution***, REA Bulletin 61-3, Rural Electrification Administration, U.S. Department of Agriculture, April 1974.

30. P. Chowdhuri and M.A. Mahaffy, "Wave Propagation in a dc Superconducting Cable Part II: Parametric Effects," ***IEEE Trans. on Power Apparatus and Systems***, Vol. PAS-100, pp. 680-690, 1981.

31. D.E. Parrish and D.J. Kvaltine, "Lightning Faults on Distribution Lines," ***IEEE Trans. on Power Delivery***, Vol. 4, pp.2179-2186, 1989.

32. IEEE Working Group Report, "Calculating the Lightning Performance of Distribution Lines," ***IEEE Trans. on Power Delivery***, Vol. 5, pp. 1408-1417, 1990.

33. IEEE Task Force Report, "Investigation and Evaluation of Lightning Protective Methods for Distribution Circuits Part I: Model Study and Analysis," ***IEEE Trans. on Power Apparatus and Systems***, Vol. PAS-88, pp. 1232-1238, 1969.

34. IEEE Task Force Report, "Investigation and Evaluation of Lightning Protection Methods for Distribution Circuits Part II: Application and Evaluation," ***IEEE Trans. on Power Apparatus and Systems***, Vol. PAS-88, pp. 1239-1247, 1969.

35. K. Uehara and G. Ohwa, "Investigation of Lightning Damages on Distribution Lines," ***IEEE Trans. on Power Apparatus and Systems***, Vol. PAS-87, pp. 1018-1025, 1968.

36. A.J. Eriksson, M.F. Stringfellow and D.V. Meal, "Lightning-Induced Overvoltages on Overhead Distribution Lines," ***IEEE Trans. on Power Apparatus and Systems***, Vol. PAS-101, pp. 960-968, 1982.

37. P. Chowdhuri, "Lightning-Induced Voltages on Multiconductor Overhead Lines," ***IEEE Trans. on Power Delivery***, Vol. 5, pp. 658-667, 1990.

38. ***Engineering and Operational Manual for Rural Electric Systems***, REA Bulletin 160-1, Rural Electrification Administration, U.S. Department of Agriculture, January 1968.

39. C.J. McMillen, C.W. Schoendube and D.W. Caverly, "Susceptibility of Distribution Transformers to Low-Voltage Side Lightning Surge Failure," ***IEEE Trans. on Power Apparatus and Systems***, Vol. PAS-101, pp. 3457-3470, 1982.

40. R.C. Dugan and S.D. Smith, "Low-Voltage-Side Current-Surge Phenomena in Single-Phase Distribution Transformer Systems," ***IEEE Trans. on Power Delivery***, Vol. 3, pp. 637-647, 1988.

41. J.L. Puri, N.C. Abi-Samra, T.J. Dionise and D.R. Smith, "Lightning Induced Failures in Distribution Transformers," ***IEEE Trans. on Power Delivery***, Vol. 3, pp. 1784-1801, 1988.

42. M. Darveniza and D.R. Mercer, "Lightning Protection of Pole Mounted Transformers," ***IEEE Trans. on Power Delivery***, Vol. 4, pp. 1087-1095, 1989.

43. W.W. Lewis: ***The Protection of Power Systems against Lightning***, Chapter 8. John Wiley & Sons, New York, 1950.

44. H.R. Armstrong, H.O. Stoelting and E.F. Veverka, "Impulse Studies on Distribution Line Construction," ***IEEE Trans. on Power Apparatus and Systems***, Vol. PAS-86, pp. 206-214, 1967.

45. M. Darveniza, G.J. Limbourn and S.A. Prentice, "Line Design and Electrical Properties of Wood," ***IEEE Trans. on Power Apparatus and Systems***, Vol. PAS-86, pp. 1344-1356, 1967.

46. M. Darveniza: ***Electrical Properties of Wood and Line Design***. University of Queensland Press, St. Lucia, Queensland, Australia, 1980.

47. E. Sarbach, "Some Aspects of Insulation Co-Ordination in H.V.D.C. Systems", ***The Brown Boveri Review***, Vol. 55, N0. 7, pp. 344-351, 1968.

48. G.D. Breuer, L. Csuros, R.W. Flugum, J. Käuferle, D. Povh and A. Schei, "HVDC Surge Diverters and Their Application for Overvoltage Protection of HVDC Schemes", ***Proc. CIGRE***, Paper No. 33-14, 1972.

49. W. Heise, U. Burger, J. Käuferle and D. Povh, "The Cabora Bassa D.C. Transmission System: Overvoltage Protection and Insulation Coordination", ***IEEE Trans. on Power Apparatus and Systems***, Vol. PAS-93, pp. 1096-1104, 1974.

50. CIGRE WG 33.05 Report, "Application Guide for Insulation Coordination and Arrester protection of HVDC Converter Stations", ***Electra***, No. 96, pp. 101-156, 1984.

51. EPRI Report, "Handbook for Insulation Coordination of HVDC Converter Stations", Report No. EL-5414, 1987.

52. H. Elahi, R.W. Flugum, S.E. Wright and D.R. Brown, "Insulation Coordination Process for HVDC Converter Stations: Preliminary and Final Designs", ***IEEE Trans. on Power Delivery***, Vol. 4, pp. 1037-1048, 1989.

53. J. Arrillaga: ***High Voltage Direct Current Transmission***. IEE Power Engineering Series 6. Peter Peregrinus Ltd., London, U.K., 1983.

54. K.R. Padiyar: ***HVDC Power Transmission Systems Technology and System Interactions***. John Wiley & Sons, New York, N.Y., 1990.

55. E.W. Kimbark, "Transient Overvoltages Caused by Monopolar Ground Fault on Bipolar DC Line: Theory and Simulation", ***IEEE Trans. on Power Apparatus and Systems***, Vol. PAS-89, pp. 584-592, 1970.

56. N.G. Hingorani, "Transient Overvoltages on a Bipolar HVDC Overhead Line Caused by DC Line Faults", ***IEEE Trans. on Power Apparatus and Systems***, Vol. PAS-89, pp. 592-610, 1970.

57. D.J. Melvold, P.C. Odam and J.J. Vithayathil, "Transient Overvoltages on an HVDC Bipolar Line during Line Faults", ***IEEE Trans. on Power Apparatus and Systems***, Vol. PAS-96, pp. 591-601, 1977.

58. J.S. Kresge, E.C. Sakshaug and S.A. Miske, Jr., "Arrester Protection of High Voltage DC Transmission System Converter Terminals", ***IEEE Trans. on Power Apparatus and Systems***, Vol. PAS-90, pp. 1555-1562, 1971.

59. M. Takanashi, T. Yokokura, S. Maruyama, J. Ozawa and S. Shirakawa, "Operating Duty Test of Surge Arresters for HVDC Transmission Systems", ***IEEE Trans. on Power Apparatus and Systems***, Vol. PAS-96, pp. 1384-1390, 1977.

60. S. Horiuchi, F. Ichikawa, A. Mizukoshi, K. Kurita and S. Shirakawa, "Power Dissipation Characteristics of Zinc-Oxide Arresters for HVDC Systems", ***IEEE Trans. on Power Delivery***, Vol. 3, pp.1666-1671, 1988.

61. D.J. Melvold, "DC Arrester Test Philosophies on Recent HVDC Projects as Used by various Suppliers", ***IEEE Trans. on Power Delivery***, Vol. 6, pp. 672-679, 1991.

CHAPTER 16

PROTECTION OF LOW-VOLTAGE SYSTEMS AGAINST TRANSIENT VOLTAGES

16.1 Introduction

Damages to low-voltage systems by transient voltages are frequent. Extensive efforts have not been undertaken because the extent and cost per occurrence of damage have not been high. However, more attention is being given to this topic recently because of the increasing application of semiconductor devices in the electrical systems, and because of the demand for higher reliability and better power quality. Semiconductor devices have been known to be very sensitive to transient voltages. This chapter will discuss the techniques for protection of low-voltage systems against transient voltage, with emphasis on semiconductor devices.

To be able to protect a system against transient voltages, one should know the following:

1. The characteristics of the ***source*** of transient voltages;
2. The means of ***transmission (or coupling)*** of the transient voltage to the system;
3. The characteristics of the ***susceptibility*** of the system; and
4. The techniques of ***protection***.

16.2 Sources of Transient Voltages in Low-Voltage Systems

The main sources of transient voltages in low-voltage systems are [1-15]:

1. Lightning;
2. Switching by mechanical switches;
3. Switching by semiconductor devices; and
4. Electrostatic discharge.

16.2.1 Lightning transients

Lightning can directly hit an overhead power line. A transient voltage wave then travels along the line. It is transferred through the transformers, and finally arrives at the low-voltage systems. Although the original voltage wave is significantly reduced when it arrives at the low-voltage system, its magnitude can still be high enough to cause damage. Lightning may also strike a grounded object near an overhead power line. This would induce a transient voltage on the nearby overhead line. Although the magnitude of the induced voltage is lower than that due to a direct lightning stroke to a line, they occur more frequently than the direct strokes. Even cloud-to-cloud discharges have caused damages to low-voltage installations. About 5 percent of the damages to low-voltage systems are caused by direct lightning strokes, about 59 percent by lightning-induced voltages and the remaining 36 percent by other lightning-related causes [3]. Lightning-related voltages exceed 5600 volts on a 120-volt overhead power line [4]. Most of the transient voltages are oscillatory; the duration of some of the transients are as short as 5 μs. Transients can frequently exceed 3000 volts in 600-volt dc transit systems [9]. Transients can also be non-oscillatory.

16.2.2 Switching transients caused by mechanical switches

Switching surges are generated when one state of an electrical circuit is changed to another state. The most obvious cases are (1) a switch closing, and (2) a switch opening. The sources of switching surges are numerous. The magnitude and the waveshape of a switching surge depend upon the type of switch which is generating the transient, and the electrical circuit which is connected to the switch. The well-known switching surge in a low-voltage system is caused by a 'showering' arc. During the opening of a mechanical switch, it restrikes several times because of insufficient dielectric recovery of the arc space. Such repeated restrikes produce 'saw-toothed' voltage transients [12]. Voltages exceeding 17 kV have been reported during the showering arc caused by a vacuum switch opening a 120-volt magnet [12].

16.2.3 Switching transients caused by semiconductor devices

Semiconductor switching devices (diodes, thyristors, transistors, etc.) are notorious for generating very fast voltage transients. Knowledge of the reverse-recovery characteristics of a semiconductor device is essential for its proper application. Some devices generate high transient voltages during reverse recovery under certain conditions; some do not produce high transients, and some seem to produce high transients under all conditions.

Semiconductor devices which worked well in a breadboard circuit might cause damage in the field because the vendor has changed the manufacturing process or the devices of the same steady-state ratings were purchased from a new vendor.

When a reverse voltage is applied to a semiconductor device while it is carrying current in the forward direction, the device cannot block the reverse voltage instantaneously. A reverse current flows through the device until the charge stored in the junction during forward conduction is neutralized by the sweeping action of the reverse current and by decaying through recombination. The amount of the stored charge depends upon:

1. The device itself;
2. The magnitude of the forward current before the application of the reverse voltage;
3. The rate of decrease of the forward current upon the application of the reverse voltage; and
4. The junction temperature.

The stored charge increases with the increase in the carrier life time of the device; it also increases with an increase in the forward current, di/dt of the commutated current and the junction temperature. The reverse recovery period is composed of two parts:

1. The storage period; and
2. The transition period.

During the storage period, high reverse current will flow through the junction and the voltage across the junction will be very small. During the transition period, the junction will exponentially recover its ability to withstand reverse voltages. During this quasi-conduction state, many free carriers will be present at the junction. If the ensuing recovery voltage across the junction is such that at a particular point the voltage across the junction is higher than the dielectric withstand capability of the recovering device, the junction will break down. The reverse recovery phenomenon is illustrated in Fig. 16.1.

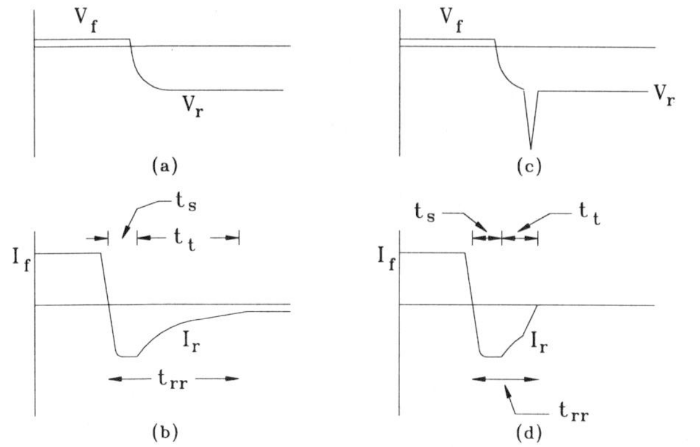

Fig. 16.1 Reverse recovery of a silicon diode.
t_{rr}=reverse recovery time; t_s=storage time; t_t=transition time.

The reverse-recovery phenomenon is influenced by the device as well as by the circuit. The reverse-recovery current goes slowly to zero in some devices, as shown in Fig. 16.1 (a) and (b). However, in some others, the reverse-recovery current may snap off, producing a high transient-voltage spike, as shown in Fig. 16.1 (c) and (d). Figure 16.2 shows the schematic of a test circuit to study the reverse recovery characteristics of semiconductor devices. Half-cycle 60-Hz forward current flows through the device under test (DUT) when the LC-resonant circuit is discharged by firing the hydrogen thyratron T_2. Reverse voltage is applied across the test device at the peak of the forward current (Fig. 16.3) by discharging the precharged capacitor, C_g, by firing the hydrogen thyratron, T_1, through the inductor, L_s. The reverse-recovery transients are shown in Figs. 16.4 to 16.6.

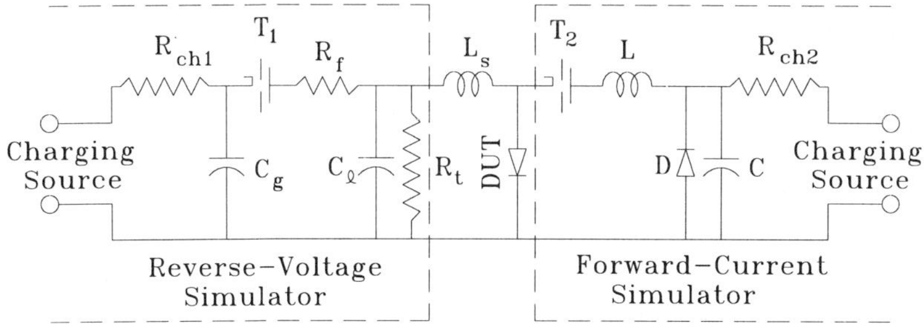

Fig. 16.2 Schematic for reverse-recovery test on semiconductor devices.

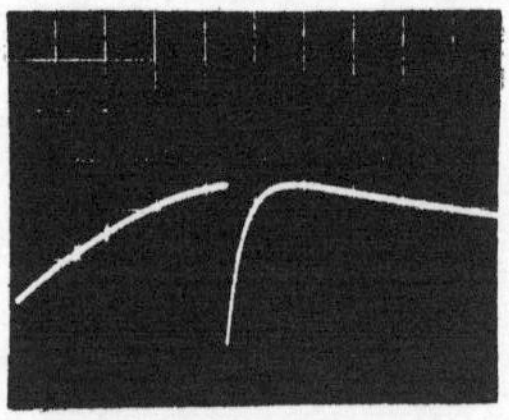

Fig. 16.3 Reverse voltage applied at peak of 60-Hz forward current.
100 A/div; 1 ms/div.

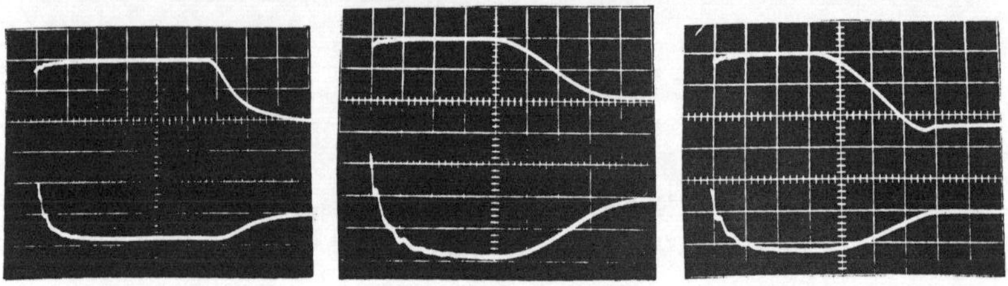

(a) (b) (c)

Fig. 16.4 Effect of reverse voltage magnitude on reverse recovery of a silicon rectifier. L_s=0

(a) upper: 100 V/div; 2 μs/div; lower: 50 A/div; 2μs/div.
(b) upper: 500 V/div; 1 μs/div; lower: 100 A/div; 1 μs/div.
(c) upper: 500 V/div; 1μs/div; lower: 200 A/div; 1 μs/div.

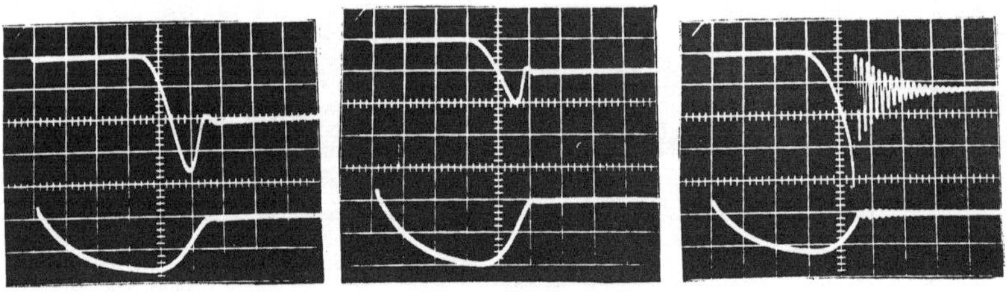

(a) (b) (c)

Fig. 16.5 Effects of reverse voltage magnitude on reverse recovery of a silicon rectifier. L_s=10 μH

(a) upper: 200 V/div; 2 μs/div; lower: 50 A/div; 2 μs/div.
(b) upper: 500 V/div; 2 μs/div; lower: 50 A/div; 2 μs/div.
(c) upper: 500 V/div; 2 μs/div; lower: 100 A/div; 2 μs/div.

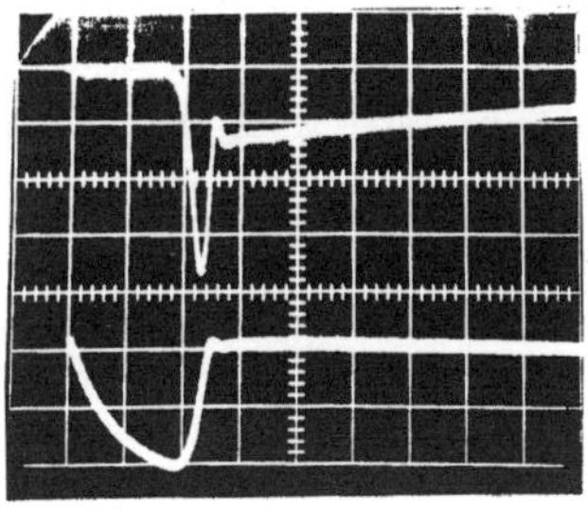

(a)

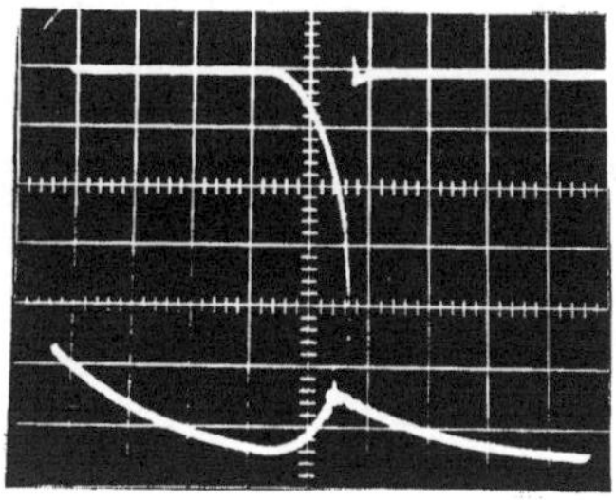

(b)

Fig. 16.6 Effects of reverse voltage magnitude on reverse recovery of a silicon rectifier. L_s=20 μH

(a) upper: 200 V/div; 5μs/div; lower: 20 A/div; 5 μs/div.
(b) upper: 500 V/div; 2μs/div; lower: 50 A/div; 2 μs/div.

In Figs. 16.4 to 16.6, the influences of the magnitude of the reverse (commutating) voltage and also the inductance in the commutating circuit can be clearly seen. The self generated commutating voltage spike can be high enough to destroy the device itself (Fig. 16.6b).

16.2.4 Transients generated by electrostatic discharges

When two materials come in contact, one tends to rob electrons from the other. As a result, the two materials acquire voltages of opposite polarities when they are eventually separated. If the two materials are non-conductors, they act like storage devices (capacitors). Sparks fly when they come in contact with a material which is charged to a different voltage level. The voltage levels could be as high as 15 to 20 kV [16]. However, the energy stored is small. The electrostatic discharge is deadly for low-power semiconductor devices which have very high input impedance, such as MOSFET devices. Plastics tend to store charge. During unwrapping of a semiconductor device from a plastic wrapper, electrostatic discharge may occur, destroying the device. One way to minimize such catastrophes is to use wrappers made of polymeric material doped with conducting agents. In addition, the vulnerable semiconductor devices should be stored and shipped with their input leads shorted together.

16.3 Transmission of Electromagnetic Transients

An electromagnetic transient can be transmitted from the source to the susceptible system by:

1. Conduction;
2. Induction; and
3. Radiation.

16.3.1 Transmission by conduction

Conduction is the most frequent means of transmission of a transient voltage. Conducted transients are also the severest transients. Transients may be conducted via the infeed line as well as by the ground. The most common means of a transient penetrating a system is via conduction over the incoming line. However, sometimes transients find 'sneak' paths to a system through the grounding system. In an extensive system, individual equipment or subsystems are separately connected to a 'ground', with the erroneous assumption that all grounds in a system have the same potential, namely zero. Heavy currents can flow through the individual grounding points during the passage of transients, thus transiently raising the individual ground points to different voltage levels. This, in turn, would transiently raise the voltage of the system components. It is the first principle in the technology of electromagnetic compatibility to have 'single-point' grounding. However, in reality, it is not always possible to achieve single-point grounding.

16.3.2 Transmission by induction

Electromagnetic transients from one circuit can be transmitted to another circuit by magnetic coupling and/or electrostatic coupling. It is known from Faraday's law that the voltage induced in a circuit is proportional to the rate of change of the magnetic flux surrounding the circuit. It can be stated mathematically as:

$$V_2 = M_{12}\frac{dI_1}{dt}, \qquad (16.1)$$

where V_2 = voltage induced on circuit 2,
I_1 = current in circuit 1, and
M_{12} = mutual inductance between circuits 1 and 2.

V_2 can be reduced by reducing the magnetic coupling (i.e., M_{12}). M_{12} can be reduced by reducing the area over which the magnetic flux is linked. If circuits 1 and 2 are lines, then M_{12} is reduced by installing each pair (infeed and return) close to each other. M_{12} can also be reduced by increasing the separation distance between circuits. IEEE has classified infeed lines into four levels, with level 1 as low-voltage, highly susceptible lines and level 4 as ac or dc power infeed lines [12]. Separation distances are recommended in the standard for the various classes of infeed lines.

Magnetic coupling can also be minimized by magnetic shielding. Circuits and subsystems are enclosed in ferromagnetic enclosures. Shielding against magnetic fields in the low-frequency region by ferromagnetic shields may be explained by stating that the ferromagnetic shield provides lower reluctance to magnetic flux than air; therefore, most of the magnetic flux is shunted by the shield. The higher the permeability of the shield, the more magnetic flux will be shunted away. However, magnetic permeability depends upon the flux density inside the shield. The higher the magnetic flux density, the lower the permeability will be. At or near saturation, the magnetic shield will be almost useless.

Electrostatic coupling can be minimized by increasing the separation distance between the noise source and the susceptible system. Electrostatic coupling can also be minimized by shielding. Electrostatic shielding is illustrated in Fig.16.7.

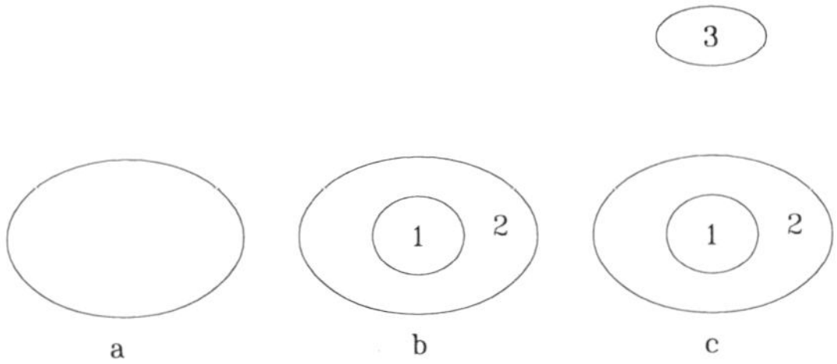

Fig. 16.7 Gauss's law extended to electrostatic shielding.

It can be proved by Gauss's law that the field inside a hollow conductor is zero if the space inside the hollow conductor contains no charge, i.e., the potential in the space inside the hollow is uniform and equal to the potential of the surface of the conductor (Fig. 16.7a). The argument still remains valid if there is an uncharged conductor in the hollow (Fig. 16.7b). If there is a third conductor outside the hollow conductor 2 (Fig. 16.7c), and the conductors 1 and 2 are uncharged, then invoking (3.1):

$$V_1 = p_{13}q_3 , \quad \text{and} \tag{16.2}$$

$$V_2 = p_{23}q_3 . \tag{16.3}$$

As $V_1=V_2$, then $p_{13}=p_{23}$. This can be extended to any number of conductors outside the hollow conductor. Now, if the conductors 1 and 2 have charges q_1 and q_2, respectively, then:

$$V_1 = p_{11}q_1 + p_{12}q_2 + V' , \quad \text{and} \tag{16.4}$$

$$V_2 = p_{12}q_1 + p_{22}q_2 + V'' , \tag{16.5}$$

$$\text{where} \quad V' = p_{13}q_3 + \cdots + p_{1n}q_n , \quad \text{and} \tag{16.6}$$

$$V'' = p_{23}q_3 + \cdots + p_{2n}q_n . \tag{16.7}$$

As $p_{i1} = p_{i2}$ for $i = 3, \cdots n$,

then $V' = V''$ and, $V_1 - V_2 = (p_{11} - p_{12})q_1 + (p_{12} - p_{22})q_2$.

In other words, the potential difference between a conductor completely surrounded by a second conductor and this second conductor is unaffected by any external fields. However, it must be pointed out that the potential of either conductor with reference to any other external potential (for example, ground) will vary by the same amount V'. Therefore, any change in the potential of the outer conductor 2 with respect to an external reference will be communicated to the inner conductor.

Next, let us suppose that the outer conductor is held at a fixed potential (Fig. 16.8).

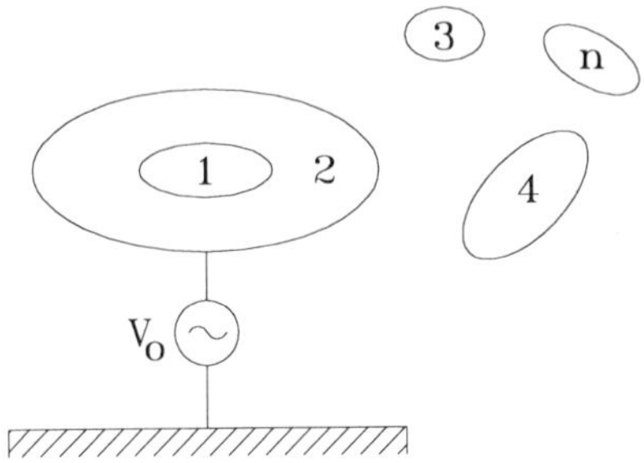

Fig. 16.8 Principle of electrostatic shielding.

From (16.4) and (16.5),

$$V_2 = V_o = p_{12}q_1 + p_{22}q_2 + V' \tag{16.8}$$

$$\text{or,} \quad V' = V_o - (p_{12}q_1 + p_{22}q_2), \quad \text{and} \tag{16.9}$$

$$V_1 = (p_{11}q_1 + p_{12}q_2) + V_o - (p_{12}q_1 + p_{22}q_2)$$

$$= (p_{11} - p_{12})q_1 - (p_{22} - p_{12})q_2 + V_o . \tag{16.10}$$

This expression for V_1 is independent of charges on the outside conductors. In particular, when conductor 2 is grounded, then $V_o = 0$, and

$$V_1 = (p_{11} - p_{12})q_1 - (p_{22} - p_{12})q_2 . \tag{16.11}$$

This is the principle of electrostatic shielding.

We can look at it another way. If $V_2 = 0$, then

$$q_1 = c_{11}V_1 + c_{13}V_3 , \tag{16.12}$$

$$q_2 = c_{12}V_1 + c_{23}V_3 , \quad \text{and} \tag{16.13}$$

$$q_3 = c_{13}V_1 + c_{33}V_3 . \tag{16.14}$$

If $q_1 = 0$, then the potential inside the hollow is the same everywhere and is equal to that of the hollow itself which is zero (because $V_2 = 0$). That is, $V_1 = 0$. Then, $c_{13}V_3 = 0$, or $c_{13} = 0$, because V_3 can have any value. Therefore, for the more general case when q_1 is not zero,

$$q_1 = c_{11}V_1 , \quad \text{and} \tag{16.15}$$

$$q_3 = c_{33}V_3 . \tag{16.16}$$

In other words, the potential of each conductor is proportional to its own charge, and is unaffected by the charge of the other conductors.

In practical cases, sometimes it is difficult to completely isolate a conductor from the others. Figure 16.9 shows two conductors above a ground plane. The voltage on conductor 2 will induce a voltage on conductor 1 by capacitive coupling:

$$V_1 = \frac{C_{12}}{C_{12} + C_{10}} V_2 = \frac{1}{1 + \dfrac{C_{10}}{C_{12}}} V_2 . \tag{16.17}$$

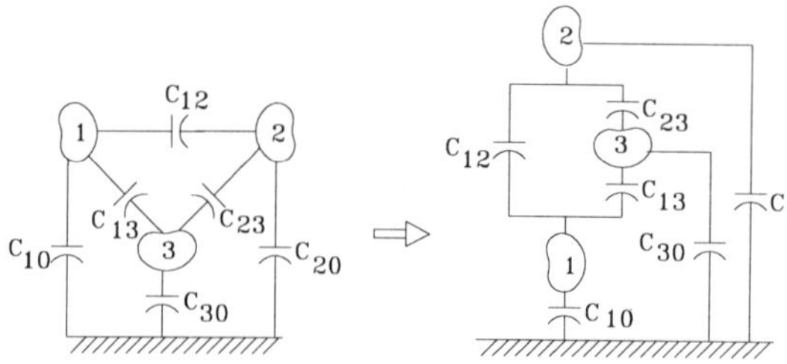

Fig. 16.9 Capacitive coupling between two conductors.

In Fig. 16.10, a third isolated conductor is brought in. The voltage on conductor 1 is given by:

$$V_1 = \frac{1}{1 + \frac{C_{10}}{C_{12e}}} V_2 , \tag{16.18}$$

where $C_{12e} = C_{12} + \frac{C_{13}C_{23}}{C_{13} + C_{23}} > C_{12}$.

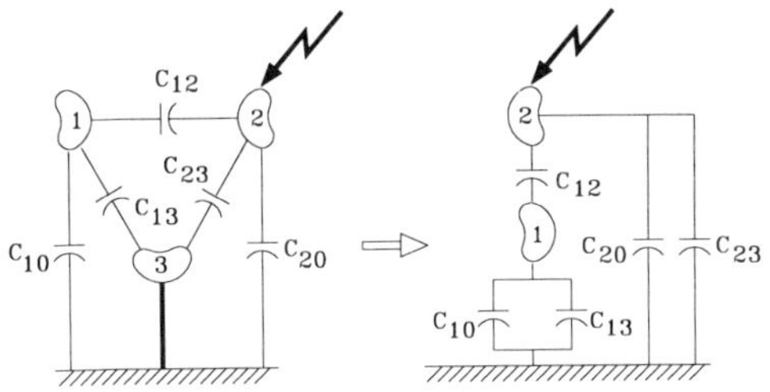

Fig. 16.10 Capacitive coupling enhanced by an isolated third conductor.

In Fig. 16.11, the third conductor is grounded. Here, V_1 is given by:

$$V_1 = \frac{1}{1 + \frac{C_{10e}}{C_{12}}} V_2 , \tag{16.19}$$

where $C_{10e} = C_{10} + C_{13} > C_{10}$, *and* $C_{12} < C_{12e}$.

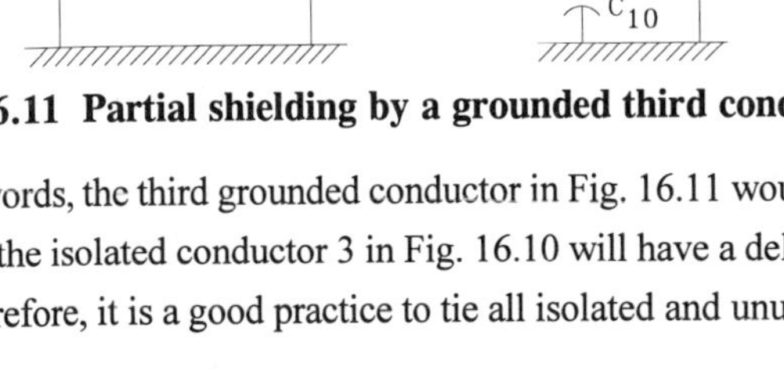

Fig. 16.11 Partial shielding by a grounded third conductor.

In other words, thc third grounded conductor in Fig. 16.11 would partially shield conductor 1, and the isolated conductor 3 in Fig. 16.10 will have a deleterious effect on conductor 1. Therefore, it is a good practice to tie all isolated and unused conductors to the ground.

16.3.3 Transmission by radiation

The electromagnetic fields from a disturbing source vary inversely as the distance of the field point from the source. These fields are generally both electric and magnetic, and they travel away from the source at a finite velocity, determined by the permittivity, ϵ, and the permeability, μ, of the medium. These electromagnetic fields induce voltages on an object (antenna), the actual magnitude depending upon the orientation, size and shape of the object, and, of course, on the strength of the radiating fields. It is not uncommon for a distant lightning stroke to induce enough voltage on the antenna of a television or a radio to cause damage to the equipment. Switching transients in a substation switchyard can cause damage to relays located in the neighboring control room.

One way to protect equipment is by shielding either the source, the receiver, or both. Naturally, a lightning bolt cannot be shielded, nor can an antenna be shielded. However, in many cases, the disturbing source can be contained inside a shield, if the source is known. Similarly, the susceptible receiver can be placed inside a shield. The other way to avoid damage is to install surge protectors at the input of the receiver. Generally, however, a radiated field penetrates a receiver via multiple paths. In that case, a surge protector will not be of much use.

The theory of shielding of electromagnetic waves is generally analyzed by the transmission line equations where attention is focussed on waves passing through the shield. A different physical picture of shielding is possible. When an electromagnetic field impinges on the walls of a shield, eddy currents are produced on the shield walls. The direction of the eddy currents is such as to produce a field which opposes the incident field and thus tends to cancel the incident field. The principal weakness in the application of the second theory from the engineering design viewpoint is that it is difficult to compute the induced currents without solving the shielding problem first. In the following, the transmission line theory will be used to explain the shielding theory.

When an electromagnetic wave traveling in a medium 1 impinges on a second medium 2, some portion of the incident wave is reflected back to medium 1 while the rest of the wave is transmitted through the medium 2. The reflection and the transmission phenomena are analogous to the transmission line theory. The reflection and the transmission coefficients are functions of the surge impedances of the two media. Incidentally, the surge impedance is also called the characteristic or wave impedance of the medium. According to the transmission line theory, a very small portion of the incident wave will penetrate through the shield if its wave impedance is small compared to that of the free space. Therefore, it would seem that a shield of low wave impedance will be a good shield. However, we must remember that the shield is imbedded in air. The transmitted wave inside the shield will be reflected back into the shield at the other boundary. We should recall that the wave at the other boundary of a medium of low surge impedance will build up to almost the initial value after multiple reflections. Moreover, the surge impedance of a metal (shield) is not a simple resistive term. It is complex and frequency-dependent. As a transient is a time-domain quantity, the surge impedance of a metal under a transient is undefinable. Therefore, it is best to neglect the effects of reflections for shielding design.

The design procedures for shielding against electromagnetic transients assume an idealized B-H curve with negligible coercive force, H_c, as shown in Fig. 16.12 [17-19].

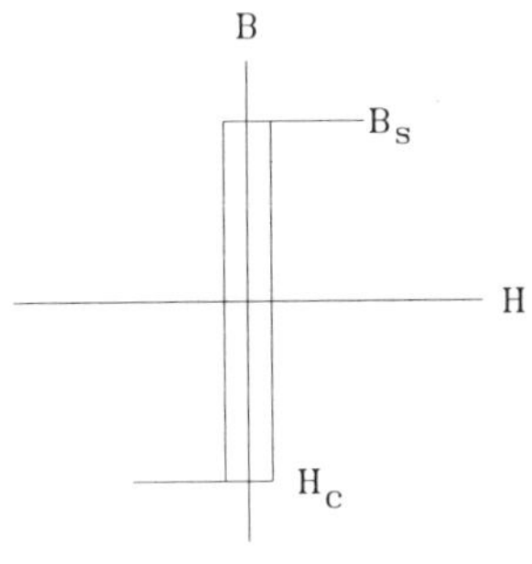

Fig. 16.12 The idealized magnetization curve.

When the electromagnetic transient impinges on the ferromagnetic shield, its surface saturates immediately because the coercive force is assumed to be negligible and a region of saturation propagates in the x-direction (Fig. 16.13). Otherwise, the shield saturates if $H_i > H_c/2$. The saturated region is shaded in Fig. 16.13. It takes a time, τ, for the saturation to spread through the whole thickness of the shield wall. This time, τ, is called the protective or diffusion time because up to this time no part of the transient can penetrate through the shield. The protective time is given by:

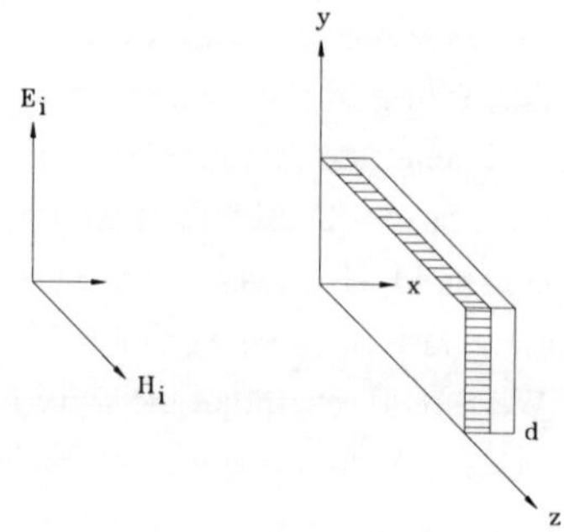

Fig. 16.13 An electromagnetic transient impinging on a ferrous shield.

$$\tau = \frac{B_{sat} d(1 + 0.5 g_s Z_w d)}{2 Z_w H_i}, \tag{16.20}$$

where B_{sat} = saturation flux density, tesla,
d = thickness of the shield wall, meter,
g_s = conductivity of shield material, mho/meter,

$$Z_w = \sqrt{\frac{\mu_o}{\epsilon_o}} \text{ ohm, and}$$

H_i = incident magnetic field, ampere-turn/meter.

For any available thickness of a ferrous material, $0.5 g_s Z_w d >> 1$. Then:

$$\tau_f \approx \frac{g_f B_{sat} d^2}{4 H_i}, \tag{16.21}$$

where τ_f and g_f pertain to the ferromagnetic shield.

It can be shown that the diffusion time for copper of the same thickness is:

$$\tau_c \approx g_c \mu_o d^2. \tag{16.22}$$

The ratio, τ_f/τ_c, is the figure of merit, F, for the performance of a ferromagnetic material impinged upon by an electromagnetic transient. It is given by:

$$F = \frac{g_f B_{sat}}{4 g_c B_i}, \tag{16.23}$$

where $B_i = \mu_o H_i$.

The above calculation is valid as long as $H_i > 0.5H_c$. If $H_i < 0.5H_c$, the ferrous material does not saturate at x=0, and it behaves much like copper. In that case, the diffusion time for the ferrous material has the same form as that for copper. The figure of merit of the ferrous material would then be:

$$F = \frac{\mu_r g_f}{g_c}, \tag{16.24}$$

where μ_r = initial permeability of the ferrous material relative to that of free space.

The shielding effectiveness of ferromagnetic materials to electromagnetic transients greatly exceeds that of copper. During the protective time, no fields greater than the coercive force leak through the ferrous shield. These are minimized by using a material with high initial permeability and low coercive force. For higher shielding effectiveness, the saturation induction should be as large as possible for the required coercive force.

From the above discussion, it may be seen that it is necessary to design all ferromagnetic shields thick enough that saturation will not penetrate through the shield under the largest electromagnetic field excitation predicted. Once the incident field has decayed, there is no further inward motion of the saturation boundary and the field in the shield gradually decays, partly as losses in the shielding material, and the rest being reradiated through the surface that was initially excited.

16.4 Susceptibility of Electrical Systems to Electromagnetic Transients

The susceptibility (withstand capability) of power apparatus was discussed in Chapter 12. The susceptibility of low-voltage apparatus and systems is basically not much different, the only exception being the semiconductor devices. We discussed the transient voltages generated by semiconductor devices from their switching action in Section 16.2.3. We will discuss the susceptibility of power semiconductor devices to electromagnetic transient voltages in this section [20-22].

In a silicon diode, no voltage will be developed across the diode if a transient voltage is applied across it in the forward-conducting direction. A voltage will be developed across it if a transient voltage is applied across it in the reverse direction while it is in the nonconducting state. If it is in conducting state, and a transient voltage is applied across it in the reverse direction, then no voltage will be developed initially; however, the transient voltage in the reverse direction will force the forward current to zero, and then develop a negative transient voltage across it. The diode may break down during this time period. Similar to the diode, a silicon controlled rectifier (thyristor) may also break down if a transient voltage is applied across it in the reverse direction either when it is conducting or when it is in its off state. In addition, a thyristor may break down in the forward direction while it is in its off state if the rate of rise of the applied transient voltage in the forward direction exceeds a threshold value (dV/dt effect). In general, this breakdown in the forward direction is not destructive for the low- and medium-power devices. For the high-power devices, the breakdown in the forward direction by high dV/dt can be destructive, because the whole cross-section of the large silicon wafer may not be fully in conduction during the interval the transient is applied. The damage is usually thermal because of the high current density in the partially conducting cross-section of the silicon wafer.

The breakdown of a nonavalanche type diode in the off state by a transient voltage is caused by dielectric failure along the surface of the semiconductor wafer. An avalanche type diode avalanches when a transient voltage in the reverse direction is applied to it during its off state. The diode breaks down by thermal runaway condition when the energy dissipated in the junction by the avalanche current exceeds its tolerable limit.

The transient-voltage breakdown levels of three types of silicon diodes are shown in Table 16.1 [21]. The types I and II were procured from two different manufacturers. Type I was an avalanche type diode while types IIA and IIB were the nonavalanche types. The types IIA and IIB belonged to the same class of diodes, differing only in their prv ratings. In other words, these two types of diodes were designed and manufactured the same way; they were segregated into different prv ratings during factory tests in accordance with their leakage current levels.

Table 16.1: Transient-Voltage Susceptibility Data of Three High-Power Diodes

Diode Type	I	IIA	IIB
Repetitive prv, volts	1200	800	1200
Transient prv, volts	-	1050	1500
Average direct current, ampere	250	250	250
Impulse breakdown[1], volts	2350[3]/2250[4]	1620[3]/1250[4]	1950[3]/1650[4]
Impulse breakdown[2], volts	1605[3]/1100[4]	767[3]/560[4]	1030[3]/440[4]

[1]: 1.5/40-μs transient voltage applied in the reverse direction at 25°C
[2]: Transient voltage applied in the reverse direction while carrying 785-ampere half-sine 17-ms current in the forward direction
[3]: Average breakdown voltage level
[4]: Lowest breakdown voltage level

It should be observed from this table that:

1. The steady-state prv rating has no relationship with the breakdown voltage level of a diode under transient voltages;
2. The avalanche type diodes can withstand higher-level transient voltages; and
3. The breakdown voltage level under a transient voltage while the diode is in its on state can be lower than its steady-state prv rating.

These trends were also observed for many other diode types [20,21]. The volt-time (time-lag) characteristic of the diode type IIB under the 1.5/40-μs voltage wave is shown in Fig. 16.14a. The volt-time characteristic of a medium-power (600-V, 25-A) diode is shown in Fig. 16.14b. Both volt-time curves show a trend towards lower breakdown levels for shorter times to breakdown. This is quite different from the volt-time characteristics of other power apparatus, where the volt-time characteristic goes up at shorter times to breakdown.

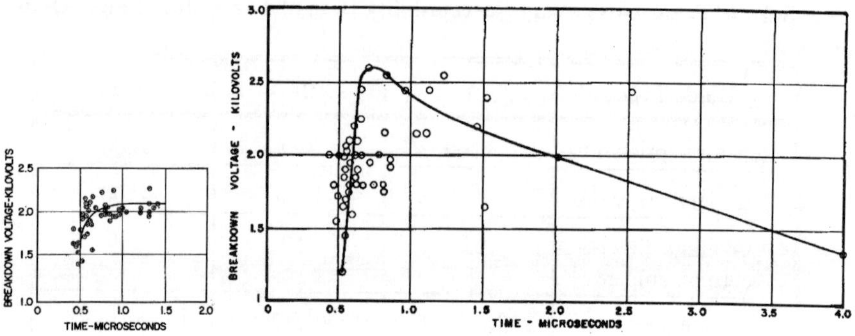

(a) (b)

Fig. 16.14 Volt-time characteristics of two types of silicon diodes [21].
(a) 600-V, 25-A diode; (b) 1200-V, 250-A diode.
(Copyright © 1973 IEEE)

The avalanche-type devices tend to clip the oncoming voltage transients, and dissipate the energy of the transient within the bulk of the semiconductor junction. However, the semiconductor junction experiences a thermal runaway condition if the oncoming transient contains enough energy to exceed the thermal capability of the semiconductor junction of the avalanche-type device. Therefore, the energy content of the oncoming transient is an important parameter for a viable transient-voltage protective scheme.

The transient-voltage susceptibilities of medium-power thyristors are shown in Table 16.2 [22]. Similar to the diodes, the prv ratings of the thyristors have no correlation to their transient-voltage breakdown voltage levels. However, unlike the diodes, the breakdown voltage level of a thyristor in the reverse direction is much higher than its prv rating even when the transient is applied while it is carrying full-load current. The dV/dt breakover level of the thyristor was found to be much higher than the specified value, at least for the thyristors tested. It is not advisable, however, to exceed the dV/dt rating of the device.

The above discussion of the experimental data shows that power semiconductor devices lack adequate specifications for their transient performance. Transient performances of devices of the same steady-state specifications are widely different. Reliable and cost effective surge protection of semiconductor-based systems will not be possible without adequate specifications on their transient performance.

Table 16.2: Transient-Voltage Susceptibility Data of Two Medium-Power Thyristors [22]

ThyristorType	IIIA	IIIB
Repetitive peak off-state voltage, volts	500	800
Non-repetitive peak off-state voltage, volts	600	960
Repetitive peak reverse voltage, volts	500	800
Non-repetitive peak reverse voltage, volts	600	960
RMS forward current, amperes	35	35
dV/dt, V/μs	100	100
Impulse breakdown[1)], volts	2596[3)]/2100[4)]	2574[3)]/2400[4)]
Impulse breakdown[2)], volts	2481[3)]/2000[4)]	-
Impulse breakdown[5)], volts	1485[3)]/1360[4)]	1469[3)]/1325[4)]

[1)]: 1.2/50-μs transient voltage applied in the reverse direction in the off state at 25°C

[2)]: Transient voltage applied in the reverse direction while carrying 49-ampere half-sine 16.67-ms current in the forward direction

[3)]: Average breakdown voltage level

[4)]: Lowest breakdown voltage level

[5)]: 1.2/50-μs transient voltage applied in the forward direction in the off state at 25°C

16.5 Insulation Coordination of Low-Voltage Electrical Systems

For proper insulation coordination, the following information about the transient-voltage environment is necessary:

1. Expected peak of the transient voltage;
2. Waveshape of voltage (front time and duration); and
3. Energy content of the transient.

The susceptibility of the equipment must also be expressed in terms of:

1. Maximum transient-voltage withstand level;
2. Waveshape related to the withstand level; and
3. Transient energy handling capacity.

Transient voltages occur at random, and no two waveshapes are the same. Therefore, it is impossible to specify a 'representative' waveshape. Nonetheless, the ANSI/IEEE Guide recommends two open-circuit voltage waveshapes and one short-circuit current waveshape [11]:

(i) 0.5-μs 100-kHz Ring Voltage Wave,
(ii) 1.2/50-μs Voltage Wave, and
(iii) 8/20-μs Current wave.

The rise time (0.5 μs) of the ring voltage wave is defined as the time difference between the 10% and 90% amplitude points on the leading edge of the waveform. After the first peak, the voltage decays while oscillating at 100 kHz. The 1.2/50-μs voltage wave and the 8/20-μs current wave are the standard transient waveforms used to test high-voltage equipment [23]. The ring voltage waveshape is recommended for indoor installations, and the other two waveshapes for outdoor and service entrance environments.

Shielding is used to protect against radiated electromagnetic transients. This was described in Section 16.3.3. For protection against conducted electromagnetic transients, surge protectors incorporating nonlinear resistors are the most suitable choice. These nonlinear resistors are widely known as metal oxide varistors (MOV). Many types of MOVs are available. A judicious selection of MOV would result in a reliable and cost effective protective scheme. Some typical characteristics of low-voltage MOVs are shown in Table 16.3. Outlines of three models of low-voltage MOVs are shown in Fig. 16.15.

Although all the models are applicable to 130-volt systems, their characteristics are different. They should be applied according to the demands of the system which is to be protected. It is obvious that a 130-volt MOV does **not** clamp the transient voltage to 130 volts. It only signifies that a 130-volt MOV can be used in a 130-volt 60-Hz power system. One should carefully note the discharge voltage of the MOV at the specified surge current, and decide whether this discharge voltage is lower than the

Table 16.3: Typical Characteristics of Metal Oxide Varistors [37]

(Courtsey of Harris Corporation)

Varistor Model	Maximum Ratings (85°)				Characteristics (25°C)			
	Continuous		Transient		Varistor Peak Voltage @ 1 mA DC	Maximum Clamping Voltage at 8x20-μs Test Current		Typical Capacitance
	RMS Voltage	DC Voltage	Energy (10x1000-μs)	Peak Current (8x20-μs)				
	Volts	Volts	Joules	Amperes	Volts	Volts	Amperes	pF
V150LA20B	150	200	80	6500	212-243	360	100	1600
V151HA40	150	200	300	30000	212-268	405	200	8000
V250LA40B	250	330	130	6500	354-413	620	100	1000
V251HA40	250	330	370	40000	354-429	630	200	5000
V480LA80B	480	640	180	6500	670-790	1160	100	550
V481HA40	480	640	650	40000	670-825	1230	200	2700
V660LA100B	660	850	250	6500	940-1100	1650	100	400
V661HA40	660	850	900	40000	940-1160	1720	200	2000

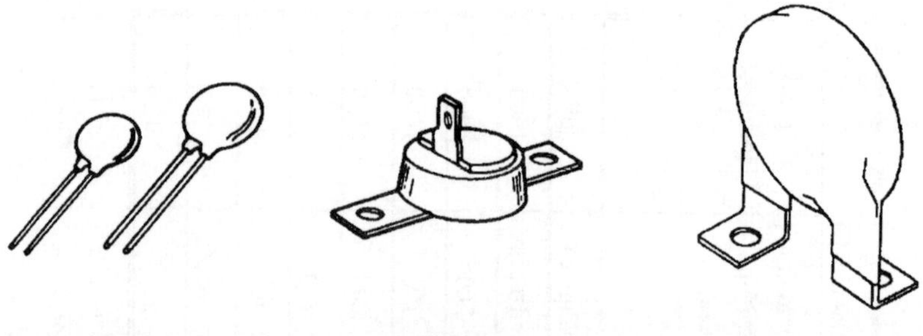

LA Series PA Series HA Series

Fig. 16.15 Low-voltage varistors of various voltage and energy ratings.
(Courtsey of Harris Corporation)

voltage withstand capability of the equipment or the system. One should also note the **energy rating** of the MOV. The MOV should be selected so that its rated energy handling capability is higher than that expected in the transient.

Selecting the proper MOV is one step in the process of insulation coordination of low-voltage systems. Proper installation of the MOV is another step. The MOV should be installed as near to the equipment as possible. Lead lengths must be the shortest possible. Many high-power semiconductor devices have pigtails which are about one foot long. It is convenient to connect the MOV between the heat sink and the tip of the pigtail. However, many power semiconductor devices have been damaged because of this 'convenience' of installation. It was mentioned in Section 16.2.3 that semiconductors have a tendency to generate very fast transients during their reverse recovery period which may damage the semiconductor p-n junction. The purpose of the MOV is to provide a low-impedance path to the internally generated transient. However, a one-foot lead between the p-n junction and the MOV would look like a very high impedance to the transient.

It was shown earlier that the transient-voltage withstand capability of a semiconductor device has no correlation with its steady-state prv rating. Installing two series-connected devices instead of one device of double the steady-state prv rating would generally provide higher transient-voltage withstand capability. Of course, the inconvenience, weight, volume and perhaps the cost of two devices with their heat sinks and voltage-sharing network must also be weighed in the final choice.

16.5.1 An example of insulation coordination in low-voltage systems

Problem: A 600-volt rms ac system will need silicon rectifiers for power rectification. Measurements in the field indicated that transient voltages as high as 10 kV with 100 joules of energy may impinge on this system at random. It was found from design considerations that one particular type of silicon rectifier would be the most suitable for this application. This type of rectifier was available at three prv (peak reverse voltage) ratings: 800 volts, 1000 volts and 1200 volts. Their design specifications showed that all three prv ratings had the same transient voltage ratings: 2.5 kV when the rectifier was not carrying any current (i.e., blocking), and 1.5 kV when the rectifier was carrying full-load current in the forward direction and the transient voltage was applied in the reverse direction. Which prv rating should be selected for this application? How many of the rectifiers should be connected in series? Should surge protectors be used? If surge protectors are needed, what should be their discharge voltage rating and their energy-handling capability?

Solution: The peak steady-state voltage of the system is 848.5 volts ($=\sqrt{2}$x600). If no surge protector is used, then enough number of rectifiers must be connected in series to withstand the expected 10-kV transient voltage. In that case, 7 (=10.0/1.5) rectifiers must be used in the series string. If surge protectors are used, obviously one single 800-V rectifier will not suffice because the system peak steady-state voltage would then exceed the steady-state voltage rating of the device. A 1000-V or a 1200-V rectifier would be sufficient to withstand the system peak steady-state voltage; however, a 660-V varistor (Table 16.3) would be required, the maximum clamping voltage of which is 1.65 kV. This is higher than the transient-voltage rating of 1.5 kV for the rectifiers. A good choice would be to connect two 800-V rectifiers in series, with a 480-V varistor (V480LA80B) across each rectifier. The transient energy rating of this varistor being 180 J, it will safely dissipate the expected transient voltage of 100 J. If, however, the energy of the expected transient were higher, e.g., 1 kJ, then the varistor V481HA40 with 650-J energy rating should be connected across each of the two series connected 800-V rectifiers.

16.6 Other Methods of Protection [12]

Low-level signals are used to control large amounts of power. These signals are sensitive to electrical noise. Lightning is a main source of such electrical noise. Heavy discharge of lightning current in the grounding mat in a switchyard can produce a potential gradient along the mat so that if the shield of the signal cable is grounded both at the transmitting end (switchyard) and at the receiving end (control room), then current will flow through the cable shield, and contaminate the signal. Single-point grounding is the standard practice under these circumstances. One has to decide which end of the cable shield to ground. The cable shield should be bonded to the 'quiet' ground. For a substation, the receiving end of the cable (i.e., the switchyard) is the 'noisy' ground. Therefore, the cable shield should be bonded to the ground at the control room. A simple device such as a common-mode choke can be very useful in many cases where two grounds are not at the same potential. A common-mode choke consists of a few turns of a coaxial cable wound on a ferrite core. This choke is connected between the signal cable and the input of the receiver. Such a choke offers a high impedance to the flow of any shield current without affecting the flow of the signal. Another method to reduce the shield current caused by the ground potential gradient is to use a 'balun' transformer. These are small high-frequency isolation transformers which impede the flow of noise currents. High transient voltages can also be induced in the 115-volt supply of the control mains. Well-shielded isolation transformers have worked well to reduce such transients in the control power supplies. Low-voltage nonlinear varistors across sensitive devices, such as solid-state devices, would prevent damage to these devices. The radiated electromagnetic fields of the lightning discharge can be strong enough to cause malfunction or even damage to sensitive equipment in the control room. Enclosing such equipment in a copper enclosure can be of great benefit. Fiber-optic cables are now being extensively used to transmit signals in highly 'noisy' environments. However, the opto-electronic devices at either end of the fiber-optic cable should be well shielded from radiated electrical noise.

16.7 References to Chapter 16

1. K. Berger, "Transient Phenomena in House Installations and Preliminary Review of the Possibilities of Protection of House Installations Against Atmospheric Overvoltages,"(*in German*), ***Bulletin SEV***, Vol. 35, No. 19, pp. 523-533, 1944.

2. K. Berger and R. Pichard, "Impulse Tests on a House Installation in Uitikon at Albis, 1945," (*in German*), ***Bulletin SEV***, Vol. 38, No. 18, pp. 543-570, 1947.

3. Von G. Lehmann and W. Naumann, "Damages on Low-Voltage Systems by Lightning Overvoltages," (*in German*), ***Bulletin SEV***, Vol. 60, No. 4, pp. 137-141, 1969.

4. F.D. Martzloff and G.J. Hahn, "Surge Voltages in Residential and Industrial Power Circuits," ***IEEE Trans. on Power Apparatus and Systems***, Vol. PAS-89, pp. 1049-1056, 1970.

5. IEEE Committee Report: "Bibliography on Surge Voltages in AC Power Circuits Rated 600 Volts and Less," ***IEEE Trans. on Power Apparatus and Systems***, Vol. PAS-89, pp. 1056-1061, 1970.

6. A.J. Baggott, "Practical Measurements and Investigations into the Causes and Effects of Voltage Disturbances on Customers Equipment," ***Proc. International Conference of Power System Disturbances***, IEE Conference Publication No. 110, Paper No. 46, London, U.K., April 1974.

7. J.H. Bull, "Susceptibility of Electronic Equipment to Voltage Spikes," ***Proc. International Conference of Power System Disturbances***, IEE Conference Publication No. 110, Paper No. 52, London, U.K., April 1974.

8. L. Lagostena, A. Mantini and M. Silvestri, "Disturbances Produced by Domestic Appliances Controlled by Thyristors. Experiments and Studies Conducted for the Purpose of Preparing Adequate Standards," ***Proc. International Conference of Power System Disturbances***, IEE Conference Publication No. 110, Paper No. 214, London, U.K., April 1974.

9. P. Chowdhuri, "Study of Transient Voltages in Transit Systems," ***IEEE Trans. on Electromagnetic Compatibility***, Vol. EMC-17, pp. 140-149, 1975.

10. H. Mitani, "Magnitude and Frequency of Transient Induced Voltages in Low-Voltage Control Circuits of Power Stations and Substations," ***IEEE Trans. on Power Apparatus and Systems***, Vol. PAS-99, pp. 1871-1878, 1980.

11. "IEEE Guide for Surge Voltages in Low-Voltage AC Power Circuits." American National Standard No. ANSI/IEEE C62.41-1991.

12. "IEEE Guide for the Installation of Electrical Equipment to Minimize Electrical Noise Inputs to Controllers from External Sources," IEEE Std 518-1982.

13. T. Bernstein, "Lightning and Power Surge Damage to Appliances," ***IEEE Trans. on Industry Applications***, Vol. IA-20, pp. 1507-1512, 1984.

14. R. Odenberg and B.J. Braskich, "Measurements of Voltage and Current Surges on the AC Power LIne in Computer and Industrial Environments," ***IEEE Trans. on Power Apparatus and Systems***, Vol. PAS-104, pp. 2681-2691, 1985.

15. J.J. Goedbloed, "Transients in Low-Voltage Supply Networks," ***IEEE Trans. on Electromagnetic Compatibility***, Vol. EMC-29, pp. 104-115, 1987.

16. W. Boxleitner, "How to Defeat Electrostatic Discharge," ***IEEE Spectrum***, pp. 36-40, August 1989.

17. F.J. Young, "Ferromagnetic Shielding Related to the Physical Properties of Iron," ***1968 IEEE Electromagnetic Compatibility Symposium Record***, IEEE Publication No. 68C12-EMC, pp. 88-95, 1968

18. R.R. Ferber and F.J. Young, "Enhancement of EMP Shielding by Ferromagnetic Saturation," ***IEEE Nuclear and Space Radiation Effects Conference***, San Diego, CA, 1970.

19. F.J. Young, "Pulse Shielding by Nonferrous and Ferromagnetic Materials," ***Proc. IEEE***, Vol. 61, pp. 404-413, 1973.

20. P. Chowdhuri, "Breakdown of P-N Junctions by Transient Voltages," ***Direct Current***, Vol. 10, No. 3, pp. 3-11, 1965.

21. P. Chowdhuri, "Transient-Voltage Characteristics of Silicon Power Rectifiers," ***IEEE Trans. on Industry Applications***, Vol. IA-9, pp. 582-592, 1973.

22. P. Chowdhuri, "Behavior of Thyristors Under Transient Voltages," IEEE/PES Winter Meeting, IEEE Publication No. 80 CH1523-0-PWR, Paper No. A 80 114-9, 1980.

23. "IEEE Standard Techniques for High-Voltage Testing", IEEE Std 4-1995.

INDEX

(Numbers refer to page numbers)